Managing Biodiversity in Agricultural Ecosystems

Managing Biodiversity
in Agricultural Ecosystems

EDITED BY D. I. JARVIS, C. PADOCH, AND H. D. COOPER

Published by Bioversity International

Columbia University Press *New York*

Schweizerische Eidgenossenschaft
Confédération suisse
Confederazione Svizzera
Confederaziun svizra

**Swiss Agency for Development
and Cooperation SDC**

IDRC ✳ CRDI

UNITED NATIONS
UNIVERSITY

CBD

Columbia University Press
Publishers Since 1893
New York   Chichester, West Sussex
Copyright © 2007 Bioversity International
All rights reserved

As of December 1, 2006 IPGRI and INIBAP operate under the name "Bioversity International."

Library of Congress Cataloging-in-Publication Data
Managing biodiversity in agricultural ecosystems / edited by D. I. Jarvis, C. Padoch, and H. D. Cooper.
    p. cm.
ISBN 13: 978-0-231-13648-8 (hard cover : alk. paper) ISBN 13: 978-0-231-51000-4 (e-book)
ISBN 10: 0-231-13648-X (hard cover : alk. paper)—ISBN 10: 0-231-51000-4 (e-book)
    1. Agrobiodiversity.   2. Agricultural ecology.   I. Jarvis, Devra I. (Devra Ivy), 1959–   II. Padoch, Christine.   III. Cooper, H. D. (H. David)
S494.5.A43M36 2007
630—dc22   2006031672
∞

This book is dedicated to our children—

Raffaella, Sofia, Charlie, and Duncan—

who connect our present world to that of the future.

Contents

Acknowledgments

The editors would like to thank the governments of Canada (IDRC, International Development Research Centre) and Switzerland (Swiss Agency for Development and Cooperation) for their generous financial support of this book.

Much of the work presented in this volume was accomplished with kind assistance from the governments of Switzerland (SDC, Swiss Agency for Development and Cooperation), The Netherlands (DGIS, Directorate-General for International Cooperation), Germany (BMZ/GTZ, Bundesministerium für Wirtschaftliche Zusammenarbeit/Deutsche Gesellschaft für Technische Zusammenarbeit), Japan (JICA), Canada (IDRC), Spain, and Peru and the Global Environmental Facility of the United Nations Environment Programme, the Secretariat of the Convention on Biological Diversity, and the Food and Agriculture Organization of the United Nations.

We thank many of our colleagues who helped at various stages of book production; our special thanks go to Steve Clement, Charles Spillane, Jean Louis Pham, Linda Collette, Julia Ndung'u-Skilton, Beate Scherf, and Paola De Santis. Several anonymous referees provided very welcome critical reviews of the chapters. We owe very special thanks to Linda Sears for her precise and rapid editing of the chapters in this volume.

Finally, our most sincere and profound thanks go to many participants whose names and affiliations do not appear in this volume. Many farmers, development workers, educators, researchers, and government officials participated in the many studies presented in this book; it is they who made this work possible.

Contributors

F. Ahkter	The Centre for Policy Research for Development Alternatives, Bangladesh
L. Arias	Centro de Investigaciones y Estudios Avanzados del Incipiente Projección Nacional, Mérida, Yucatán, Mexico
W. Ayalew	International Livestock Research Institute, Nairobi, Kenya
J. Bajracharya	Agriculture Botany Division, Nepal Agriculture Research Council, Khumaltar, Lalitpur, Nepal
D. Balma	Direction de la Recherche Scientifique, Ouagadougou, Burkina Faso
D. Bartley	FAO Inland Water Resources and Aquaculture Service, Rome, Italy
D. E. Bennack	Instituto de Ecología, Xalapa, Veracruz, Mexico
H. Brookfield	Australian National University, ACT 0200, Australia
A. H. D. Brown	Centre for Plant Biodiversity Research, CSIRO Plant Industry, Canberra, Australia

G. G. Brown	Soil Invertebrate Laboratory, Embrapa Soybean, Londrina, PR, Brazil
L. Brussaard	Wageningen University, Soil Quality Section, Wageningen, The Netherlands
S. Bunning	Land and Plant Nutrient Management Service (AGLL), Food and Agriculture Organization of the United Nations, Rome, Italy
L. A. Burgos-May	Centro de Investigaciones y Estudios Avanzados del Incipiente Projección Nacional, Mérida, Yucatán, Mexico
T. C. Camacho-Villa	Centro de Investigaciones y Estudios Avanzados del Incipiente Projección Nacional, Mérida, Yucatán, Mexico, and Wageningen University and Research Center, Participatory Approaches Studies, Wageningen, The Netherlands
M. Ceroni	Department of Botany and Gund Institute for Ecological Economics, University of Vermont, USA
J. L. Chavez-Servia	Centro Interdisciplinario de Investigación para el Desarrollo Integral Regional–Instituto Politecnico Nacional, Oaxaca, Mexico
L. Collado	Consorcio para el Desarrollo Sostenible de Ucayali, Pucallpa, Perú
H. D. Cooper	Secretariat, Convention on Biological Diversity, Montreal, Quebec, Canada
R. Costanza	Rubenstein School of Environment and Natural Resources and Gund Institute for Ecological Economics, University of Vermont, USA
M. Dijmadoum	Fédération National des Groupements Naam, Ouahigouya, Burkina Faso
A. G. Drucker	School of Environmental Research, Charles Darwin University, Australia

M. R. Finckh — Department of Ecological Plant Protection, University of Kassel, Wutzenhausen, Ecological Agricultural Science, Germany

B. M. Freitas — Departamento de Zootecnia, Universidade Federal do Ceará, Fortaleza, Brazil

D. Gauchan — Nepal Agricultural Research Council, Kathmandu, Nepal

B. Gemmill — African Pollination Initiative, Nairobi, Kenya

J. P. Gibson — Institute for Genetics and Bioinformatics Homestead, University of New England, Armidale NWS 2351, Australia

M. Halwart — Inland Water Resources and Aquaculture Service, Food and Agriculture Organization of the United Nations, Rome, Italy

O. Hanotte — International Livestock Research Institute, Nairobi, Kenya

T. Hodgkin — International Plant Genetic Resources Institute, Maccarese, Rome, Italy

I. Hoffmann — Animal Production Service, Food and Agriculture Organization of the United Nations, Rome, Italy

N. N. Hue — Vietnamese Agricultural Science Institute, Hanoi, Vietnam

V. Imbruce — New York Botanical Garden, Bronx, NY, USA

D. I. Jarvis — International Plant Genetic Resources Institute, Maccarese, Rome, Italy

T. Johns — Centre for Indigenous Peoples' Nutrition and Environment and School of Dietetics and Human Nutrition, McGill University, Ste. Anne de Bellevue, Quebec, Canada

D. Karamura — International Network for the Improvement of Banana and Plantain, Kampala, Uganda

E. Karamura	International Network for the Improvement of Banana and Plantain, Kampala, Uganda
P. G. Kevan	Department of Environmental Biology, University of Guelph, Guelph, Ontario, Canada
L. Latournerie	Instituto Tecnológico Agropecuario de Condal (SIGA-ITA2), Mérida-Motul, Condal, Yucatán, Mexico
D. Lope	Fundación Kan Uak, A.C. Mérida, Yucatán, Mexico, and Wageningen University and Research Center, Bio-Cultural Diversity Studies, Wageningen, The Netherlands
S. Liu	Rubenstein School of Environment and Natural Resources and Gund Institute for Ecological Economics, University of Vermont, USA
I. Mar	Institute for Agrobotany, Tapioszele, Hungary
A. Montáñez	Adriana Montañez, Universidad de Montevideo, Uruguay
A. Ochieng	University of Nairobi, Department of Botany, Nairobi, Kenya
J. Ochoa	Estación Experimental, Santa Catalina, Quito, Ecuador
C. Padoch	The New York Botanical Garden, Bronx, New York
U. Partap	International Centre for Integrated Mountain Development, Kathmandu, Nepal
M. Pinedo-Vasquez	Center for Environmental Research and Conservation, Columbia University, New York, NY, USA
R. Rana	Local Initiatives for Biodiversity, Research and Development, Pokhara, Nepal
V. R. Rao	International Plant Genetic Resources Institute, Regional Office for Asia, Pacific, and Oceania, Serdang, Malaysia

K. Rerkasem	Faculty of Agriculture, Chiang Mai University, Chiang Mai, Thailand
D. Rijal	Local Initiatives for Biodiversity, Research and Development, Pokhara, Nepal, and Noragric, Norwegian University of Life Sciences, Aas, Norway
M. Sadiki	Institut Agronomique et Vétérinaire Hassan II, Département d'Agronomie et d'Amélioration des Plantes, Rabat, Morocco
M. Sawadogo	University of Ouagadougou, Unité de Formation et de Recherche en Science de la Vie et de la Terre, Ouagadougou, Burkina Faso
M. Smale	International Plant Genetic Resources Institute, Rome, Italy, and International Food Policy Research Institute, Washington DC, USA
B. Sthapit	International Plant Genetic Resources Institute, Regional Office for Asia, Pacific, and Oceania, Pokhara, Nepal
A. Subedi	Intermediate Technology Group for Development, Kathmandu, Nepal
M. J. Swift	Institut de Recherche et Développement, Centre de Montpellier, Montpellier, France
M. B. Thomas	Centre for Plant Biodiversity Research, CSIRO Entomology, Canberra, Australia
P. Trutmann	International Integrated Pest Management, International Programs, Cornell University, Ithaca, NY, USA
J. Tuxill	Joint Program in Economic Botany, Yale School of Forestry and Environmental Studies and the New York Botanical Garden, New Haven, CT, USA
R. Valdivia	Centro de Investigación de Recursos Naturales y Medio Ambiente, Puno, Perú

Y. Y. Wang   Yunnan Agricultural University, Kunming, Yunnan, P.R. China

A. Wilby   Department of Agricultural Sciences and NERC Centre for Population Biology, Imperial College, Wye, Kent, UK

D. Williams   USDA, Foreign Agricultural Service, International Cooperation and Development, Research and Scientific Exchanges Division, Washington, DC, USA

V. A. Wojcik   Environmental Science Policy and Management, University of California, Berkeley, CA, USA

J. H. Zhou   Yunnan Agricultural University, Kunming, Yunnan, P.R. China

Y. Y. Zhu   Yunnan Agricultural University, Kunming, Yunnan, P.R. China

Managing Biodiversity in Agricultural Ecosystems

1 ❦ Biodiversity, Agriculture, and Ecosystem Services

D. I. JARVIS, C. PADOCH, AND H. D. COOPER

Biodiversity in agricultural ecosystems provides our food and the means to produce it. The variety of plants and animals that constitute the food we eat are obvious parts of agricultural biodiversity. Less visible—but equally important—are the myriad of soil organisms, pollinators, and natural enemies of pests and diseases that provide essential regulating services that support agricultural production. Every day, farmers are managing these and other aspects of biological diversity in agricultural ecosystems in order to produce food and other products and to sustain their livelihoods. Biodiversity in agricultural ecosystems also contributes to generating other ecosystem services such as watershed protection and carbon sequestration. Besides having this functional significance, maintenance of biodiversity in agricultural ecosystems may be considered important in its own right. Indeed, the extent of agriculture is now so large, any strategy for biodiversity conservation must address biodiversity in these largely anthropogenic systems. Moreover, biodiversity in agricultural landscapes has powerful cultural significance, partly because of the interplay with historic landscapes associated with agriculture, and partly because many people come into contact with wild biodiversity in and around farmland.

This book examines these various aspects of agricultural biodiversity. A number of chapters examine crop genetic resources (chapters 1, 2, 3, 10, 11, and 16) and livestock genetic resources (chapters 4, 5, and 17). Other chapters examine aquatic biodiversity (chapter 6), pollinator diversity (chapter 7), and soil biodiversity (chapter 8). Three chapters (9, 10, and 11) examine various aspects of the relationship between diversity and the management of pests and diseases. Chapters 12 and 13 explore farmer management of diversity in the wider context of spatial complexity and environmental and economic change. Chapter 14 looks at the contribution of diversity to diet, nutrition, and human health. Chapters 15 through 17 explore the value of genetic resources and of the ecosystem services provided by biodiversity in agricultural ecosystems.

This introductory chapter sets the scene for the subsequent chapters. After reviewing recent efforts to address agricultural biodiversity in the academic community and international policy fora, the multiple dimensions of biodiversity in agricultural ecosystems are surveyed. Subsequent sections examine the value of ecosystems services provided by biodiversity, the functions of biodiversity, and how these are influenced by management. The chapter concludes with a brief consideration of the future of biodiversity in agricultural ecosystems.

Recent and Current Initiatives to Address Agricultural Biodiversity

The importance to agriculture of crop, livestock, and aquatic genetic resources has long been recognized, but only in the last decade or so has the global community acknowledged the significance of the full range of agricultural biodiversity in the functioning of agricultural ecosystems. In the international policy arena, agricultural biodiversity was addressed for the first time in a comprehensive manner by the Conference of the Parties of the Convention on Biological Diversity (CBD) in 1996. The CBD program of work on agricultural biodiversity, which was subsequently developed and adopted in 2000, recognizes the multiple dimensions of agricultural biodiversity and the range of goods and services provided. In adopting the program of work, the Conference of the Parties recognized the contribution of farmers and indigenous and local communities to the conservation and sustainable use of agricultural biodiversity and the importance of agricultural biodiversity to their livelihoods. Within the framework of the convention's program of work on agricultural biodiversity, specific initiatives on pollinators, soil biodiversity, and biodiversity for food and nutrition have been launched.

This new spotlight on agricultural biodiversity is a response to a broad consensus that global rates of agricultural biodiversity loss are increasing. Estimates from the World Watch List of Domestic Animal Diversity note that 35% of mammalian breeds and 63% of avian breeds are at risk of extinction and that one breed is lost every week. The *State of the World's Plant*

Genetic Resources for Food and Agriculture (PGRFA) describes as "substantial" the loss in diversity of plant genetic resources for food and agriculture, including the disappearance of species, plant varieties, and gene complexes (FAO 1998). Every continent except Antarctica has reports of pollinator declines in at least one region or country. Numbers of honeybee colonies have plummeted in Europe and North America, and the related Himalayan cliff bee (*Apis laboriosa*) has experienced significant declines (Ingram et al. 1996). Other pollinator taxa are also the focus of monitoring concerns, with strong evidence of declines in mammalian and bird pollinators. Globally, at least 45 species of bats, 36 species of nonflying mammals, 26 species of hummingbirds, 7 species of sunbirds, and 70 species of passerine birds are considered threatened or extinct (Kearns et al. 1998).

The broad consensus on amplified rates of biodiversity loss in agricultural systems, with the need to have better quantification of these rates of change, has spurred an increasing number of international, national, and local actions on agricultural biodiversity management over the last few years. The International Plant Genetic Resources Institute (IPGRI) global on-farm conservation project (Jarvis and Hodgkin 2000; Jarvis et al. 2000); the People, Land Management and Environmental Change (PLEC) Project (Brookfield 2001; Brookfield et al. 2002); the Community Biodiversity Development and Conservation (CBDC) Programme; the Centro Internacional de Agricultura Tropical (CIAT), Tropical Soil Biology and Fertility Institute (TSBF), and Global Environmental Facility Below Ground Biodiversity (BGBD) Project; the Global Pollinator Project supported by FAO; and Operational Programme on Agricultural Biodiversity and projects supported under the Global Environment Facility (GEF) are a few prominent examples. Many case studies carried out under these and other initiatives were reviewed at the international symposium "Managing Biodiversity in Agricultural Ecosystems," held in 2001 in Montreal on the margins of the meeting of the Scientific Subsidiary Body to the CBD.

This book builds on case studies presented at the Montreal symposium. Whereas conventional approaches to agricultural biodiversity focus on its components as static things, many of the chapters in this book emphasize instead the dynamic aspects of agricultural biodiversity and the interactions between its components. Researchers with backgrounds and interests in the social and environmental sciences have also brought new perspectives and approaches to the field. They seek to understand the processes and linkages, the dynamism and practices that are essential to the way biodiversity has long been and continues to be managed in farming systems, agricultural communities, and the broader societies.

Multiple Dimensions of Agricultural Biodiversity

Agricultural biodiversity includes all components of biological diversity relevant to the production of goods in agricultural systems: the variety and variability of plants, animals, and microorganisms at genetic, species, and ecosystem levels that are necessary to sustain key functions, structures, and processes in the agroecosystem. Thus it includes crops, trees, and other associated plants, fish and livestock, and interacting species of pollinators, symbionts, pests, parasites, predators, and competitors.

Cultivated systems contain *planned biodiversity*, that is, the diversity of plants sown as crops and animals raised as livestock. Together with crop wild relatives, this diversity comprises the genetic resources of food agriculture. However, *agricultural biodiversity* is a broader term that also encompasses the associated biodiversity that supports agricultural production through nutrient cycling, pest control, and pollination (Wood and Lenne 1999) and through multiple products. Biodiversity that provides broader ecosystem services such as watershed protection may also be considered part of agricultural biodiversity (Aarnink et al. 1999; CBD 2000; Cromwell et al. 2001).

This volume takes a broad and inclusive approach and attempts to point to emerging issues in research on biodiversity in agricultural ecosystems. Chapters 2 to 7 focus primarily on diversity among crops, livestock, and fish that constitute much of the planned biodiversity in agricultural systems. In addition to domesticated crops and livestock, *managed* and *wild* biodiversity provides a diverse range of useful plant and animal species, including leafy vegetables, fruits and nuts, fungi, wild game insects and other arthropods, and fish (including mollusks and crustaceans as well as finfish) (Pimbert 1999; Koziell and Saunders 2001; also see Halwart and Bartley, chapter 7). These sources of food remain particularly important for the poor and landless (Ahkter in box 13.2, chapter 13) and are especially important during times of famine and insecurity or conflict where normal food supplies are disrupted and local or displaced populations have limited access to other forms of nutrition (Scoones et al. 1992; Johns, chapter 15). Even at normal times such associated biodiversity—including "weeds"— often is important in complementing staple foods to provide a balanced diet.

Some indigenous and traditional communities use 200 or more species for food (Kuhnlein et al. 2001; Johns and Sthapit 2004; Johns, chapter 15).

Diversity at species and genetic levels comprises the total variation present in a population or species in any given location. Genetic diversity can be manifested in different phenotypes and their different uses. It can be characterized by three different facets: the number of different entities (e.g., the number of varieties used per crop and the number of alleles at a given locus), the evenness of the distribution of these entities, and the extent of the difference between the entities. Crop genetic diversity can be measured at varying scales as well (from countries or large agroecosystems to local communities, farms, and plots), and indicators of genetic diversity are scale dependent. These issues are examined for crops by Brown and Hodgkin (chapter 2) and Sadiki et al. (chapter 3), for livestock by Gibson et al. (chapter 5), and for aquatic diversity in rice ecosystems by Halwart and Bartley (chapter 7). These chapters are complemented by case studies that illustrate how farmers name and manage units of diversity in their agricultural systems for crops (Sadiki et al., chapter 3; Hodgkin et al., chapter 4), animals (Hoffmann, chapter 6), and aquatic resources (Halwart and Bartley, chapter 7).

Chapters 8 to 10 focus on the essential role of *associated biodiversity* in supporting crop production (see also Swift et al. 1996; Pimbert 1999; Cromwell et al. 2001). Earthworms and other soil fauna and microorganisms, together with the roots of plants and trees, maintain soil structure and ensure nutrient cycling (Brown et al., chapter 9). Pests and diseases are kept in check by parasites, predators, and disease-control organisms and by genetic resistances in crop plants themselves (Wilby and Thomas, chapter 10; Jarvis et al., chapter 11; Zhu et al., chapter 12), and insect pollinators contribute to the cross-fertilization of outcrossing crop plants (Kevan and Wojcik, chapter 8). It is not only the organisms that directly provide services supporting agricultural production but also other components of food webs, such as alternative forage plants for pollinators (including those in small patches of uncultivated lands within agricultural landscapes) and alternative prey for natural enemies of agricultural pests. This has been shown in Javanese rice fields, where complex food webs ensure that the natural enemies of crop pests such as insects, spiders, and other arthropods have alternative food sources when pest populations are low, providing stability to this natural pest management system (Settle et al. 1996).

The multiple dimensions of biodiversity in cultivated systems make it difficult to categorize production systems as a whole into high or low biodiversity, especially when spatial and temporal scales are also included. In chapter 11, Jarvis et al. discuss whether crop genetic diversity is a benefit in reducing disease in time or whether it could be a hazard, given the potential emergence of pathogen super-races. They present case studies of resistant local genotypes used by farmers, use of resistance in intraspecific variety mixtures, and breeding programs that have selected for and used genotypes resistant to pests and pathogens to reduce crop vulnerability. The authors note the challenge of developing criteria that determine when and where genetic diversity can play or is playing a role in managing pest and disease.

Although academic research on agricultural biodiversity typically has focused on specific components (e.g., crops, pests, livestock), farmers manage whole systems as well as their separate parts. Built on long histories of adaptation, innovation and change, and rich bases of knowledge and practice, biodiversity management is not easily bounded or described. In chapter 7, Halwart and Bartley explain how farmers integrate the management of fish into their agricultural systems. In chapter 13, Brookfield and Padoch discuss approaches to understanding management of agricultural biodiversity by farmers over larger and more complex spatial and temporal scales. They argue that farmers often manage biodiversity in heterogeneous landscapes using a range of technologies. The authors use the term *agrodiversity* to describe the integration of biodiversity with the technological and institutional diversity typical of small-scale production. The concept of agrodiversity is also the core of chapter 14. In this chapter Rerkasem and Pinedo-Vasquez discuss a set of examples of how small-scale farmers manage biodiversity to solve emerging problems. Emphasizing the complexity, dynamism, and *hybrid* nature of their examples, the authors revise and update conventional views of traditional knowledge and practice to better reflect the realities of smallholder production.

Ecosystem Services and Their Value

Biodiversity in agricultural ecosystems underpins the provision of a range of goods and services from these ecosystems (Millenium Ecosystem Assessment 2000). The value of biodiversity can be expressed in economic terms because people and societies derive benefit (or utility) from the use of the ecosystem services it provides. The concept of total economic value, which includes current use value, option value (insurance value plus exploration value), and existence value or human preference for the existence of the resource unrelated to any use, is widely used by economists to identify various types of value from biodiversity (Orians et al. 1990; Pearce and Moran 1994, Swanson 1996). In addition, biodiversity goods and services often have either public or mixed private and public properties. The economic value of such goods is not well captured by market prices because they are not traded (Brown 1990). For example, the combinations of seed types grown by farmers produce a harvest from which they derive private benefits through food consumption, sales, or other utility. When they are considered as genotypes, however, the pattern of seed types across an agricultural landscape contributes to the crop genetic diversity from which not only these farmers but also people residing elsewhere and in the future may derive public benefit (Smale 2005). Because farmers' decisions on the use and management of crop varieties in their fields can result in loss of potentially valuable alleles, their choices have intergenerational and interregional consequences. Economic theory predicts that as long as agricultural biodiversity is a good, farmers as a group will underproduce it relative to the social optimum, and institutional interventions are necessary to close the gap (Sandler 1999).

In chapter 15, Johns gives empirical evidence of the value of agricultural biodiversity to dietary diversity, nutrition, and health. Gauchan and Smale (chapter 16) and Drucker (chapter 17) describe case studies that illustrate crop and animal diversity (variation within and between crops and breeds, respectively) values to farmers in ways not captured in analysis of market prices. Indeed, much of the value of crop and livestock variation is related to the potential for future adaptation or crop improvement and to ecosystem services such as erosion prevention and disease control. As discussed in chapters 16 and 17, different sectors of society perceive these values in different ways (see also Smale 2005). Chapter 16 compares geneticists' and farmers' values, identifying the factors that influence whether farmers will continue to grow (i.e., find valuable) the rice landraces that plant breeders and conservationists consider to be important for future adaptation or crop improvement. Chapter 17 discusses how declines in indigenous breeds may reflect the lack of availability of indigenous breeding stock rather than farmer net returns.

Although the worth of biodiversity in providing food is most widely appreciated, other values derived from biodiversity can be highly significant (Ceroni et al., chapter 18). The value of biodiversity and related ecosystems usually is calculated at the margin, that is, for assessing the value of changes in ecosystem services resulting from management decisions or other human actions or for assessing the value of the biodiversity of or service provided by an area that is small compared with the total area. Despite the existence of various valuation methods to estimate the different values of biodiversity, only ecosystem goods (or *provisioning ecosystem services*) are routinely valued (Ceroni et al., chapter 18). Most supporting and regulating services are not valued at all because they bear the characteristics of public goods and are not traded in markets.

Interactions Between Components of Biodiversity and Management by Farmers

Although our understanding of the relationship between biodiversity and ecosystem functioning is incomplete, several points can be stated with a high degree of certainty. First, species composition may be more important than absolute numbers of species. A high diversity of functional guilds is more important from a functional perspective than species richness itself (Brown et al., chapter 9). For example the range of functional guilds of predators of pests is key to effective natural pest control (Wilby and Thomas, chapter 10). Second, genetic diversity within populations is important for continued adaptation to changing conditions and farmers' needs through evolution and, ultimately, for the continued provision of ecosystem goods and services (see Brown and Hodgkin, chapter 2; Sadiki et al., chapter 3; Hodgkin et al., chapter 4; Hoffmann, chapter 6; Halwart and Bartley, chapter 7; Jarvis et al., chapter 11). And third, diversity within and between habitats and at the landscape level is also important in multiple ways (Brookfield and Padoch, chapter 13; Rerkasem and Pinedo-Vasquez, chapter 14). Diversity at the landscape level may include the diversity of plants needed to provide crop pollinators with alternative forage sources and nesting sites or to provide the alternative food sources for the natural enemies of crop pests (Kevan and Wojcik, chapter 8; Wilby and Thomas, chapter 10).

Many of the case studies of small-scale management described throughout the book feature exploitation of what are conventionally viewed as environments unsuited or marginal for agricultural production. It is in such environments (steep, infertile, flood-prone, dry, or distant) that many small farmers and much agricultural biodiversity continue to be found. In these circumstances, management of high levels of diversity can become a central part of the livelihood management strategies of farmers and pastoralists and survival of their communities (Brookfield and Padoch, chapter 13; Rerkasem and Pinedo-Vasquez, chapter 14). Agricultural biodiversity helps guarantee some level of resilience, with the capacity to absorb shocks while maintaining function. Smallholder farmers and the social and ecological environments in which they operate are continually exposed to many changes. When sudden change occurs, those most resilient have the capacity to renew, reorganize, and even prosper (Folke et al. 2002). In a system that has lost its resilience, adaptation to change is difficult at best, and therefore even small changes are potentially disastrous. Inability to cope with risks, stresses, and shocks, be they political, economic, or environmental, undermines and threatens the livelihoods of small-scale farmers.

Future of Agricultural Biodiversity

It is commonly said that globalization and the drive to higher agricultural productivity are the enemies of agricultural biodiversity. The spread of Green Revolution hybrid seeds and technologies, new diets, and laws on intellectual property, and seed and variety release, registration and certification, as well as access restrictions worldwide have all had negative impacts on diversity. The effects of these modernization and globalization trends have been neither simple nor linear, however. New opportunities to manage agricultural biodiversity and threats are provided by modern technologies and the globalization of markets. In some cases these tend to favor further specialization and uniformity in agricultural systems; some services provided by on-farm agricultural biodiversity are replaced in part by external inputs such as fertilizers, pesticides, and improved varieties. Inappropriate or excessive use of some inputs often reduces biodiversity in agricultural ecosystems (thus compromising future productivity) and in other ecosystems. As many of the chapters of this book suggest, alternative approaches that make use of agricultural biodiversity to provide these services can result in benefits for both productivity and biodiversity conservation. In order to identify management practices, technologies, and policies that promote the positive and mitigate the negative impacts of agriculture on biodiversity, enhance productivity, and increase the capacity to sustain livelihoods, we will need an improved understanding of the links, interactions, and associations between different components of agricultural biodiversity and the ways in which they can contribute to stability, resilience, and productivity in different kinds of production systems. As the creators and custodians of most of the world's agricultural biodiversity, farmers must be fully engaged in these efforts.

References

Aarnink, W., S. Bunning, L. Collette, and P. Mulvany, eds. 1999. *Sustaining Agricultural Biodiversity and Agro-Ecosystem Functions: Opportunities, Incentives and Approaches for the Conservation and Sustainable Use of Agricultural Biodiversity in Agro-Ecosystems and Production Systems*. Rome: FAO.

Brookfield, H. 2001. *Exploring Agrodiversity*. New York: Columbia University Press.

Brookfield, H., C. Padoch, H. Parsons, and M. Stocking. 2002. *Cultivating Biodiversity: The Understanding, Analysis and Use of Agrodiversity*. London: ITDG Publishing.

Brown, G. M. 1990. Valuing genetic resources. In G. H. Orians, G. M. Brown, W. E. Kunin, and J. E. Swierzbinski, eds., *Preservation and Valuation of Biological Resources*, 203–226. Seattle: University of Washington Press.

CBD (Convention on Biological Diversity). 2000. *Programme of Work on Agricultural Biodiversity*. Decision V/5 of the Conference of the Parties to the Convention on Biological Diversity, May 2000, Nairobi: Convention on Biological Diversity.

CBD (Convention on Biological Diversity). 2003. *Monitoring and Indicators: Designing National-Level Monitoring Programmes and Indicators*. Montreal: Convention on Biological Diversity.

Cromwell, E., D. Cooper, and P. Mulvany. 2001. Agricultural biodiversity and livelihoods: Issues and entry points for development agencies. In I. Koziell and J. Saunders, eds., *Living Off Biodiversity: Exploring Livelihoods and Biodiversity Issues in Natural Resources Management*, 75–112. London: International Institute for Environment and Development.

FAO (Food and Agriculture Organization of the United Nations). 1998. *The State of the World's Plant Genetic Resources for Food and Agriculture*. Rome: FAO.

Folke, C., S. Carpenter, T. Elmqvist, L. Gunderson, C. S. Holling, and B. Walker. 2002. *Resilience and Sustainable Development: Building Adaptive Capacity in a World of Transformation*. Scientific background paper on resilience for the World Summit on Sustainable Development, on behalf of the Environmental Advisory Council to the Swedish government. Available at www.un.org/events/wssd.

Hilton-Taylor, C., ed. 2000. *IUCN Red List of Threatened Species*. Gland, Switzerland: IUCN.

Ingram, M., G. C. Nabhan, and S. Buchmann. 1996. Impending pollination crisis threatens biodiversity and agriculture. *Tropinet* 7:1.

Jarvis, D. I. and T. Hodgkin. 2000. Farmer decision-making and genetic diversity: Linking multidisciplinary research to implementation on-farm. In S. B. Brush, ed., *Genes in the Field: On-Farm Conservation of Crop Diversity*, 261–278. Boca Raton, FL: Lewis Publishers.

Jarvis, D. I., L. Myer, H. Klemick, L. Guarino, M. Smale, A. H. D. Brown, M. Sadiki, B. Sthapit, and T. Hodgkin. 2000. *A Training Guide for In Situ Conservation On-Farm*, Version 1. Rome: International Plant Genetic Resources Institute.

Jarvis, D. I., D. Nares, T. Hodgkin, and V. Zoes. 2004. On-farm management of crop genetic diversity and the Convention on Biological Diversity programme of work on agricultural biodiversity. *Plant Genetic Resources Newsletter* 138:5–17.

Johns, T. and B. R. Sthapit. 2004. Biocultural diversity in the sustainability of developing country food systems. *Food and Nutrition Bulletin* 25:143–155.

Kearns, C. A., D. W. Inouye, and N. M. Waser. 1998. Endangered mutualisms: The conservation of plant–pollinator interactions. *Annual Review of Ecological Systems* 29:83–112.

Koziell, I. and J. Saunders, eds. 2001. *Living Off Biodiversity: Exploring Livelihoods and Biodiversity Issues in Natural Resources Management*. London: International Institute for Environment and Development.

Kuhnlein, H. V., O. Receveur, and H. M. Chan. 2001. Traditional food systems research with Canadian indigenous peoples. *International Journal of Circumpolar Health* 60:112–122.

Magurran, A. E. 2004. *Measuring Biological Diversity*. Oxford: Blackwell.

Millenium Ecosystem Assessment. 2005. Ecosystems and Human Wellbeing. Vol 1: Status and Trends. Washington, DC: Island Press.

Noss, R. F. 1990. Indicators for monitoring biodiversity: A hierarchical approach. *Conservation Biology* 4:355–364.

Orians, G. H., G. M. Brown, W. E. Kunin, and J. E. Swierzbinski, eds. 1990. *Preservation and Valuation of Biological Resources*. Seattle: University of Washington Press.

Pearce, D. and D. Moran. 1994. *The Economic Value of Biodiversity*. London: Earthscan.

Pimbert, M. 1999. *Sustaining the Multiple Functions of Agricultural Biodiversity*. Background paper for the FAO/Netherlands Conference on the Multifunctional Character of Agriculture and Land. Rome: FAO.

Sandler, T. 1999. Intergenerational public goods: Strategies, efficiency, and institutions. In I. Kaul, I. Grunberg, and M. A. Stein, eds., *Global Public Goods*, 20–50.

Oxford, UK: United Nations Development Programme and Oxford University Press.

Scoones, I., M. Melnyk, and J. N. Pretty. 1992. *The Hidden Harvest: Wild Foods and Agricultural Systems—A Literature Review and Annotated Bibliography.* London: International Institute for Environment and Development.

Settle, W. H., H. A. Ariawan, E. T. Cayahana, W. Hakim, A. L. Hindayana, P. Lestari, and A. S. Pajarningsih and Sartanto. 1996. Managing tropical rice pests through conservation of generalist natural enemies and alternative prey. *Ecology* 77:1975–1988.

Smale, M. 2005. Concepts, metrics and plan of the book. *Valuing Crop Biodiversity: On-Farm Genetic Resources and Economic Change.* Wallingford, UK: CAB International.

Swanson, T. 1996. Global values of biological diversity: The public interest in the conservation of plant genetic resources for agriculture. *Plant Genetic Resources Newsletter* 105:1–7.

Swift, M. J., J. Vandermeer, P. S. Ramakrishnan, J. M. Anderson, C. K. Ong, and B. A. Hawkins. 1996. Biodiversity and agroecosystem function. In H. A. Mooney, J. H. Cushman, E. Medina, O. E. Sala, and E.-D. Schulze, eds., *Functional Roles of Biodiversity: A Global Perspective.* Chichester: Wiley, SCOPE/UNEP.

Wood, D. and J. M. Lenne. 1999. Why agrobiodiversity? In D. Wood and J. M. Lenne, eds., *Agrobiodiversity: Characterization, Utilization and Management,* 1–14. Wallingford, UK: CAB International.

2 ❦ Measuring, Managing, and Maintaining Crop Genetic Diversity On Farm

A. H. D. BROWN AND T. HODGKIN

The great challenge now facing the global agricultural community is how to develop and improve the productivity of agricultural ecosystems to alleviate poverty and ensure food security in a sustainable fashion. For meeting short-term needs and achieving long-term sustainability, it is universally recognized that plant genetic diversity is essential.

Management of biodiversity is complex and synthetic, involving all levels of diversity (ecosystem, species, gene, and environment), and depends on a variety of disciplines (genetics, farming systems, social sciences). Does genetic diversity itself merit any special focus or concern amid these disciplines? We contend that it does.

If so, then we need a framework of knowledge for managing agrobiodiversity at the gene level, *in situ*, sustainably, and that framework must take account of its conservation and use. This chapter discusses the conservation of plant genetic diversity in production systems, describing how different kinds of genetic information can inform the task of managing genetic diversity and deriving actions and indicators for progress. Three categories of plant species make up plant biodiversity in the rural landscape:

- The plant species that are deliberately cropped or tended and harvested for food, fiber, fuel, fodder, timber, medicine, decoration, or other uses
- At the other extreme, wild species that occur in natural communities and that benefit the agricultural environment by providing protection, shade, and groundwater regulation

- Between these extremes, the wild related species of domesticates that can interbreed with and contribute to the genepool of their crop cousins, that survive autonomously, that share many of the pests and diseases of crops, and that sometimes are eaten to relieve famine

Of the three categories, the main focus in this chapter is on the first.

New Perspectives on Genetic Diversity

Human appreciation of genetic diversity in plants has a long history (Frankel et al. 1995). Traditionally, farmers have manipulated, selected, and used the differences they perceived between and within the plant species that sustained them and their families. These differences are in morphology, productivity, reliability, quality, pest resistance, and the like, including variation that may not be apparent to the untrained eye. Now we have entered the era of molecular biology. It provides us with new tools and the means to understand genetic diversity at its fundamental level in new ways. This section sketches some emerging perspectives on genetic diversity and relates them to more established studies on the agromorphological variation in crop species.

Molecular Diversity

Genetic diversity arises primarily as changes in the linear sequence of nucleotides in DNA. Changes can occur in the sequence in the coding region of genes or in the spacer regions between and within genes. Changes happen also in the number of copies of genes, the linkages of several genes, or indeed in whole chromosomes. A fraction of these changes translates into protein variation, marker polymorphisms, characters, and morphological variation in agronomic characters, and ultimately into varieties with different names.

To manage diversity effectively, we need to measure it and understand its extent and distribution. Efforts to measure variation have ranged from the evaluation of plant phenotypes using morphological characters to the use of molecular genetic markers. More recently, three of the major new tools of molecular biology are providing new perspectives on crop genetic diversity and opening new ways to manage plant genetic resources: single nucleotide polymorphisms (SNPs), phylogenetic analyses, and functional genomics.

They have emerged as research tools because of the increasing ability to obtain DNA sequence data on larger numbers of samples.

SINGLE NUCLEOTIDE POLYMORPHISM

Table 2.1 summarizes some recent estimates of diversity at the DNA level in crop plants or wild relatives as the probabilities per base pair of difference between two sequences in samples from various collections. These estimates of SNP are preliminary and at the level of species because population data are still lacking. The *richness statistic* of diversity K is the average number of polymorphic sites per base pair, and the *evenness statistic* θ corresponds roughly to heterozygosity. Alternatively, one can think of its inverse as the average number of base pairs lying between each SNP when two randomly chosen sequences are compared.

These and similar estimates show that genetic diversity is extensive at the DNA level. The estimates also stress wide differences in amounts of diversity for different parts of the gene or between gene and spacer regions in the

Table 2.1. Recent nucleotide diversity studies.

Species	Sample	Gene(s)	K (bp)*	θ (bp)*	Sequence per Individual (kb)
Zea mays (maize)[a]	9 inbreds, 16 landraces	21 loci	0.036	0.010	14.4
Hordeum spontaneum (wild barley)[b]	25 widespread accessions	*Adh1*	0.01	0.003	1.4
		Adh2	0.02	0.005	2.0
		Adh3	0.06	0.015	1.8
Triticum aestivum (bread wheat)[c]	<8 varieties	Restriction fragment length poly morphism probes	0.004	—	2.4
Glycine max (soybean)[d]	25 genotypes: coding	115 loci	0.002	0.00053	29
	Noncoding		0.005	0.00125	48

Sources: [a]Tenaillon et al. (2001), [b]Lin et al. (2002), [c]Bryan et al. (1999), [d]Zhu et al. (2003).
*The *richness statistic* of diversity K is the average number of polymorphic sites per base pair, and the *evenness statistic* θ corresponds roughly to heterozygosity.

genome. Molecular diversity in the alcohol dehydrogenase system in wild barley illustrates the tendency to accumulate extra DNA diversity in the less important parts of the genome. The main alcohol dehydrogenase (*Adh1*) has about half the diversity in the minor *Adh2* locus (table 2.1). The third locus (*Adh3*), which is silent in a major lineage of wild barley, appears to be the least crucial to function and harbors the most diversity. In a sample of wheat cultivars, the low diversity estimates appear to reflect the restrictions on diversity in the highly selected genepools of modern varieties and a possible bottleneck in the restricted number of origins of hexaploid wheat.

Breeding system is also a key variable. Charlesworth and Pannell (2001) have recently reviewed molecular diversity estimates from natural plant populations and emphasized the importance of breeding systems. Table 2.1 gives some data on maize to compare with wheat and wild barley estimates, and as expected maize has at least twice the values of inbreeders. This difference between outcrossing and inbreeding species is much more evident at the population level than at the species level (Hamrick and Godt 1997).

Much of this nucleotide sequence diversity would not be functionally expressed, and the question arises as to what purposes it could serve in the management of agricultural biodiverssity. Such selectively neutral diversity is ideal for measuring lineages and comparative relationships between individuals, populations, and species, obtaining evidence of recent bottlenecks in population size, documenting gene flow, recombination, seed supply, and variety identification.

PHYLOGENY AND COALESCENCE

The second outcome of the growing body of DNA sequence data and the growing capacity to generate samples of sequences from populations is more accurate phylogenies and the addition of an evolutionary time dimension to sequence diversity analysis (Clegg 1997). Once this technology becomes more widely available, it will be ideal for tracking in time the movement of genes and populations. Understanding relatedness helps in improving conservation decisions, developing core collections, searching for new characters such as new resistances, and choosing parents for plant breeding.

For example, a phylogeny of the alleles in wild barley samples at the *Adh3* locus separates accessions into two distinct lineages, which according to the molecular clock diverged some 3 million years ago (Lin et al. 2002).

One cluster had populations from the northern and western half of the Fertile Crescent (Israel, Jordan, Turkey, Syria, and Iraq). The second cluster partially overlaps and stretches east (Iraq, Iran, Turkmenistan, and Afghanistan). This result opens up the question of whether the divergence applies to the other parts of the genome and the extent of incorporation of the two *Adh3* lineages into the genepool of cultivated barley.

Molecular phylogenies also bring a new perspective for appraising biodiversity for conservation (Brown and Brubaker 2000). In the perennial subgenus *Glycine,* whose species are wild relatives of soybean, phylogenies based on organellar (chloroplast) sequences and on nuclear single-gene and multigene families have led to new insights into species relationships and the origin of polyploidy lineages. Diversity measures that incorporate distinctiveness can then be used to assess the effectiveness of a network of nature reserves in conserving the entire genepool of the subgenus. For assessing diversity on farm, distinctiveness measures assist in pointing to areas that need survey and more intensive effort.

FUNCTIONAL GENOMICS

With so much of the nucleotide diversity within a species located in the non-expressed part of the genome, how will we track down the small fraction of genetic diversity that is functionally important? The new technique of microarrays in the field of genomics provides a new approach (Aharoni and Vorst 2001; Peacock and Chaudhury 2002). Genomics is the study of all the genes in an organism at once. Microarrays (DNA chips) allow us to lay out the genome of a plant in a spatial array. For example, the *Arabidopsis* genome can be accommodated as 100,000 droplets on a single microscope slide, which can be replicated and used as a reference array many times.

The reference array can then be screened against two messenger RNA populations from two contrasting sources. The approach derives its great power by being fundamentally comparative, distinguishing the genes that have responded to a specific stress from those that have not. Differential expression between stress-tolerant and stress-sensitive genotypes could arise from genetic differences in the control regions that regulate these indicator sequences or from differences in the structural genes themselves. In *Arabidopsis,* significant overlap occurs between the genes expressed in response to different kinds of stress (E. Klok and E. Dennis, pers. comm., 2003). Thus expression of the same 34 genes changed with low oxygen and with wounding, and 5 genes responded to all of three stresses (hypoxia, wounding, and drought). Determining such genes in *Arabidopsis* could give us a powerful tool for screening populations for adaptedness to stress in crop plants. Thus genomic approaches and the use of microarrays offer the chance to link differential expression at the DNA level with adaptive divergence.

Adaptedness of Variation in Landraces

Microarray technology is a new, promising way to uncover the adaptively important genetic diversity in populations at the molecular level, still largely untried on a substantial scale. It is already apparent from more established procedures that landraces are reservoirs of adaptive variation. Teshome et al. (2001) recently reviewed the published research on variation in landraces of cereals and pulses in their centers of origin (table 2.2). The review was concerned with studies of the influence of human, biotic, and abiotic factors that maintain genetic diversity and population differentiation in traditional cultivars. There were many descriptive reports that measured variation for genetic markers or morphology. However, fewer reports sought to analyze the function of the diversity and the key factors that maintain it. Furthermore, most of the studies examined divergence be-

Table 2.2. Number of studies that report divergence between cereal or grain legume landraces for genetic markers (isozyme, DNA polymorphism) or morphological characters (e.g., agronomic and plant traits, quality, yield).

Kind of Diversifying Factors	Genetic Markers	Morphological Characters
Geographic separation at different levels (between countries, regions, or localities)	12	19
Biotic interactions (diseases and pests)	0	7
Abiotic gradients and mosaics (altitude, climate, soil, field size)	7	14
Abiotic stress at extremes of waterlogging, aridity, heat, cold, salinity	2	8
Farmers' selection criteria	1	3
Total (42 population and 31 genebank samples)	22	51

Source: Teshome et al. (2001).

tween populations; fewer have focused on the variation within individual populations. Despite these shortcomings, the growing body of evidence indicates that landraces are adapted to special features of their environment and represent a store of diversity that techniques such as microarrays could identify further.

Among the recent studies of landrace adaptedness, based on thorough fresh sampling of original populations rather than conserved material in genebanks, is that by Weltzien and Fischbeck (1990), who demonstrated the superior performance of barley landraces in their marginal, arid Near East environments. To identify the major factors affecting sorghum landrace diversity, Teshome et al. (1999) studied samples from North Shewa and South Welo, Ethiopia. Systematic sampling of more than 200 fields found 64 farmer-named varieties with an average of 10 different landraces per field. In this example, where each field had a mixture of landraces, each named landrace formed a countable unit of genetic diversity. Diversity statistics that measure richness and evenness are readily computable from morphotype frequencies. Multiple regression between landrace richness and an array of variables at field level found that higher diversity was found in fields at intermediate altitudes, fields with soils of low pH and with a lower clay content, and fields where farmers used more selection criteria in choosing the landraces they grew. Chapter 4, for example, presents evidence from case studies that exemplify this adaptation.

The study of morphological characters and population performance in benign and adverse environments seems to be worlds apart from estimates of DNA diversity and its patterns. (We set aside functional genomics studies and microarrays, which are technologies that may bridge the gap between molecular and morphological diversity.) If allozyme studies are included, then a large literature has arisen from various treatments of this relationship in all kinds of plant populations and is too extensive to review here. There are far fewer studies of diversity in landraces of crops as such. Today the scale and intensity of sampling for DNA sequence data typically are very different from those of morphological studies, but this is bound to change as projects aim to detect "linkage disequilibrium" between markers and characters in collections (Rafalski 2002).

Ideally we need information at both the molecular and the morphological level for a complete understanding of adaptive traits and their joint interpretation and analysis in terms of environment and human management. The strengths of the DNA sequence data are that they tell of evolutionary processes (population sizes, connections, shared ancestry, recombination), whereas those of adaptive traits are direct measures of crop improvement and benefit that relate directly to farmers' needs.

Indicators for Managing Genetic Diversity In Situ

In monitoring genetic diversity on farm, we need indicators. An indicator is a significant physical, chemical, biological, social, or economic variable that is measurable in a defined way for management purposes. Table 2.3 lists suggested indicators for monitoring and managing agricultural bio-diversity in situ in two groups (domesticated and wild species) and adds indicators for considering the links between in situ and ex situ activities (Brown and Brubaker 2002). The first group are plants that are or may be cultivated by humans. This includes domesticates that depend on humans for survival and wild species that are directly used by farmers such as plants that are the source of traditional medicines and other diverse cultural uses. The second group is the remaining plant species growing naturally in the agroecosystem and not directly used. (We omit consideration of indicators for ex situ strategies here. Such indicators are discussed elsewhere [Brown and Brubaker 2002], and the focus here is on in situ diversity.)

Domesticated Plants and Harvested Wild Species

Brown and Brubaker (2002) suggested the number of distinct landraces of each crop in an area as a primary indicator, together with some measure of prevalence, or the area devoted to them as a proportion of the available area in a region. Although this is straightforward in principle, experience in the International Plant Genetic Resources Institute (IPGRI) in situ project has pointed to several practical difficulties in assembling such data. Researchers are concerned with the recognition and naming of landraces: how they differ between crop species and between cultures and how much difference occurs between populations in different villages for landraces with the same name, in both time and space. A certain level of imprecision is inevitable and sometimes may be desirable, allowing some flexibility. Analyses such as those of Teshome et al. (1999) of farmers' recognition of sorghum landraces in Ethiopia and by Sadiki et al. (chapter 3) of faba bean landraces in Morocco have shown

Table 2.3. Indicators proposed for monitoring.

Proposed Indicator	Validity and Interpretive Issues	Lowest Applicable Level or Unit	Able to Combine to Higher Levels*
Domesticated plants and harvested wild species			
Number, frequency, and area of distinct landraces or harvested wild populations	Are names reliable? How does a specific landrace vary genetically in space and time?	Field or parcel	++
Environmental amplitude of area devoted to each crop	Does genetic diversity relate to abiotic and biotic environmental diversity, and on what scale? What is the relationship between occurrence and productivity?	Region	++
Number, durability, and evolution of farmer management and selection criteria	Do diverse criteria and diverse uses lead to genetic diversity?	Farm	+
Security of traditional knowledge	At what levels are there relationships between diversity and knowledge?	Administrative district	–
Wild species and crop relatives			
Species presence in designated areas that cover environmental range	What are the relative geographic locations, management, and benefit-sharing policies of the specified areas?	Natural resource administrative district	+
Population numbers and sizes	How does census size relate to durability? What are minimum viable sizes?	Metapopulation (valley)	++
Gene diversity, population divergence, and distribution	What is the relationship between genetic information and strategy?	Population	++

Table 2.3. continued

Proposed Indicator	Validity and Interpretive Issues	Lowest Applicable Level or Unit	Able to Combine to Higher Levels*
Links between in situ and ex situ activities			
Ex situ backup samples for vulnerable in situ populations; con served in situ sites for recalcitrant species	Scale of sampling, replenishment, and use strategies.	Individual collections	++
Cooperative links between ex situ institutions and farming communities	Information and seed exchange protocols, benefit-sharing, technology transfer.	National programs	–

Source: Modified from Brown and Brubaker (2002).
*Refers to whether the value for an indicator at higher levels (e.g., village level) can be derived from its value at lower level (e.g., farm level) by appropriate averaging.

that traditional knowledge is remarkably reliable. Presumably, the knowledge, recognition, and naming of farmers' crop diversity is crucial to their subsistence.

If appropriate baseline information is available, the percentage area occupied by traditional landraces is likely to be an important indicator of change in on-farm genetic diversity in any area. Surveys of individual landrace occurrence and frequency can yield huge amounts of data. However, these variables can be combined into summary measures. One example is the simple classification of rice landraces in Nepal as to frequency (present on few or many farms) and area planted (in large stands or as few plants in small fields or gardens) into four classes (table 2.4). Just as comparison between three regions is possible (in this example, the intermediate Kaski site has a higher component of rare, restricted landraces), so can one compare trends over time and assess differences in vulnerability, usage patterns, conservation strategy, and participatory plant breeding (PPB) options for each class.

Turning to medicinal, fuel, and other species harvested or grazed directly from the wild, the census of population numbers and sizes is an essential tool. Local communities have a direct interest in implementing

Table 2.4. Number and distribution of rice landraces in three in situ study sites, Nepal.

	Bara (80 m asl)	Kaski (650–1,200 m asl)	Jumla (2,200–3,000 m asl)
Mean area per landrace (ha)	0.95	1.17	0.91
Total number of landraces	33	63	23
Kinds of landraces			
Large area, many households	9	9	4
Large area, few households	2	3	0
Small area, many households	3	3	3
Small area, few households	19	48	16

Source: Data by Joshi et al., as summarized in Jarvis et al. (2000:83–85).

conservation plans to maintain these plants. Yet immediate need—particularly in harsh times—brings overexploitation. Addressing the numerical decline in populations of both highly prized and neglected or underused species is an obvious focus for conservation strategies, measurable by such indicators.

In theory, population size should relate to genotype richness: bigger populations or bigger samples should include more genotypes. If this relationship is general, then size (crop) area could be a quick way to gauge richness in a region, where it is impossible to investigate genetically.

A step beyond mapping landrace cultivation is relating such maps to climate, topography, and soil maps to measure the diversity of the environments that they occupy in sum. Examples of this kind of indicator used for the erosion of natural vegetation are the maps of increasing land clearing over time in the cereal belt of western Australia. The integrative tools of geographic information systems (GISs) (Guarino et al. 2002) offer the ability to estimate patterns of diversity and to monitor changes in the area devoted to landraces and changes in the distribution and size of populations of useful wild species so as to determine whether specific habitats are losing diversity.

Such information will be more useful with supporting research on the link between environmental divergence and genetic diversity. This link is not always straightforward and should be a focus of research. As Teshome et al. (2001) pointed out, one landrace may have high adaptability and yield well in many different environments, in many kinds of habitat. Its widespread use could be a deliberate choice of farmers from known performance rather than the unavailability of other varieties. In such a case, the wide adaptability of that population is of great value and not to be discounted because of the apparent lack of landrace richness. Conclusions also will be firmer if GIS data based on landrace occurrence in the region are related to performance. For example, if the landrace crop is mapped in a marginal area early in the growing season but that crop subsequently fails, then the early mapping of its presence is no longer evidence of its sustainable maintenance in that area.

As mentioned earlier, investigating the basis of the farmer's choices is a clear route to understanding how diversity is being maintained. Uses are selective forces, and, as with names, there may be much variation between farmers and between years as to the purpose of growing some landraces. Taba's (1997) survey showed that in Argentina, some farmers grew 13 of the 16 landraces of maize primarily for grain (table 2.5). But others grew the same 13 plus the 3 extra, for a total of 31 additional primary uses, 24 secondary uses, and 13 tertiary uses. Overall, these diverse uses suggest a multiniche model of diversifying selection that fosters adaptive diversity (Crow and Kimura 1970:262; Gillespie 1998:71). Aside from different culinary uses, we can add as uses the choice of genotypes that farmers make for specific environmental reasons (e.g. varieties that are known to cope with stressed patches of farmland or varieties chosen for waterlogged parcels).

One problem is that multiple use of a set of crop landraces in itself may not ensure diversifying selection. For example, if a particular new variety serves several purposes well, it could become widely planted and push out more specialized types. However, an overall rapid decline in the value of

Table 2.5. Number of different specific primary, secondary, and tertiary uses and culinary purposes specified for the landraces of maize grown in various countries.

Country	Number of Landraces	Number Used for Grain	Additional Specific Primary Uses	Secondary Uses	Tertiary Uses
Argentina	16	13	31	24	13
Bolivia	42	All	8	10	2
Chile	13	3	5	4	2
Mexico	12	All	5	11	3

Source: Summarized from Taba (1997).

statistics that measure kinds of use is likely to signal a loss of diversifying selection, the prelude to a loss of diversity.

Clearly, simply counting different selection criteria is only indicative. The presumed relationship between the diversity of reasons behind farmers' choices and actual genetic diversity must be tested. At present this can be attempted at the named landrace level.

Finally, the maintenance of genetic diversity on farm is more likely if mechanisms are in place to stop the erosion of traditional knowledge and to share the benefits that follow from exploitation of that diversity in exotic locales. A variety of approaches are needed to determine what traditional knowledge is being maintained and by whom. These approaches may provide the basis for indicators of the security of traditional knowledge. The processes that affect traditional knowledge are hard to measure; indeed, we are just beginning to address these concerns. A further problem is the separation in time between a farmer's decision to grow diverse populations today and the reaping of some remotely possible benefit from using that material elsewhere in the future. Today's benefit springs from yesterday's decision and is probably only a weak incentive for continued planting of diversity today. For such reasons, this indicator is likely to be most applicable at national or regional levels.

Wild Species and Crop Relatives in Agricultural Areas

The previous section discussed indicators for managing on-farm diversity in domesticated species. However, as noted at the beginning of the chapter, managing agricultural biodiversity in agricultural areas also includes wild species. The need to include wild crop relatives in our discussion arises from their several links with cultivated species. There are ecological linkages when agriculture directly leads to wild habitat damage or loss. Wild relatives often are the weeds of farmers' fields. Crops and their relatives share beneficial insects and microbes, pests, and diseases, leading to complex coevolutionary linkages. In addition, wild relatives may serve as sources of new and useful genes (Jarvis and Hodgkin 1999). Therefore it is important to extend conservation concerns to the wild species in agricultural systems. Indeed, wild species may be indicators of serious changes in production systems. Therefore indicators of genetic management of wild plant species are needed.

Monitoring the situation for wild plant species in agricultural areas is challenging. The major questions include whether certain species merit priority and whether populations in reserved areas or regions inhospitable to agriculture make up for vulnerable or decimated populations in rural areas. As to species priority, Brown and Brubaker (2002) argued that wild crop relatives provide an appropriate focus because they can be used as flagships and because they are hosts for the same pests and diseases as their related crop. The future of populations of such relatives that occur beside farmers' fields may be tenuous unless farmers deliberately foster them. This certainly occurs for the wild progenitors of certain crops such as corn but is unlikely to happen for more distant relatives. These feral populations do not lend themselves easily to specific management for conservation purposes, hence the importance of wild populations in specified reserved areas (such as the Sierra de Manantlan Biosphere Reserve, Mexico, designated for *Zea diploperennis* but managed also as a productive area).

Because conservation agencies are assembling data on the threatened status of many flora, it is possible to extract broad-scale information for species related to crops. For example, Brown and Brubaker (2002) summarized the conservation status of the wild species of crop genera that are native to Australia. This list revealed two main features: More than half of the crop-related taxa thought to be at risk were classed as "too poorly known" to assess their status, and only about 20% of crop relatives at risk could be confirmed as occurring in protected areas. It forms a challenge for government conservation policy to improve such measures.

However, we recognize that lists of species present on such broad geographic scales provide only a rough overview over substantial areas. They convey no details as to how precarious such species are or whether the conserved areas adequately represent the species. A reliable interpretation is contingent on information about management plans and benefit sharing in the rural setting (e.g., whether farming practices such as herbicide use threaten populations of wild relatives, whether farmers have access to the benefits of such populations). Broad-scale indicators alone are inexact and insensitive to change. Furthermore, it is possible for national programs to improve such statistics numerically but still mask genetic erosion through loss of habitat in the rural landscape.

As with domesticates, the listing of population numbers and their size distribution is a more refined dataset, possible to implement at lower scales than species presence and therefore more revealing of trends. Rocha et al. (2002) provided a detailed example for wild lima bean populations in Costa Rica. Rare and endangered species are amenable to this approach, as is evident in the population biology literature on such species. Such data may hide the following concerns: How does size actually relate to persistence beside farmers' fields, and how does it relate to genetic diversity? How do we assess the minimum sizes, population numbers, and distribution needed? Combining within similar species may be straightforward, but how are population numbers and sizes of a woody perennial to be combined with those of an ephemeral? Seed bank dynamics also must be monitored.

Wild relatives continue to attract studies of their population genetic structure using both the newer molecular methods and tests of disease resistance and environmental stress tolerance. Such data can be used to infer the proportion of diversity (allele richness or heterozygosity) conserved by various strategies or to follow trends over time. Perhaps more fundamentally, they reveal major features of the genetic system, such as cryptic polyploidy, breeding system variation, or impoverishment of self-incompatibility alleles in *Rutidosis leptorrhynchoides* (Young et al. 2000), which population sizes alone may not display. Also, under mating systems of predominant self-fertilization or apomixis, populations are likely to differ in their level of polymorphism. Newly founded, widely dispersed, colonial populations are likely to be less variable than populations at the center of origin. In the context of rural landscapes, genetic studies on populations of wild relatives sampled from farmers' fields would be instructive as to comparative responses to shared biotic and abiotic pressures.

Links Between In Situ and Ex Situ Actions

Programs for conserving and using genetic resources in situ probably will involve actions ex situ if they are to be resilient to fluctuations and upheavals in the field. Extreme examples of the need for action in both domains are catastrophes such as wars that lead to a loss of varieties in situ or major losses of viability occurring in genebank samples. Thus there is a need to monitor the links between in situ (on farm) and ex situ (large genebanks) tactics for coordinated action. The question is how the activities will interrelate and what additional indicators are needed to measure progress toward maintaining diversity.

Sample complementarity is one level that could be assessed. Of the populations in situ known to be at risk, what fraction has backup samples in ex situ collections? Likewise, of the samples that are recalcitrant to seed banking, lose viability quickly, or are expensive to regenerate, how many have secure in situ sources?

Joint strategies place a premium on collaboration between institutions and farming communities. If they are to endure, these links must grow. Mechanisms for the exchange of information and technology and agreed plans for seed exchange and benefit sharing will cement these links.

Research Challenges and Development Opportunities

The better we understand the forces acting on variation in situ in farmers' fields, the better will be our monitoring and management. Figure 2.1 is a diagram of the key features of a scheme to relate research and development efforts. The focus is on understanding and making use of the evolutionary forces that affect on-farm diversity. In the two central boxes are the evolutionary forces that act in situ at the level of population genetic structure. On the left are the forces that are nonselective structural population processes

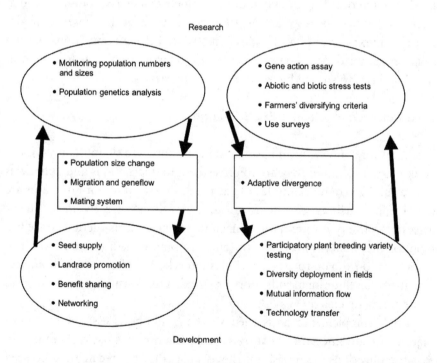

FIGURE 2.1. Research and development opportunities in relation to population genetic processes. Research foci *(upper left and right)* and options for management and development action *(lower left and right)* that target the two groups of forces affecting diversity on farm.

(e.g., population size and fluctuation, migration, mating system, and re-combination) and act on the whole genome, whereas on the right are the selective forces (including environmental forces and farmers' choices) that relate to gene function. The division is to some extent arbitrary because selection—the principal agent of adaptive evolution—acts through differ-ential survivorship, reproduction, and recombination. Yet the separation helps to arrange the research and management options and opportunities and to display their relationships.

Research thrusts are listed in the upper two balloons, divided accord-ing to the group of forces operating and the genetic changes they elicit. In the case of population forces on the left half of the diagram, the research yields data on numbers of populations and allele frequencies of marker genes. The related development actions are shown in the lower balloon (e.g., seed supply systems, landrace promotion schemes). They are strate-gies that aim to counter unwanted changes in population numbers and gene flow. The experience of studying or conducting these actions can be used to guide further research.

The right half of the diagram focuses on the adaptive divergence and functional diversity of landrace populations. Research strategies include analyzing gene action in microarrays, testing germplasm responses to biotic and abiotic stress, and surveying farmer selection criteria in rela-tion to diversity. The development actions that promote diversity and are relevant here are participatory variety testing and plant breeding and improvements in the deployment of diversity in fields and the flow of in-formation and technology to improve farmer selection.

The division between the two kinds of forces and attendant activities is perhaps strained because interactions and links abound. For example, suc-cessful participatory breeding schemes open up issues of landrace promo-tion, benefit sharing, seed supply, and spread of their products (e.g., the adoption of the rice cultivar Kalinga III in northwest India, asanalyzed by Witcombe et al. 1999). Yet when we consider how these actions will affect diversity, it is helpful to classify them according to whether their main ef-fect is on the diversity of the whole genome through population structure or on specific genes through selection regimes.

How might development action guide further research? One example is the issue of the impact of PPB on on-farm diversity. The question of whether PPB leads to an increase or a decrease in diversity is an important one for which there are too few data. Early results for rice cultivars in Nepal (Sthapit and Joshi 1998) show an encouraging rapid increase in the number of landraces for farmers in marginal upland areas that arise from PPB. However, the generalizability of this result, as well as indirect effects on local agrobiodiversity and its management, warrant investigation.

A multiplicity of uses is one well-known key factor that increases the likelihood of conservation of diverse local varieties. Many examples include sorghum in Ethiopia (Teshome et al. 1999) and maize in Latin America (Taba 1997) (table 2.5). Diversity of uses therefore is a key topic for research and documentation. Such information provides a baseline of indigenous knowledge that can reinforce diversifying selection regimes by promoting selection for different uses. It can be the basis of molecular genetic studies of the diversity of use-related traits.

Conclusion

Genetic diversity merits special focus in the management of agrobiodiversity because this is the very resource we are attempting to husband. Despite the challenging complexity of measuring it in an agricultural setting, we must know whether genetic erosion is effectively being slowed or whether it is gaining pace.

Genetic diversity is important from two perspectives: the population structural aspects, which are reflected in marker gene monitoring and show the history and present presumed health of the system; and the functional side, providing current adaptation to environmental diversity and extremes and providing the raw variation for future needs.

We cannot view genetic diversity as an amorphous, undifferentiated entity for which it suffices to have a certain quantity. We must identify change that matters. How do those concerned with on-farm conservation handle change, and what sort of indicators would be best for differentiating change that matters (erosion or loss) from change that is just a feature of any dynamic farm system? Indicators for monitoring the management of genetic diversity should track both population genetic structure and functional diversity.

Holistic approaches to agrobiodiversity conservation and development rightly stress an ecosystem perspective, extended to include human community betterment. Such approaches account for genetic erosion and species endangerment in general terms. However, they run the risk of assuming that developing the agroecosystem more sustainably will automatically stall the loss of genetic diversity and maintain underused species. Indeed, some researchers ask, why track genetic diversity at all? This question is particularly pertinent, given the increasing power of modern molecular technology to reveal our genetic heritage at its most detailed level. This chapter sought to clarify the task of conserving plant genetic diversity and suggests how we might monitor progress toward better conservation outcomes. Strategies must address both species that are cropped or harvested and wild species that occur in the ecosystem. In several cases, the wild species deserving particular attention are evolutionary relatives of crop species. Recent data on SNPs emphasize the extent of diversity at the gene level and great differences between species. An increasing number of new estimates of genetic diversity, and their patterns of variation between farms and of changes over time, confront the manager. Crucial indicators can be swamped easily. The nature of population divergence is a key characteristic encompassing issues ranging from the recognition and naming of landraces to the focus of farmers' selection criteria.

Acknowledgments

This work is a result of the IPGRI-supported global project "Strengthening the Scientific Basis of In Situ Conservation of Agricultural Biodiversity On-Farm." The authors would like to thank the governments of Switzerland (Swiss Agency for Development and Cooperation), the Netherlands (Directorate-General for International Cooperation), Germany (Bundesministerium für Wirtschaftliche Zusammenarbeit/Deutsche Gesellschaft für Technische Zusammenarbeit), Canada (International Development Research Centre), Japan (Japan International Cooperation Agency), Spain, and Peru for their financial support.

References

Aharoni, A. and O. Vorst. 2001. DNA microarrays for functional plant genomics. *Plant Molecular Biology* 48:99–118.

Brown, A. H. D. and C. L. Brubaker. 2000. Genetics and the conservation and use of Australian wild relatives of crops. *Australian Journal of Botany* 48:297–303.

Brown, A. H. D. and C. L. Brubaker. 2002. Indicators for sustainable management of plant genetic resources: How well are we doing? In J. M. M. Engels, V. Ramanatha Rao, A. H. D. Brown, and M. T. Jackson, eds., *Managing Plant Genetic Diversity*, 249–262. Wallingford, UK: CAB International.

Bryan, G. J., P. Stephenson, A. Collins, J. Kirby, J. B. Smith, and M. D. Gale. 1999. Low levels of DNA sequence variation among adapted genotype of hexaploid wheat. *Theoretical and Applied Genetics* 99:192–198.

Charlesworth, D. and J. R. Pannell. 2001. Mating systems and population genetics structure in the light of coalescent theory. In J. Silverton and J. Antonovics, eds., *Integrating Ecology and Evolution in a Spatial Context,* 73–95. Oxford, UK: Blackwell.

Clegg, M. T. 1997. Plant genetic diversity and the struggle to measure selection. *Journal of Heredity* 88:1–7.

Crow, J. F. and M. Kimura. 1970. *An Introduction to Population Genetics Theory.* New York: Harper & Row.

Frankel, O. H., A. H. D. Brown, and J. J. Burdon. 1995. *The Conservation of Plant Biodiversity.* Cambridge, UK: Cambridge University Press.

Gillespie, J. H. 1998. *Population Genetics: A Concise Guide.* Baltimore: Johns Hopkins University Press.

Guarino, L., A. Jarvis, R. J. Hijmans, and N. Maxted. 2002. Geographic information systems (GIS) and the conservation and use of plant genetic resources. In J. M. M. Engels, V. R. Rao, A. H. D. Brown, and M. T. Jackson, eds., *Managing Plant Genetic Diversity,* 387–404. Wallingford, UK: CAB International.

Hamrick, J. L. and M. J. W. Godt. 1997. Allozyme diversity in cultivated crops. *Crop Science* 37:26–30.

Jarvis, D. I. and T. Hodgkin. 1999. Wild relatives and crop cultivars: Detecting natural introgression and farmer selection of new genetic combinations in agroecosystems. *Molecular Ecology* 8:S159–173.

Jarvis, D. I., L. Myer, H. Klemick, L. Guarino, M. Smale, A. H. D. Brown, M. Sadiki, B. Sthapit, and T. Hodgkin. 2000. *A Training Guide for In Situ Conservation On-Farm.* Rome: IPGRI.

Lin, J. Z., P. L. Morrell, and M. T. Clegg. 2002. The influence of linkage and inbreeding on patterns of nucleotide sequence diversity at duplicate alcohol dehydrogenase loci in wild barley (*Hordeum vulgare* ssp. *spontaneum*). *Genetics* 162:2007–2015.

Peacock, J. and A. Chaudhury. 2002. The impact of gene technologies on the use of genetic resources. In J. M. M. Engels, V. R. Rao, A. H. D. Brown, and M. T. Jackson, eds., *Managing Plant Genetic Diversity,* 33–42. Wallingford, UK: CAB International.

Rafalski, J. A. 2002. Novel genetic mapping tools in plants: SNPs and LD-based approaches. *Plant Science* 162:329–333.

Rocha, O. J., J. Degreef, D. Barrantes, E. Castro, G. Macaya, and L. Guarino. 2002. Metapopulation dynamics of lima bean (*Phaseolus lunatus* L.) in the central valley of Costa Rica. In J. M. M. Engels, V. R. Rao, A. H. D. Brown, and M. T. Jackson, eds., *Managing Plant Genetic Diversity,* 205–215. Wallingford, UK: CAB International.

Sthapit, B. R. and K. D. Joshi. 1998. Participatory plant breeding for in situ conservation of crop genetic resources: A case study of high altitude rice in Nepal. In T. Partap and B. Sthapit, eds., *Managing Agrobiodiversity,* 311–328. Kathmandu, Nepal: International Centre for Integrated Mountain Development.

Taba, S. 1997. Maize. In D. Fuccillo, L. Sears, and P. Stapleton, eds., *Biodiversity in Trust,* 213–226. Cambridge, UK: Cambridge University Press.

Tenaillon, M. I., M. C. Sawkins, A. D. Kong, R. L. Gaut, J. F. Doebley, and B. S. Gaut. 2001. Patterns of DNA sequence polymorphism along chromosome 1 of maize (*Zea mays* ssp. *mays* L.). *Proceedings of the National Academy of Science USA* 98:9161–9166.

Teshome, A., A. H. D. Brown, and T. Hodgkin. 2001. Diversity in landraces of cereal and legume crops. *Plant Breeding Reviews* 21:221–261.

Teshome, A., L. Fahrig, J. K. Torrance, J. D. Lambert, J. T. Arnason, and B. R. Baum. 1999. Maintenance of sorghum (*Sorghum bicolor,* Poaceae) landrace diversity by farmers' selection in Ethiopia. *Economic Botany* 53:69–78.

Weltzien, E. and G. Fischbeck. 1990. Performance and variability of local barley landraces in Near-Eastern environments. *Plant Breeding* 104:58–67.

Witcombe, J. R., R. Petre, S. Jones, and A. Joshi. 1999. Farmer participatory crop improvement. IV. The spread and impact of a rice variety identified by participatory varietal selection. *Experimental Agriculture* 35:471–487.

Young, A. G., A. H. D. Brown, B. G. Murray, P. H. Thrall, and C. H. Miller. 2000. Genetic erosion, restricted mating and reduced viability in fragmented populations of the endangered grassland herb *Rutidosis leptorrhynchoides.* In A. G. Young and G. M. Clarke, eds., *Genetics, Demography and Viability of Fragmented Populations,* 334–359. Cambridge, UK: Cambridge University Press.

Zhu, Y. L., Q. J. Song, D. L. Hyten, C. P. VanTassel, L. K. Matukumalli, D. R. Grimm, S. M. Hyatt, E. W. Fickus, N. D. Young, and P. B. Cregan. 2003. Single-nucleotide polymorphisms in soybean. *Genetics* 163:1123–1134.

3 🌱 Variety Names

An Entry Point to Crop Genetic Diversity and Distribution in Agroecosystems?

M. SADIKI, D. JARVIS, D. RIJAL, J. BAJRACHARYA, N. N. HUE,
T. C. CAMACHO-VILLA, L. A. BURGOS-MAY, M. SAWADOGO, D. BALMA,
D. LOPE, L. ARIAS, I. MAR, D. KARAMURA, D. WILLIAMS,
J. L. CHAVEZ-SERVIA, B. STHAPIT, AND V. R. RAO

The names farmers give to their traditional varieties or landraces are fundamental to their very essence and use. Harlan (1975) discusses how landraces are recognizable morphologically, farmers have names for them, and different landraces are understood to differ in adaptation to soil type, time of seeding, date of maturity, height, nutritive value, use, and other properties. Many studies have pointed out how farmers recognize and name populations of the crops they grow according to their agromorphological, ecological–adaptive, quality, and use characteristics (Boster 1985; Quiros et al. 1990; Bellon and Brush 1994; Teshome et al. 1997; Schneider 1999; Soleri and Cleveland 2001). Yet, in contrast to this wealth of literature on how farmers name their varieties, there has been little systematic study on the consistency among farmers within and between villages on these names and the descriptive traits they give to their varieties. Even less attention has been given to whether the units of diversity that farmers manage actually have specific names attached to them that can be compared (Jarvis et al. 2000).

Only recently have studies begun to investigate whether landrace names can actually be used as a basis for arriving at estimates of local crop diversity on farm. Moreover, the question still arises of whether named varieties are the identifiable units of diversity that farmers manage. Whether these identified farmer's units of diversity management are clearly genetically distinct and form genetically identifiable populations at agromorphological, biochemical, or molecular levels is a concern in ensuring the availability of useful diversity. If farmer-named varieties are not genetically distinct, then the usefulness of names as a means to identify and quantify the diversity in agricultural ecosystems is limited. In contrast, if farmer-named varieties are genetically distinct, then structured sampling based on names can be used to assess the amount of diversity on farm, and genetic relationships can be established between varieties. Additionally, information on distinctiveness of farmers' varieties bearing similar names would help in preserving and using this diversity because it might be lost if such varieties are discarded because they bear the same names.

The distribution of variety names within and between communities and regions furnishes the raw data for estimating richness and evenness of diversity across the landscape. Diversity levels may vary dramatically from one local crop variety to another. Identifying which varieties or traits are rare or common in any given population or landscape helps us understand how diversity is distributed across the agricultural landscapes. Farmers' understanding and beliefs of the production spaces they manage influence the patterns of diversity. If diversity is an answer to support farmers in their agricultural needs, then identifying areas of high diversity is important. This will include identifying areas where local crop varieties have adaptive capacity to particular environmental parameters. It is in these areas that diversity has a role to play in the use of sustainable agriculture.

In chapter 2, Brown and Hodgkin discussed recent molecular advances to quantify the amount of diversity on farm. This chapter presents recent empirical studies on farmers' fields and laboratory assessments of genetic distinctiveness, consistency, and distribution of the local crop varieties maintained on farm.

Names as an Indicator of Diversity

Farmers use many of the phenotypic features of plants to identify and select their crop varieties. These agromorphological criteria may take a wide range of forms and usually are linked to the genetic makeup of a crop. They are used by farmers to distinguish and name crop varieties and often are the basis for farmers' selection of planting seed. In assessing crop diversity maintained on farm, it may be important to distinguish the names farmers give to their varieties from the agromorphological traits they use to identify and select varieties, the traits they use to value varieties, and the traits they use to select seeds or propagules for the next generation. For instance, a farmer may identify a named variety of maize by its color, leaf shape, and region of origin, value it for its cooking quality, and select in the progeny for pure-to-type seeds. Work on taro (table 3.1) illustrates how the literal meaning of names can differ from the descriptors used by farmers to describe their named taro varieties (*Colocasia esculenta*) in Begnas, Nepal. The household survey revealed that the farmers used at least 15 key descriptors to distinguish their taro varieties.

In Hungary, Mar and Holly (2000) reported that farmers' names for local common bean varieties were correlated to some of the crop's agromorphological traits, especially seed color (e.g., *fehérbab* = white bean, *feketebab* = black bean, *barnabab* = brown bean), and in several cases referred to the traditional use of these local varieties (*menyecskebab* = young wife's bean). Studies carried out in the Yucatán, Mexico, to understand how farmers name and manage maize varieties showed that growth cycle was the primary trait for distinguishing varieties, followed by the form and color of the cob and grain. Key traits used to distinguish populations within named varieties were the form of the husk, plant height, and grain size. Husk form ranked very high as a distinguishing trait because the husk protects the grain from pests during storage. Insufficient husk around the larger cobs of improved varieties has limited the acceptance of improved varieties in the area (table 3.2; Arias et al. 2000; Chavez-Servia et al. 2000; Arias 2004; Burgos-May et al. 2004; Latournerie Moreno et al. 2005).

Farmers in Uganda give names to banana cultivars using one or more of the characters expressed in their locality and also the characters important to them and other consumers. Karamura (2004) classified highland banana clones using characteristics important to both farmers and consumers into five clusters that share characters. For example, clones in one cluster sucker profusely, mature quickly, and produce soft-textured food; in another cluster the clones are slow in production of suckers, take a long time to mature, and produce hard-textured food. The characters used by the farmers—particularly those expressed in their locality—are consistent to a large extent, though not completely. The characters important to farmers and consumers are based on long-term experience, and selection for similar traits could have been practiced over generations.

Studies on maize in Mexico and faba bean in Morocco show that farmers may emphasize traits to distinguish populations within varieties that are different from traits used to distinguish between varieties. For both maize and faba bean, morphological traits are important for

Table 3.1. Description of farmer-named cultivars of taro (*Colocasia esculenta*), Begnas, Nepal.

Local Name	Botanical Name	Literal Meaning of Farmers' Descriptors	Distinguishing Morphological Characteristics
Bhaishi khutte	Var. *esculenta*	Multiple corms like buffalo footprints, annual; unbranched corm; many buds, few cormels; cup-shaped leaf; morphotype similar to *hattipow, panchamukhe seto, panchamukhe*	Flat and multicorm types, slow and late leaf senescence, white bud
Chhattre	Var. *antiquorum*	Leaf shaped like umbrella; long and green leaf color; red bud with round corm	Dumbbell corms with pink bud, pink skin, and conical cormels
Chhaure	*C. esculenta*	Puppies; multicormel types like puppies	Long cormels with red bud; round corm
Dhudhe karkalo	*C. esculenta*	Milky white petiole, bud, and sap color and thick plant; no corm but profuse root system; round leaf; adapted to home gardens	Multicorm type, no cormel, cylindrical corm, and cup-shaped leaf
Gante	*C. esculenta*	Short; petiole black, branching corm, and large cormel; red bud, petiole, and sheath; round, small corm	Dumbbell-shaped corm with round corms
Hattipow	Var. *esculenta*	Corm shaped like elephant foot; tall and thick plants, whitish and broad leaves; large multicorms with depressed bud; light green petiole; rough (*jerro*) leaf; few cormels; adapted to open field	Flat and multitype corm, slow and late leaf senescence with white bud

Table 3.1 Continues next page

Table 3.1. continued

Local Name	Botanical Name	Literal Meaning of Farmers' Descriptors	Distinguishing Morphological Characteristics
Kaat	*Var. esculenta*	Easy cooking in Gurung dialect; soft and round leaf shape, excellent cooking quality, similar to *rato panchamukhe*	Red buds; many buds
Khari chhoto	*C. esculenta*	Short corm; pink petiole; long corm size, similar to *thagne khari*, *thangne*, *khari pindalu*	Corm grown upright; taro covered by feathery sheath
Khari pindalu	*C. esculenta*	Cylindrical corm	
Khujure	*Var. antiquorum*	Multicormels; with many small cormels; itchy corms; many cormels with white buds, white petiole; dark purple petiole junction; purple leaf margin; round leaf	Round corm and cormels with white buds; corms are acrid
Khujure kalo	*Var. antiquorum*	Black multicormels; petiole black, corm and cormels both edible; nonitching; white bud; many cormels; petiole purple (*kalo*); long leaf	Branching corm, round with white bud
Khujure seto	*Var. antiquorum*	White multicormels; petiole black, corm and cormels itching; many cormels, plenty of black petiole, purple leaf margin	Branching, corm and cormels round and very small with white bud

Panchamukhe	Var. esculenta	Five-faced white corm; tall and thick plants, whitish and broad leaves; looks like hattipow; many buds with depressed buds; large corm size; rough leaf (jarro) with tall plant; thick vein	Flat and multicorm type; without cormels; white bud and slow and late senescence
Panchamukhe seto	Var. esculenta	Five-faced white corm; tall and thick plants, whitish and broad leaves; multicorm types with white bud color; light green petiole; similar to panchamukhe, bhaishi khutte, hattipow	Clustered corm without cormels; white bud; slow and late senescence
Rato or raate	Var. antiquorum	Red; purple petiole	Red seminal roots with white buds, curbed peduncle at petiole junction
Rato mukhe	Var. antiquorum	Red corm; red bud with large and round corm and cormels	Base of petiole pink; leaf peduncle curbed, thick leaf blade; red roots
Rato thado	Var. antiquorum	Red corm grown upright; white buds; tall plant	Purple point at dorsal side of petiole junction; upright growth of corm
Satmukhe	Var. esculenta	Seven-faced corm; morphotypes similar to khari chhoto, thado mukhe, thagne	Multicorm type covered by feather-like stuff (bhutla)
Thado	Var. esculenta	Upright growth of the corm	Cylindrical corm, nonbranching
Thagne khari	Var. esculenta	Old clothes; feather-like tissues covering the corms (thagne)	Purple point at petiole junction, large cormels

Sources: Rijal et al. (2003); focus group discussions, Begnas, 2001.

Table 3.2. Traits used by farmers to distinguish maize varieties in Yaxcaba, Yucatán, Mexico.

Farmer-Named Varieties*	Ear				Cob (Pith)			Kernel		Stalk			Cycle
	Size	Shape	Color	Husk	Size	Color	Flex	Color	Shape	Height	Thickness	Color	Months
Xnuk nal kannal	Large	Long	Color	Thick, color	Thick, long, thin	Color		Yellow	Large	Tall	Thick	Purple	3.5–4.5
Xnuk nal saknal	Large	Thick	Color	Thick, long,	Thin, thick thin	Color		White	Large	Tall	Thick		3.5–44
Xbe ub	Large		Color	Many leaves	Thick, thin	Color	Thick, thin	Purple-black	Large, small	Tall			2.5–3.5
X-mejen kannal	Small	Round	Color	Thick	Small			Yellow	Small, hard	Short, tall			2–2.5
X-mejen nal, saknal	Small, regular		Color	Thick	Small, color			White	Large, shape	Tall, short			2–3
X-tup nal	Small with point		Color	Thick with many leaves	Thick				Small	Short			2–2.5
Ts'iit bakal	Small, large			Thin with points	Thin, long					Tall			3–3.5
Nal xoy	Large		Color	Spines, thick	Thick				Small	Tall			3–3.5

Source: Morales-Valderrama and Quiñones-Vega (2000), data analyzed by Claudia Ezyguirre, 2002.
kannal=yellow; *saknal*=white; *xhe ub*=purple.

distinguishing between varieties, whereas adaptive and use traits are used more commonly to distinguish populations of a single named landrace (table 3.3; see also Cazarez-Sanchez 2004 and Cazarez-Sanchez and Duch-Gary 2004 for nutritional and physical qualities of maize landraces and their association with specific dishes). Interesting is that precocity—or time to harvest—is extremely important in distinguishing maize varieties in the Yucatán, where shorter growth cycle is important for avoiding the drought periods, whereas in Morocco, the cycle is used only to distinguish populations within a variety.

Table 3.3. Comparison of some traits used by farmers to distinguish between and within varieties of faba bean in Morocco and maize in the Yucatán, Mexico.

	Morphology					Growth Cycle	Adaptation		Use	
	Pod Shape, Length, and Form (fb)	Seed Color	Seed Size	Leaf Shape (fb)	Husk Thickness (mz)	Time to Harvest	Resistance to Drought	Resistance to Sun	Cooking Ability or Ease	Ease of Grain Removal from Cob
Faba Bean										
Distinction between varieties	X	X	X							
Distinction within varieties (between seed lots)				X			X	X	X	
Maize										
Distinction between varieties	X	X	X			X				
Distinction within varieties (between seed lots)				X	X			X	X	X

Sources: Arias et al. (2000); Sadiki et al. (2001); Morales-Valderrama and Quiñones-Vega (2000).

One farmer's name for a crop variety may well be the same name that other farmers in the village give to the same variety, but as spatial scales increase these names may no longer be consistent with those in the next village. Evidence in Ethiopia has revealed different names for the same variety, reflecting emphasis on different qualities by different farmers or communities. An example is durum wheat in Ethiopia: In some villages a variety is called "white," whereas in others the same variety is called "early" (Tanto 2001). Tesfaye and Ludders (2003) found similar evidence in Ethiopia for ensete, a clonally propagated crop, for which a few landraces assumed different names at different locations.

Even within a village, different farmers may have different names for the same crop varieties. Sawadogo et al. (2005) measured the consistency of farmers within sites in Burkina Faso in naming sorghum varieties (table 3.4). Variety names are linked to plant morphology (height, shape, color, grain size, color and opening of glumes for cereals), agronomic behavior (growth cycle, flowering dates), adaptation to the environment (tolerance to drought, resistance to pests, disease, and birds, soil adaptation), and uses (freshly consumed, cooking quality, taste). Certain differences of variety names in the same village or community reflect differences in the languages used to name the variety. For example, variety *pokmi-*

Table 3.4. Consistency of local farmers' sorghum variety names, Burkina Faso.

Most Common Name Given to the Variety	Farmers Recognizing the Variety Under the Common Name in the Site Where Widely Grown (%)		Other Names Given to the Variety by Farmers in One or Both Sites (%)		
Name 1	Thiougou Site (6 villages)	Tougouri Site (6 villages)	Name 2	Name 3	Name 4
Kurbuli	100	5.55	0	0	0
Zugilga	0	100	33.34	0	0
Zuwoko	72.22	0	77.8	27.77	22.22
Fibmiugu	83.4	0	77.8	22.22	16.66
Z. fibsablega	100	16.70	83.4	0	0
Gambré	100	0	0	0	0
Z. wabugu	0	94.44	100	5.55	0
Balingpelga	0	100	100	0	0
Pokmiugu	5.55	77.8	94.44	22.22	0
Pisyobe	0	50	27.77	22.22	0
Zuzeda	0	72.22	27.77	22.22	0

Source: Sawadogo et al. (2005).

ugu of Thiougou is *fibmiugu* in Tougouri. *Pok* is *fiba* and means "glume" in Mooré, so *fibmiugu* is *pokmiugu*. Some varieties (*kurbuli* and *gambré*) are known in only one site, whereas varieties *zuwoko* and *de fibmiugu* were found and used in only one site but were known by farmers from the other site by three different names (*pokmiugu, banigpelega, fibmiugu*).

Agronomic Traits as an Indicator of Diversity

The level at which on-farm crop genetic diversity is assessed will provide different information on the quantity and type of diversity maintained. Counting the number of named varieties to assess variety richness over specified spatial and temporal scales has been used to give an indication of diversity maintained on farm for many major crops, including potatoes (Quiros et al. 1990; Brush et al. 1995; Zimmerer 2003), corn (Bellon and Taylor 1993; Bellon and Brush 1994; Louette et al. 1997), beans (Martin and Adams 1987; Voss 1992), cassava (Boster 1985), and sorghum (Teshome et al. 1997; see also chapter 2). Yet the question remains whether the use of local crop variety names correctly estimates local crop varietal diversity because farmers may not be consistent in naming and describing local crop varieties (Jarvis et al. 2004).

In Morocco, Sadiki et al. (2001, 2002) showed that farmers in different villages use different names to designate faba bean varieties described by the same set of seed and pod traits. Names and farmers' descriptions of local faba bean varieties in northern Morocco were collected together with seed samples from 185 randomly selected farms in 15 villages belonging to five communities of three provinces. The farmers were asked to list the names and describe the local types of faba bean varieties they know and grow. Characteristics of each cultivar were listed along with distinctive traits according to each farmer's statement. The consistency among farmers for naming the local varieties of faba bean was assessed by the percentage of farmers recognizing the same variety by the same name and description.

Table 3.5a shows that some varieties have different names, such as *foul sbaï lahmar, foul roumi,* and *lakbir lahmar* but are described by the same traits by farmers. In other instances, varieties such as *moutouassate labiade* are described differently by different farmers. Finally, other cases

Table 3.5a. Names and descriptions of farmers' varieties of faba bean cited and described during field surveys among the types they heard about, they know, they have seen, or they grow in Morocco.

Variety Name	Variety Code	Pod Length	No. of Seeds per Pod	Seed Size (mg/seed)	Seed Color	Pod Shape
Foul sbaï labiade	A	Long	7	Large (>1.5)	Light yellow	Flattened
Foul sbaï lahmar	B	Long	6–7	Large	Brown	Flattened
Foul roumi	C	Long	6–7	Large	Brown	Flattened
Lakbir lahmar	D	Long	6–7	Large	Brown	Flattened
R'baï labiade	E	Medium	4–5	Large	Light yellow	Flattened
R'baï laghlid lahmar	F	Medium	4–5	Large	Brown	Flattened
Khmassi laghlide khdar	G	Medium	4–5	Large	Green	Flattened
Foul beldi lou l'khal	H	Medium	4–5	Large	Violet	Flattened
T'lati laghlide beldi	I	Short	3	Large	Dark brown	Flattened
Beldi moutouassate labiade	J	Medium	4–5	Medium (0.8–1.5)	Light yellow	Flattened
Foul beldi aadi	K	Medium	4–5	Medium	Light yellow	Flattened
Moutouassate labiade	L	Medium	4–5	Medium	Light yellow	Cylindrical
Foul lahmar moutouassate	M	Medium	4–5	Medium	Light brown	Flattened
Foul moutouassate lou l'khal	N	Medium	4–5	Medium	Violet	Cylindrical
Moutouassate labiade	O	Medium	4–5	Medium	Brown	Cylindrical
Beldi moutouassate lakhdar	P	Medium	4–5	Medium	Green	Flattened
Beldi (A)	Q	Short	3	Medium	Light gray	Flattened

Table 3.5a. continued

Variety Name	Variety Code	Pod Length	No. of Seeds per Pod	Seed Size (mg/seed)	Seed Color	Pod Shape
Beldi (B)	R	Short	3	Medium	Light gray	Cylindrical
Beldi (C)	S	Short	3	Medium	Violet	Flattened
Foul r'guigue lahmar	T	Short	3	Small (<0.8)	Brown	Cylindrical
Filt labiade	U	Short	3	Small	Light gray	Cylindrical
Fouila baldia khadra	V	Short	3	Small (<0.8)	Green	Cylindrical
Foul bouzid s'ghir	W	Short	3	Small (<0.8)	Violet	Cylindrical
Lou l'khal s'ghir	X	Short	3	Small (<0.8)	Black	Cylindrical

were found in which the varieties were not given specific names but were designated by a generic name *beldi*, although farmers were able to distinguish different units within this *beldi* category without giving precise names. Consistency in names of faba bean varieties was noted among farmers of eight Moroccan villages in three different communities using a nonparametric correlation coefficient for pairs of villages based on chi-square (table 3.5b). An example is given for one faba bean variety; which shows that consistency decreases as geographic distance increases. Sadiki (unpublished data) also compared the consistency of variety names with sets of traits farmers used to describe varieties and found that sets of traits to describe a variety had much higher consistency over geographic areas than variety names. Consistency of variety names among farmers is highest between close villages (villages of the same community). The consistency index (correlation coefficient) decreases as geographic distance between villages increases, significantly more rapidly for names than for traits (figure 3.1), indicating that sets of agromorphological traits have the potential to be more consistent over geographic space than names.

In Hungary, Mar et al. (2004) found that, for common bean, in some cases farmers would distinguish different common bean varieties

Table 3.5b. Consistency of classification of faba bean variety *foul sbaï labiade* among farmers of 8 villages in 4 different communities computed as correlation coefficient for pairs of villages based on chi-square.

Province	Community	Village	Sidi Senoun	Bouajoul	Hadarine	El Jir	Aïn Barda	Ghiata-Al-Gharbia	Bou Rhoufa
Taounate	Ourtzagh	Ain Kchir	0.70	0.53	0.46	0.18	-0.03	0.06	-0.21
	Ourtzagh	Sidi Senoun		0.53	0.23	0.24	0.22	0.14	0.04
	Ghafsai	Bouajoul			0.63	0.11	0.07	0.30	0.09
	Ghafsai	Hadarine				0.23	-0.07	0.24	0.09
	Tammadit	El Jir					0.60	0.11	0.13
	Tammadit	Aïn Barda						0.07	0.23
Taza	Oued Amlil	Ghiata-Al-Gharbia							0.63

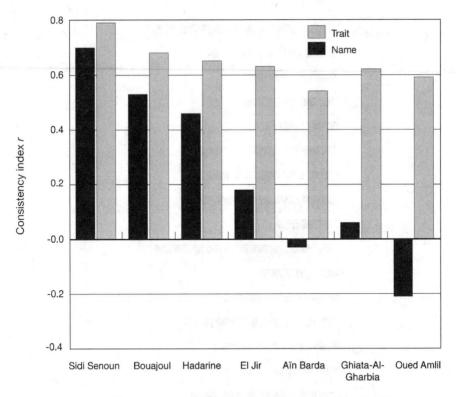

FIGURE 3.1. Comparison of consistency of names with consistency of traits among villages for the faba bean variety *foul sbaï labiade* based on consistency index (r). Coefficient of correlation between r (consistency index) and d (distance in km from Ain Kchir to other 7 villages) for names and traits = −0.537 and −0.173, respectively; degree of significance of correlation for names and traits = 0.002 and 0.280, respectively (M. Sadiki, M. Arbaoui, L. G. Houti, and D. Jarvis, unpublished data, 2004).

by agromorphological traits but gave only the generic name *beans*. This was similar to the barley-naming system in Morocco, where the majority of local varieties were called *beldi,* meaning "local," as opposed to modern introduced varieties. Nevertheless, farmers clearly identify and separate varieties based on seed, ear, and plant characteristics, straw yield and quality for animal feed, and often quality of flour (Rh'rib et al. 2002).

In contrast to faba bean, studies on durum wheat names in Morocco indicated that the farmers designate broad categories comprising different varieties or entities (Taghouti and Saidi 2002). This metaclassification

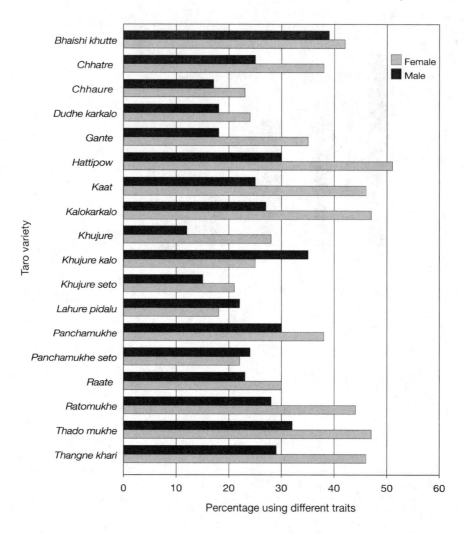

FIGURE 3.2. Descriptions of farmer-named taro cultivars by male and female farmers (percentage of men and women using different traits for each variety) (Deepak Rijal, unpublished data, 2004). The difference of frequency traits used by women to men was tested by Wilcoxon ($Z=-5.696$).

is based on ear characteristics, particularly color (black or white). Within each category, varieties share the same broad name but are distinguished by farmers by other traits, such as flour quality and plant height. In another case, alfalfa varieties in Morocco are generally named after their geographic origin. The names of alfalfa varieties derived from the same ecosite are generic and recall adaptation to local soil and climatic conditions. Two main groups are distinguished: *demante* (mountain area) and *rich* (oasis area). These two groups differ from each other in growth habit, growth speed after cutting, and winter hardiness. Within each group, farmers separate varieties on the basis of plant agronomic and morphological traits (Bouzeggaren et al. 2002).

If sets of traits are the unifying unit for recognizing varieties, do all farmers recognize the same local cultivar using the same traits? Can divergence of gender interests be correlated with the naming of diversity units and diversity patterns on farm? Investigations in the naming of local taro varieties in Nepal, shown in figure 3.2, indicate that women are more consistent than men in the traits they use to describe particular taro varieties. Farmers characterized 18 taro landraces against 24 descriptors related to corm (type, shape, size, growth), cormel (number, size), leaf (shape, size, texture, color), petiole (color, sheath color, number), plant height (short, medium, tall), and root system (profuse). Compared with male farmers, female farmers used the same descriptors more often across landraces and were more reliable in recognizing specific descriptors than male counterparts when asked to characterize landraces. Men often used only corm and shoot characteristics to distinguish varieties, whereas women used cormel, leaf form, and size and growth habits as additional descriptors.

A similar study was conducted in Vietnam, where study results showed that the consistency level among women (80.57–98.5%) was slightly higher than among men (78.2–94%) but not statistically significant in naming and describing 47 taro cultivars grown in seven sites across Vietnam (Canh et al. 2003; Hue et al. 2003).

Farmer Variety Names and Genetic Distinctiveness

The names or traits farmers use to distinguish their varieties may be consistent within and between villages, yet this does not address the question of the extent to which these farmer-named units are genetically distinct or at what level—agromorphological, biochemical, molecular—this distinction might be found.

Cluster analyses were performed on agromorphological data of named rice varieties from three sites at different elevations in Nepal (Bara, < 100 m; Kaski/Begnas, 600–1,400 m; and Jumla, 2,200–3,000 m) to assess the distinctiveness of these named varieties at the agromorphological level

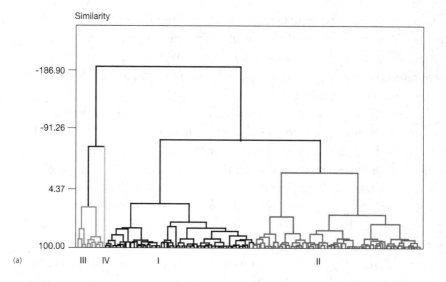

(a)

III IV   I   II

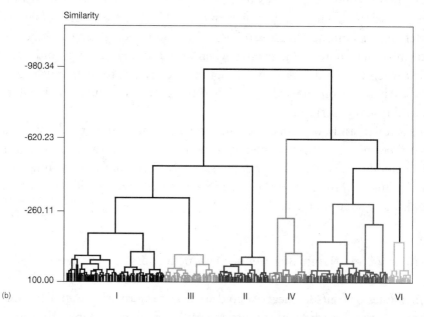

(b)

I   III   II   IV   V   VI

FIGURE 3.3. Dendrogram of rice accessions from three sites and check modern varieties, Nepal (Bajracharya 2003; Bajracharya et al. 2006). (a) Jumla site: Diversity of names but little genetic (morphological) diversity in landraces Figure 3(a) adapted from Bajracharya et al. 2006; awns and stigma color were distinguishing traits. (b) Kaski site: Identically named landrace populations clustered together, showing a high degree of consistency in names and agromorphological descriptions; a wide range of agromorphological variability was encountered; leaf and grain morphological traits were important. (c) Bara site: Clear clustering by farmers' units of diversity and agroecosystems encountered, showing high consistency in the names and agromorphological descriptions; quantitative traits explained 60% of total variation in the principal component analysis; growth duration traits were important.

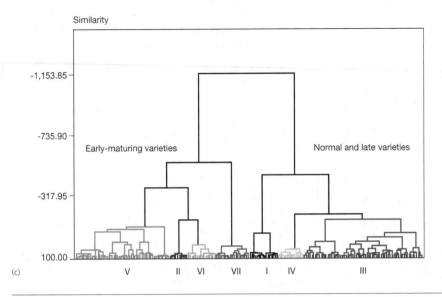

Similarity

-1,153.85

-735.90

Early-maturing varieties   Normal and late varieties

-317.95

100.00

(c)

V   II VI   VII   I   IV   III

FIGURE 3.3. *continued*

(Bajracharya 2003; Bajracharya et al. 2006; figure 3.3). At the two lower-elevation sites identically named landrace populations clustered together, showing a high degree of consistency in names and agromorphological descriptions, whereas in the high-elevation site of Jumla, although there were numerous names, little morphological diversity of measured traits was found.

Evaluation of taro morphology in Vietnam also revealed variation within similar names (Hai et al. 2003) and between the different named taro landraces (Hue et al. 2003; Tuyen et al. 2003). Work on rice cultivars in India and in the Cagayan Valley of the Philippines showed that samples with the same name often had a quite different genetic constitution at the biochemical and molecular levels (Pham et al. 1999; Sebastian et al. 2001).

Similar analysis of the genetic differences between farmers' faba bean varieties in Morocco—agromorphological evaluation using 10 consistently described varieties among the 14 grown in the region—revealed a large amount of phenotypic diversity in these variety types for most of the characteristics studied (Sadiki et al. 2002). Hierarchical cluster analysis and multivariate discriminant analysis revealed that the seed lots bearing the same name generally clustered together. These results agree with phenotypic description of the types by farmers, indicating that for these populations, the farmer-named units are distinct, and the traits have

Table 3.6. Percentage of accessions from each variety classified into classes defined by discriminant functions based on posterior probability of membership.

Original Classes (Farmer-Named Factorial Discriminant Analysis, (FDA)-Based Varieties) Name	Code	Classes (Clusters)*									
		1	2	3	4	5	6	7	8	9	10
Lakbir labmar	D	100	0	0	0	0	0	0	0	0	0
R'baï labiade	E	0	100	0	0	0	0	0	0	0	0
Foul beldi lou l'kbal	H	0	0	86	0	0	0	14	0	0	0
T'lati laghlide beldi	I	0	0	0	71	0	0	29	0	0	0
Beldi moutouassate labiade	J	0	0	0	14	86	0	0	0	0	0
Foul labmar moutouassate	M	0	0	0	0	0	100	0	0	0	0
Foul moutouassate lou l'kbal	N	0	0	0	0	0	0	100	0	0	0
Filt labiade	U	0	0	0	0	0	0	0	100	0	0
Foul r'guigue labmar	T	0	0	0	0	0	0	0	0	100	0
Foul bouzid s'ghir	W	0	0	0	0	0	0	0	0	0	100

Source: Sadiki et al., unpublished data.
*94% of the accessions were classified correctly by FDA-based analysis into their original varieties.

a genetic basis. Preliminary molecular marker–based genetic distances data using rapid amplification of polymorphic DNA confirmed that the differences between the seed lots of different types exceeded the differences between seed lots bearing the same name (Belqadi 2003; Benchekchou 2004). In Morocco the accession ranking pattern established for 10 local varieties based on phenotypic traits is very consistent with the farmers' descriptors of faba bean varieties (Belqadi 2003). Table 3.6 shows that 94% of the 70 accessions analyzed were correctly classified in their variety types based on similarities of agromorphological traits. Therefore, the phenotypic clustering pattern closely agrees with farmers' descriptions of the local varieties. The distinction of the varieties based on the phenotypic characters corresponds to the farmers' perceptions in designating the varieties. For these 10 varieties, farmers' diversity units coincide with the units of measured phenotypic diversity.

Acceptable Variability in Names and Genetic Distinctiveness

How large is the variability associated with a "consistent" name? In Uganda, new banana clones may come into an area without their original names and be given new names, just as different ethnic groups may change a name after they receive a new clone (Karamura and Karamura 1994). In order to identify possible duplication of 192 named banana variants, Karamura (1999) first calculated a pairwise estimation of dissimilarity for pairs of ramets of the same plant based on 61 morphological characters. The pairs of ramets from the same plant had a distance coefficient that ranged from 0.044 to 0.147. This was used as a basis to look at landraces sharing the same name but not the same ramets to determine whether the samples were in fact genetically different varieties. Seventy-nine clones came out as distinct from the study of 192 named variants. Eighteen pairs of accessions with similar names had dissimilarity values below 0.1, giving them similar values to ramets that came from the same plant. Four pairs had similarity values falling outside the range of ramets and thus could be considered different clones.

Farmers' Units of Diversity Management

Discussions on farmers' units of diversity management (FUDs) in contrast to simply using a variety's name raise several questions: Is this unit as unique as it is dynamic, and may it change over time? Would joining the variety named with the informant who named it make the unit unique? Names and traits to describe named varieties have the potential to differ not only over spatial scales but also over time. A name may stay the same, but the traits used to describe the named variety may change. Likewise, the traits farmers use to describe a variety may stay the same, but the name associated with these sets of traits may change over time as new farmers adopt and grow the materials. Farmers may also modify generic names by adding on a new descriptor to the name as new traits appear in their populations (see Pandey et al. 2003 for an example in sponge gourd and Rijal et al. 2003 for an example in taro). Name changes may also depend on how the traditional knowledge of these varieties changes over generations and how attitudes and perceptions toward these varieties change over time. Basic to all these possibilities is the concept that here the genetic makeup of the landrace populations is not static but continues to evolve over time (Brown 2000). As farmers select seeds or plants for the next generation, the traits used in selecting for the next cycle of planting may remain consistent or may change, which might result in changes in the genetic structure of the plant (see chapter 4).

It could also be that the traits most important to the farmer to distinguish a variety are not the genetically distinct ones used by the researcher to distinguish varieties. In Nigeria, Busso et al. (2000) found that farmer management practices of pearl millet, a cross-pollinated crop, resulted in more differences between farmers than between the same-name variety grown by different farmers. Thus individual landraces with different names grown by a single farmer were more similar in their genetic structure than same-name landraces of different farmers. In this case the traits farmers use to distinguish the different named varieties did not lead to genetic identity at the molecular level. Similarly for maize in Mexico (another cross-pollinated species) Pressoir and Berthaud (2004) found that the high variation in flowering range and in anthesis silking interval between populations suggested that the pattern of population structure for these agromorphological traits could be very different from that described with molecular markers.

Despite the changes that occur to the variety over time, the name of the variety in the household or village may continue as in the past, or the custodian of the variety may decide that what is now available (compared with the past) is entirely different from what it used to be (as seen in chapter 4 in the case of banana) and thus change the name. What is key, however, is that named populations—irrespective of genetic differences—will be treated in a particular way, and this in turn will affect or even control the future genetic structure of populations on farm (Brown and Brubaker 2002).

There could be differences in the way in which diversity is partitioned depending on the biological characteristics of the crop. Hamrick and Godt (1997) summarized the effect of breeding system on variation within and between crop populations, revealing that selfers (inbreeders) showed twice as much population differentiation as outcrossers (outbreeders). Thus differences between varieties would be expected to be less prominent in cross-pollinated crops than in self-pollinated ones. On the other hand, farmer-named varieties may be on a finer scale for inbreeders than for outbreeders. For the outbreeding alfalfa, the generic name *local* might apply to the material grown in a whole village, whereas for sorghum, a partially outcrossing crop, several named landraces may be grown in the same plot. Once outbreeding landraces such as maize are named, managed as separate units, distinguished by morphological heritable characters such as seed color, and divergent in flowering time, they may accumulate very significant genetic divergence over time.

Is it possible to categorize named varieties into functional groups? Is a metaclassification of names possible through analyzing farmers' perceptions of functional groups? Xu et al. (2001) showed that although different ethnic groups in southwestern China had different names for crops according to local languages, the major morphological types were clearly distinguished across the different ethnic groups. Gauchan et al. (2003; see also chapter 16) categorized named varieties into those with high diversity levels, those with particular adaptive traits, and those that are rare, noting that different types of households are more likely to maintain varieties in one category than in another.

Spatial Diversity Patterns and Variety Names

The distribution of variety names within and between communities and regions may give indications of richness and evenness patterns of genetic diversity on farm. Methods to characterize the amount and distribution of crop cultivars were developed in Nepal based on average area and number of households growing each cultivar (Sthapit et al. 2000). Local cultivars were categorized into groups of cultivars that occupied large or small areas (based on average area) and cultivars that were grown by many and few households (based on number of households). This four-cell analysis (FCA) method has been used in a variety of ways. Rana and colleagues (see Rana

2004) calculated a mean area in hectares per household for each variety grown in a village to determine whether a variety should be considered as grown in a large or small area at the household level. This method resulted in a problem because households with more agricultural land have the capacity to plant larger areas of any one variety, whereas households with less land could plant only a small area (with one or more varieties). To rectify this problem, the information was reanalyzed using percentage area covered within the village by any one variety compared with the percentage of farmers in the village growing the variety. The results are shown in figure 3.4.

The varieties in the upper right-hand corner of figure 3.4c are grown by many farmers and cover a significant percentage of the village area dedicated to rice agriculture. There are also a significant number of varieties grown by very few farmers that in total cover a small percentage of the rice-growing areas. From the graph it is evident that, for the majority of the varieties grown, area covered increases as the number of farmers growing a variety increases; the area covered by these varieties also increases. The varieties that fall outside of the main trend are noteworthy, such as the two points in the lower right-hand corner of figure 3.4c, which are grown by many farmers but in such small areas that the total percentage coverage does not increase at the same rate as that of the other varieties. These two varieties—*rato anadi* and *seto anadi*—are glutinous rice varieties grown most commonly in irrigated areas or in *dhab* (persistent waterlogged areas). They are valued as local cuisine during festivities. Farmers tend to grow these two landraces, which have special religious and cultural significance, in small areas for their own household needs.

Hue et al. (2003) noted a second limitation to the FCA methods just described when they are used for taro in Vietnam. The average area recorded under taro varied greatly depending on the agroecological conditions and market fluctuations. In study sites, it ranged from 28 to 3,600 m^2, whereas the average number of farmers growing each taro cultivar ranged from 1 to 25; thus definitions of "large and small area" and "many and few farmers" were relative and differed from village to village. Moreover, an analysis of diversity distribution pattern showed that villages with a large taro area were not necessarily rich in taro diversity. In general, two to three cultivars of local taro were grown in large areas by many households; thus they can be defined as commonly widespread cultivars at the location. These common and widespread cultivars have high demand in the market because of their quality traits. However, many cultivars (four to nine) were still managed by few households in small plots.

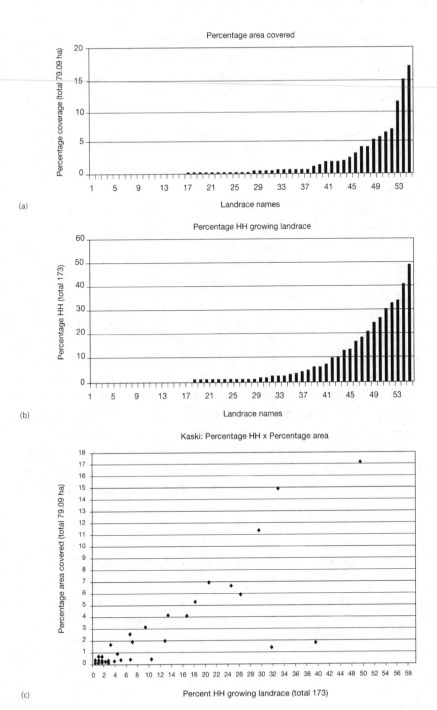

(a)

(b)

(c)

FIGURE 3.4. (a) Reanalysis of rice data based on (b) percentage of area covered (79.09 ha total) and (c) percentage of farmers (HH=household; 173 total) growing the variety in Kaski, the mid-hill ecosite (raw data from Sthapit et al. 2000).

Table 3.7 shows the amount of taro diversity in terms of the number of named varieties. The midland, mountain, and south coastal sandy ecosystems are rich in numbers of taro varieties. Genotype diversity indices for taro diversity were computed using the Simpson index to compare variety of evenness of taro (i.e., the frequency of farmers growing each taro cultivar at village level). Table 3.7 also illustrates the comparison between sites for diversity indices for taro. The highest diversity index was found in Sa pa site, followed by Da bac, Phu vang, and Nho quan. Lowest diversity was in Tra cu.

In Morocco, the importance of local varieties was assessed as percentage of area grown to each local variety on farm and at a given geographic area. Distribution of local varieties in space was assessed by frequency of farmers using each variety. It was found that the number of faba bean varieties grown in the same season by each farmer was not correlated to farm size (Sadiki et al. 2005).

Grum et al. (2003) showed that instead of calculating actual areas, researchers together with farmers and local development workers could use the FCA method to give farmers an opportunity to discuss their perceptions on where a variety should sit within the four cells and whether they consider it rare or common, widespread or local. Grum and colleagues used this method in Sub-Saharan Africa to discuss farmers' perceptions on rice, yam, sorghum, millet, and cowpea. The method raised significant awareness in farmers and extension workers on the extent and distribution of local crop varieties. It is now being used in Benin and Zimbabwe at the university as a tool to assess diversity on farm (M. Grum,

Table 3.7. Amount of taro diversity in different ecological regions, Vietnam, 2003.

Ecological Site	No. of Varieties	Range of Planting Area/ Cultivar (m²)	Average No. Varieties/ Household	Diversity Index (H')
Sa pa	12	28–907	2–4	0.847
Da bac	10	25–360	2–3	0.800
Nho quan	9	40–1,810	1–3	0.680
Phu vang	9	50–241	2–4	0.730
Nghia hung	4	36–216	1–2	0.378
Tra cu	3	50–310	1–2	0.340

Source: Hue et al. (2003).

pers. comm., 2003). The method is also being used to understand farmers' rationale for allocating land area for each cultivar, to identify common and rare cultivars, and to monitor local crop diversity for conservation actions in Nepal, Mozambique, Sri Lanka, and Malaysia (B. Sthapit, pers. comm., 2003).

In Uganda, Mulumba et al. (2004) used this FCA method to identify and understand the best practices for the conservation of rare banana landraces in Uganda's semiarid area of Lwengo. Using this method, researchers recorded a total of 66 varieties of bananas in the subcounty. For the 19 varieties farmers considered to be rare cultivars, a total of 21 management practices were identified. The principal component analysis showed that of the 21 practices, 9 were very critical for the survival of rare landraces.

Karamura et al. (2004) used the same analysis to understand farmers' management strategies to determine which of those strategies contribute to pest reduction while retaining diversity. Their results indicated that communities at the study site keep accessing and selecting new banana genotypes to give an average number of 13 genotypes per farm. Of the diversity found on this site, approximately 45% was maintained in small quantities and on very small acreage for various reasons. Although the farmers identified more than 20 community banana-based management practices as essential for maintaining the possible maximum diversity of cultivars, successful management of banana groves relied on careful implementation of a number of selected, integrated, community banana-based practices (including isolated planting, loosening soil, continuous relocation, transplanting, and manuring).

Spatial allocation of varieties to different production spaces and land uses links ecological knowledge of the environment with cultural practices. Box 3.1 describes an example in which allocation of varieties is linked to gendered production spaces in Mexico. Farmers often cannot take risks in optimizing their allocation of varieties to particular production spaces. It could happen that varieties that have been passed down through generations as being niche specific may in fact be more widely adapted or even do better in different production spaces outside their home environments because adaptations are complex. Their assessment with selected traits may be incomplete, as was found in Nepal (box 3.2).

Using the FCA methods helps to reveal the extent of differences between individual landraces that is not captured by assessing only the presence or absence of landraces found on farm. What is common to the work

In Yucatán, Mexico, continuous cultivation of crop diversity relies heavily on the interaction between the main production spaces within traditional agricultural systems: home gardens, agricultural fields (milpas), and village plots.[1] Gender relations are manifested through production spaces that are related

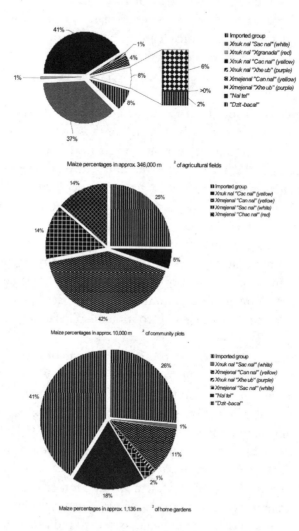

BOX FIGURE 3.1. (a) Amount and distribution of maize varieties in the research population. (b) Amount and distribution of squash varieties in the research population (figure adapted from Lope 2004).

Box 3.1 *continud*

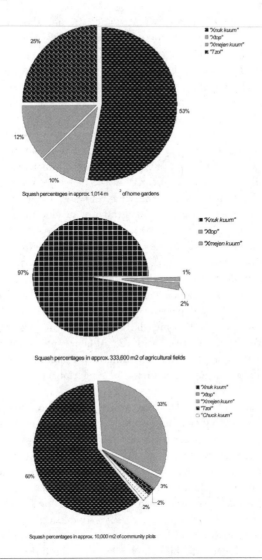

Squash percentages in approx. 1,014 m² of home gardens

Squash percentages in approx. 333,600 m2 of agricultural fields

Squash percentages in approx. 10,000 m2 of community plots

(a)

BOX FIGURE 3.1. *continued*

more to one sex than the other, which in turn are related to the gender division of labor and gender-specific knowledge and therefore likely to reflect different cropping patterns and contain different varieties. Men are exclusively responsible

Box 3.1 continues to next page

Box 3.1 continued

for cultivating crops in milpas, where women are not allowed to go without men present, and women participate in field labor only when additional labor is needed, such as for harvest. On the other hand, women perform most of the labor and make most of the decisions in home gardens and can work in them alone. Home gardens are considered primarily a female domain, whereas agricultural fields are a male domain, and gendered codes of behavior were found to be strongest in these two traditional spaces. In the case of the village plots, such boundaries are more fluid because both men and women explicitly decide what cultivars to grow in these spaces and in what amounts. Women can go alone to village plots and work in them alone as well; men also work in village plots (especially those that are located on the fringes of the community), and both may make decisions regarding what and how to plant. Box figure 3.1 summarizes the varieties and amounts grown by production space.

The results indicate that it is difficult to characterize maize or squash as being specifically a women's or men's crop, that is, as having an affiliation exclusively with one gender. Instead, in the study area it was found that the traditional production spaces, agricultural fields (men's space) and home gardens (women's space), are interdependent in terms of varietal selection and maintenance and that this interdependence is an outcome of both the similar and different reasons men and women have for cultivating a given cultivar in a given production space, coupled with the influence men and women exert on each other for varietal selection according to production space, which is exerted physically and as an explicit or subtle negotiation.

1. Mayan agricultural fields, or milpas, are intercropped with maize, beans, and squash using swidden techniques, with no mechanization, and all production is rain-fed. Spaces for other horticultural crops also are found in milpas but usually are separate from maize and its associated crops. Field size may vary from a few mecates (20 × 20 m, a unit local farmers use to measure their milpas) to 4–5 ha. Home gardens contain a greater amount of interspecies diversity, where species are used principally for food, medicines, fodder, fuel, and ornamentals. In addition to these two traditional spaces, village plots (*terrenos*) consist of old home gardens or seasonal residence sites (*ranchos*) that are no longer inhabited and community land that has been distributed to individual families for future use according to village spatial planning and population growth. Among the families studied in this research, these plots have an average size of 40 × 60 m² and are used in a way that reflects at times patterns found in agricultural fields and at times those found in home gardens.

Source: Lope (2004).

Box 3.2 Comparison of Rice Cultivars for Grain Yield in a Reciprocal Transplant Experiment, Kaski, Nepal (1,150 m)

Farmers often cannot take risks to evaluate where their allocation of varieties to particular production spaces is optimized. In Nepal, the relative performance of rice cultivars under varied moisture regimes was examined to determine whether relative performance of landraces differs when they are tested under varied moisture and fertility regimesw and whether different rice ecosystems necessitate specific variety adaptation (box table 3.2). Reciprocal planting was done in three different moisture regimes of rice ecosystems: *ghaiya* (upland rice ecosystem), *tari* (rain-fed rice ecosystem), and *sinchit* (irrigated rice ecosystem).

Results showed that variety by ecosystem interactions were significant and that adaptation was variety specific (box figure 3.2). Varieties from *tari* and *ghaiya* environments produced the highest yield in their home ecosystems, whereas among varieties from the *sinchit* rice ecosystem, only *rato anadi* and *khumal 4* had their highest yield in this home ecosystem. The varieties *kalo jhinuwa* and *ekle* had significantly higher yields outside their home environment in the *tari* rice ecosystem (under rain-fed conditions).

The ranking of sites based on yield of individual varieties showed that *mansara, kathe gurdi, kalo jhinuwa,* and *ekle* cultivars had better yield in *tari*, followed by *sinchit*, then *ghaiya*. *Rato aanadi* performed similarly in *ghaiya* and

BOX TABLE 3.2. Rice cultivars and moisture regimes.

Moisture Regimes (Different Agroecosystems)	Test Cultivars Native to Different Moisture Regimes	Distinct Traits and Values
Ghaiya (upland)	*Rato ghaiya*	Good straw value, nutrient demanding
Tari (rain-fed)	*Mansara*	Grown in marginal areas, low-fertility environments, poor eating quality
	Kathe gurdi	Early maturing
Sinchit (irrigated)	*Kalo jhinuwa*	Aromatic and fine rice
	Ekle	Popular, high yielding
	Rato aanadi	Sticky rice
	Khumal 4 (check)	Improved variety

Source: D. Rijal, unpublished data, 2004.

Box 3.2 continues to next page

Box 3.2 continued

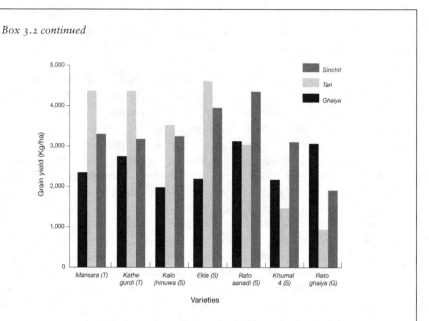

BOX FIGURE 3.2. Comparison of rice cultivars for grain yield in a reciprocal transplant experiment, Kaski, Nepal (1,150 m) (Rijal, unpublished data).

tari but less than in *sinchit*, and *rato ghaiya* had better yield in *sinchit* than in *tari* but significantly less than in its original ecosystem (*ghaiya*). The check *khumal* 4 had higher yield under irrigation (*sinchit*), followed by yield in *ghaiya*.

Source: D. Rijal, unpublished data.

presented here and in other published studies is the existence of a small number of highly abundant landraces that are grown throughout a region and a much larger number of moderately common varieties, together with a substantial amount of rare varieties grown by only one or two households (Boster 1985; Zimmerer and Douches 1991; Pham et al. 1999; Tesfaye and Ludders 2003).

The FCA method takes a similar approach to that proposed by Marshall and Brown (1975) and Brown (1978) for sampling alleles during germplasm collection. Marshall and Brown have argued that alleles that deserve priority in sampling are those that have a restricted or localized occurrence but high frequency, and this sampling technique has been used extensively in plant genetic resource collection. But how does this method apply at the level of landraces? If many farmers are growing a landrace over many areas, it could be considered widespread. This would lead us to concentrate on selecting landraces that are grown by many farmers over small areas as the priority for conservation. However, it must be noted that because this particular landrace is grown by many farmers in many small areas, the threat of loss is not very high.

A landrace that is grown by only a few farmers and in restricted areas could be said to have highly localized distribution (few farmers and small areas). Then this landrace could be considered unique and may be under great threat; thus it becomes important for ex situ conservation because on-farm conservation of such unique material may not be cost effective. This is also an entry point for linking on-farm conservation and ex situ conservation. If on-farm conservation of this material is to be contemplated, then more information on its survival over many years is needed.

If a landrace is grown by few farmers on large areas, it deserves some attention for on-farm conservation because such a landrace might have adaptive gene complexes and potential for specific adaptation. In addition, the chances of its survival on farm is ensured and conservation becomes more cost effective, along with the opportunities for its continued evolution. Finally, landraces that are grown by many farmers and in large areas probably are more recent in origin and not threatened with loss. These may be candidates for on-farm conservation in the near future, after the conservation of more important ones is ensured.

Representativeness of Local Varieties for Regional Diversity

In this chapter, some evidence has been presented to show that the richness of farmer varieties or number of farmer varieties may not necessarily increase at the same rate as the amount of diversity (allelic richness). It could be that the genetic diversity contained in a few varieties in some villages is similar to the amount of genetic diversity contained in villages with many varieties, or it could be that some villages contain the majority of the characters for much larger regions.

How representative of the diversity in the region is one site? In order to determine how well maize diversity was represented by a single study site—Yaxcaba, in the Province of Yucatán, Yucatán Peninsula—Chavez and

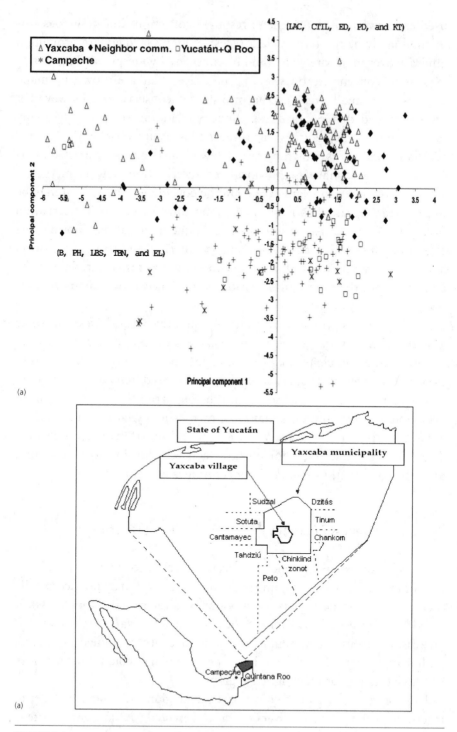

(a)

FIGURE 3.5

colleagues (Chavez et al. 2000; Camacho-Villa and Chavez-Servia 2004) compared agromorphological diversity of 314 maize varieties from all three provinces of the Yucatán Peninsula with that of 15 maize landraces from Yaxcaba. Spatial distribution in the first two principal components showed that maize samples from Yaxcaba possess highly variable characters. The diversity, measured by agromorphological traits, in the 15 landraces from Yaxcaba covered almost the whole agromorphological variation present in the entire Yucatán Peninsula (figure 3.5).

Similarly, agromorphological characterization conducted on 312 faba bean accessions representing different local varieties collected throughout major faba bean–growing areas in Morocco showed that most diversity was covered in the material that originated from two northern provinces (Belqadi 2003). There seems to be some geographic association with genetic diversity, and it would be interesting to explore whether this could be traced to differences in farmers' practices in these different provinces.

Conclusion

A key result of the examination of the relationship between farmer-named varieties and genetic distinctiveness across countries and crops has been

---

FIGURE 3.5. (a) Scatter plot of the first two principal components showing the dispersion of 314 maize populations from Yucatán Peninsula. In 1999 maize populations were collected in the three states of the Yucatán Peninsula: Yucatán State, Quintana Roo State, and Campeche State. Of the 314 populations, 182 were collected in the village of Yaxcabá and its neighboring communities, which are located in the geographic and cultural heart of the Yucatán Peninsula. The populations were characterized for 34 morphological and phenological traits. Axis 1 (principal component 1) was determined basically by the blossom (B), plant height (PH), length of branching space (LBS), total number of tassel branches (TBN), and ear length (EL). Axis 2 (principal component 2), according with its eigenvectors, was determined by leaves above the cob (LAC), central spike internode length (CTIL), ear diameter (ED), pith diameter (PD), and kernel texture (KT). The main morphological characters that describe the difference between maize populations of Peninsula Yucatán are related to reproductive traits such as tassel (length, branches, internodes), ear (shape, length, diameter, pith), and kernel (thickness, texture). As the graph shows, the maize populations from Yaxcabá and its neighboring communities cover almost the entire spread of morphological diversity along the second principal component axis, which was determined by leaves above the cob (LAC), central spike internode length (CTIL), ear diameter (ED), pith diameter (PD), and kernel texture (KT). The blossom (B), plant height (PH), length of branching space (LBS), total number of tassel branches (TBN), and ear length (EL) in the first principal component axis established the differences between Quintana Roo and Campeche State landraces and Yucatán State landraces (Chavez-Servia et al. 2000; Chavez-Servia, Camacho, and Burgos-May, unpublished data). Neighbor comm.=neighbor municipalities; Q. Roo=Quintana Roo; Yaxcaba=target municipality. (b) Map of the origin of maize samples under study. The states of Yucatán, Campeche, and Quintana Roo make up the Yucatán Peninsula.

recognition of the fact that farmers' characterization of the units of crop diversity they manage may range from a simple application of a generic crop name, such as *bean*, to all varieties of the crop, even if the different populations are managed differently, to a location-specific name modified by an accompanying set of traits. This recognition that the name may or may not represent the level of diversity of farmer management has helped to refine methods to understand how farmers manage diversity on farm. If a name clearly reflects the level of diversity of a landrace managed by the farmer, then this name could be used as a unit for conservation. When the name is not consistent with the unit managed by the farmer, then other parameters must be added to accurately define the unit for conservation. If clearly distinct landraces are all called "local," then the name of the village or household would enter the equation to define these materials.

It is possible that some of the rare named varieties in a village or region are selections from some of the common varieties and that the common varieties contain all the diversity found in the rare varieties. Analyzing this problem would entail examining a set of rare varieties and comparing them with common ones. These questions are necessary to understand the link between the farmer-recognized unit—his or her named variety— and the amount of genetic diversity in the system he or she manages. It is not known whether common varieties used throughout a village or region tend to be more variable than less common varieties. It could be that the differences are related much more to character differences (and possibly to the distribution of diversity), and all varieties have about the same allelic richness. If the locally common varieties maintained by farmers have the greatest number of locally common alleles, or if the rare varieties that farmers maintain are, in fact, selections from more common ones, then the question to ask would be whether the maintenance of common varieties on farm is sufficient. If locally common varieties are the varieties that appear to be particularly important for any farmers for certain specific characteristics, one might expect them to have a high proportion of locally common alleles of adaptive significance. Supporting the maintenance of these locally common varieties on farm therefore would be particularly important both for farmers' continued use today and for humanity's potential use in the future.

Acknowledgments

This work is a result of the International Plant Genetic Resources Institute–supported global project "Strengthening the Scientific Basis of *In Situ* Conservation of Agricultural Biodiversity On-Farm." The authors would like to thank the governments of Switzerland (Swiss Agency for Development and Cooperation), the Netherlands (Directorate-General for International Cooperation), Germany (Bundesministerium für Wirtschaftliche Zusammenarbeit/Deutsche Gesellschaft für Technische Zusammenarbeit), Canada (International Development Research Centre), Japan (Japan International Cooperation Agency), Spain, and Peru for their financial support, Dr. A. H. D. Brown for his review of the manuscript, and Dr. Jean Louis Pham for his help in synthesizing ideas on variety names as an entry point to diversity on farm.

References

Arias, L. 2004. *Diversidad genetica y conservación in situ de los maices locales de Yucatan, México* [Genetic Diversity and In Situ Conservation of Local Maize in Yucatán, Mexico]. PhD thesis, Instituto Tecnologico de Merida, Yucatán, México.

Arias, L., J. Chávez, B. Cob, L. Burgos, and J. Canul. 2000. Agro-morphological characters and farmer perceptions: Data collection and analysis. Mexico. In D. Jarvis, B. Sthapit, and L. Sears, eds., *Conserving Agricultural Biodiversity In Situ: A Scientific Basis for Sustainable Agriculture*, 95–100. Rome: IPGRI.

Bajracharya, J. 2003. *Genetic Diversity Study in Landraces of Rice (*Oryza sativa L.*) by Agro-morphological Characters and Microsatellite* DNA *Markers*. PhD thesis, University of Wales, Bangor, UK.

Bajracharya, J., K. A. Steele, D. I. Jarvis, B. R. Sthapit, and J. R. Witcombe. 2006. Rice landrace diversity in Nepal. Variability of agro-morphological traits and SSR markers in landraces from a high altitude site. *Field Crop Research* 95:327–335.

Bellon, M. R. and S. B. Brush. 1994. Keepers of the maize in Chiapas, Mexico. *Economic Botany* 48:196–209.

Bellon, M. R. and J. E. Taylor. 1993. "Folk" soil taxonomy and the partial adoption of new seed varieties. *Economic Development and Cultural Change* 41(4):763–786.

Belqadi, L. 2003. *Diversité et ressources génétiques de* Vicia faba L. *au Maroc: Variabilité, conservation ex situ et in situ et valorisation.* Doctorat de thèse ès-sciences agronomiques, Institut Agronomique et Vétérinaire Hassan II, Rabat, Morocco.

Benchekchou, Z. 2004. *Analyse de la structure de la diversité génétique de la fève in situ en relation avec sa gestion à la ferme: Contribution au développement des bases scientifiques pour la conservation in situ de la fève au Maroc.* Mémoire de 3ème cycle du diplôme d'ingénieur d'état en agronomie, Option: Amélioration Génétique des Plantes. Rabat, Morocco: Institut Agronomique et Vétérinaire Hassan II.

Boster, J. S. 1985. Selection for perceptual distinctiveness: Evidence from Aguaruna cultivars of *Manihot esculenta. Economic Botany* 39(3):310–325.

Bouzeggaren, A., A. Birouk, S. Kerfal, H. Hmama, and D. Jarvis. 2002. Conservation in situ de la biodiversité des populations noyaux de luzerne locale au Maroc. In A. Birouk, M. Sadiki, F. Nassif, S. Saidi, H. Mellas, A. Bammoune, and D. Jarvis, eds., *La conservation in situ de la biodiversité agricole: Un défi pour une agriculture durable.* Rome: IPGRI.

Brown, A. H. D. 1978. Isozymes, plant population genetics structure and genetic conservation. *Theoretical and Applied Genetics* 52:145–157.

Brown, A. H. D. 2000. The genetic structure of crop landraces and the challenge to conserve them in situ on farms. In S. B. Brush, ed., *Genes in the Field,* 29–48. Ottawa, Canada: IDRC/IPGRI/Lewis Publishers.

Brown, A. H. D. and C. Brubaker. 2002. Indicators for sustainable management of plant genetic resources: How well are we doing? In J. M. M. Engles, V. R. Rao, A. H. D. Brown, and M. T. Jackson, eds., *Managing Plant Genetic Diversity,* 249–262. Rome: IPGRI and Wallingford, UK: CABI.

Brush, S., R. Kesselli, R. Ortega, P. Cisneros, K. Zimmerer, and C. Quiros. 1995. Potato diversity in the Andean Center of Crop Domestication. *Conservation Biology* 9(5):1189–1198.

Burgos-May, L. A., J. L. Chavez-Servia, and J. Ortiz-Cereceres. 2004. Variabilidad morfologica de maices criollos de la peninsula de Yucatan, Mexico. In J. L. Chavez-Servia, J. Tuxill, and D. I. Jarvis, eds., *Manejo de la diversidad de los cultivos en los agroecosistemas tradicionales,* 58–66. Cali, Colombia: Instituto Internacional de Recursos Fitogeneticos.

Busso, C. S., K. M. Devos, G. Ross, M. Mortimore, W. M. Adams, M. J. Ambrose, S. Alldrick, and M. D. Gale. 2000. Genetic diversity within and among landraces of pearl millet (*Pennisetum glaucum*) under farmer management in West Africa. *Genetic Resources and Crop Evolution* 60:1–8.

Camacho-Villa, C. and J. L. Chavez-Servia. 2004. Diversidad morfologica del maiz criollo de la region centro de Yucatan, Mexico. In J. L. Chavez-Servia, J. Tuxill, and D. I. Jarvis, eds., *Manejo de la diversidad de los cultivos en los agroecosistemas tradicionales*, 47–57. Cali, Colombia: Instituto Internacional de Recursos Fitogeneticos.

Canh, N. T., T. V. On, N. V. Trung, C. A. Tiep, and H. V. Lam. 2003. Preliminary study of genetic diversity in rice landraces in Ban Khoang Commune, Sa Pa District. In H. D. Tuan, N. N. Hue, B. R. Sthapit and D. I. Jarvis, eds., *On-Farm Management of Agricultural Biodiversity in Vietnam*. Proceedings of a symposium, December 6–12, 2001, Hanoi, Vietnam. Rome: IPGRI.

Cazarez-Sanchez, E. 2004. *Diversidad genetica y su relacion con la tecnologia de alimentos tradicionales*. MS thesis, Colegio de Postgraduados, Montecillos, Texcoco, Mexico.

Cazarez-Sanchez, E. and J. Duch-Gary. 2004. La diversidad genetica de variedades locales de maiz, frijol, calabaza y chile, y su relación con caracteristicas culinarias. In J. L. Chavez-Servia, J. Tuxill, and D. I. Jarvis, eds., *Manejo de la diversidad de los cultivos en los agroecosistemas tradicionales*, 250–255. Cali, Colombia: Instituto Internacional de Recursos Fitogeneticos.

Chavez-Servia, J. L., L. Burgos-May, J. Canul-Ku, T. C. Camacho, J. Vidal-Cob, and L. M. Arias-Reyes. 2000. Analisis de la diversidad en un proyecto de conservacion in situ en Mexico [Diversity analysis of an in situ conservation project in Mexico]. In *Proceedings of the XII Scientific Seminar*, November 14–17, 2000, Havana, Cuba.

Gauchan, D., M. Smale, and P. Chaudhary. 2003. *(Market-Based) Incentives for Conserving Diversity on Farms: The Case of Rice Landraces in Central Terai, Nepal*. Paper presented at the fourth Biocon Workshop, August 28–29, 2003, Venice, Italy.

Grum, M., E. A. Gyasi, C. Osei, and G. Kranjac-Berisavljevie. 2003. *Evaluation of Best Practices for Landrace Conservation: Farmer Evaluation*. Paper presented at Sub-Saharan Africa Meeting, Nairobi, 2003.

Hai, V. M., H. Q. Tin, and N. N. De. 2003. Agromorphological variation of Mon Sap taro populations in the Mekong Delta, Vietnam: Role of on-farm conservation. In H. D. Tuan, N. N. Hue, B. R. Sthapit, and D. I. Jarvis, eds., *On-Farm Management of Agricultural Biodiversity in Vietnam*. Proceedings of a symposium, December 6–12, 2001, Hanoi, Vietnam. Rome: IPGRI.

Hamrick, J. L. and M. J. W. Godt. 1997. Allozyme diversity in cultivated crops. *Crop Science* 37:26–30.

Harlan, J. R. 1975. Our vanishing genetic resources. *Science* 188:618–621.

Hue, N., L. Trinh, N. Ha, B. Sthapit, and D. Jarvis. 2003. Taro cultivar diversity in three ecosites of North Vietnam. In H. D. Tuan, N. N. Hue, B. R. Sthapit, and D. I. Jarvis, eds., *On-Farm Management of Agricultural Biodiversity in Vietnam*.

Proceedings of a Symposium, December 6–12, 2001, Hanoi, Vietnam. Rome: IPGRI.

Jarvis, D., L. Myer, H. Klemick, L. Guarino, M. Smale, A. H. D. Brown, M. Sadiki, B. Sthapit, and T. Hodgkin. 2000. *A Training Guide for In Situ Conservation On-Farm.* Version 1. Rome: IPGRI.

Jarvis, D. I., V. Zoes, D. Nares, and T. Hodgkin. 2004. On-farm management of crop genetic diversity and the Convention on Biological Diversity's Programme of Work on Agricultural Biodiversity. *Plant Genetic Resources Newsletter* 138:5–17.

Karamura, D. A. 1999. *Numerical Taxonomic Studies of the East African Highland Bananas (Musa AAA–East Africa) in Uganda.* Montpellier, France: INIBAP.

Karamura, D. 2004. *Estimation of Distinct Clones in the Uganda National Banana Germplasm Collection.* Presentation at "Workshop on Data Variables and Structure to Answer Questions That Support the Conservation and Use of Crop Genetic Diversity On-Farm," September 20–24, 2004, Rome.

Karamura, D. and E. Karamura. 1994. *A Provisional Checklist of Banana Cultivars in Uganda.* Kampala, Uganda: National Agricultural Research Organization (NARO) and INIBAP.

Karamura, D., E. Karamura, J. Wasswa, B. Kayiwa, A. Kalanzi, and C. Nkwiine. 2004. *Analysis of Community Banana Based Management Practices: A Farmers' Perspective Towards Maintaining Diversity.* Presentation at "Workshop on Data Variables and Structure to Answer Questions That Support the Conservation and Use of Crop Genetic Diversity On-Farm," September 20–24, 2004, Rome.

Latournerie Moreno, L., J. Tuxill, E. Yupit Moo, L. Arias Reyes, J. Crisotbal Alejo, and D. I. Jarvis. 2006. Traditional maize storage methods of Mayan farmers in Yucatan, Mexico: Implication for seed selection and crop diversity. *Biodiversity and Conservation,* 15(5): 1771–1795.

Lope, D. 2004. *Gender Relations as a Basis for Varietal Selection in Production Spaces in Yucatan, Mexico.* M.S. thesis, Wageningen University, The Netherlands.

Louette, D., A. Charrier, and J. Berthaud. 1997. In situ conservation of maize in Mexico: Genetic diversity and maize seed management in a traditional community. *Economic Botany* 51:20–38.

Mar, I. and L. Holly. 2000. Hungary. Adding benefits. In D. Jarvis, B. Sthapit, and L. Sears, eds., *Conserving Agricultural Biodiversity In Situ: A Scientific Basis for Sustainable Agriculture,* 194–198. Rome: IPGRI.

Mar, I., A. Simon, and A. Gyovai. 2004. *Data Variables on Percent Coverage, Number of Farmers, Measurements of Richness and Evenness in Maize and Beans in Hungary.* Presentation at "Workshop on Data Variables and Structure to Answer

Questions That Support the Conservation and Use of Crop Genetic Diversity On-Farm," September, 20–24, Rome.

Marshall, D. R. and A. H. D. Brown. 1975. Optimal sampling strategies in genetic conservation. In O. H. Frankel and J. G. Hawkes, eds., *Crop Genetic Resources for Today and Tomorrow*, 53–80. Cambridge: Cambridge University Press.

Martin, G. B. and M. W. Adams. 1987. Landraces of *Phaseolus vulgaris* (Fabacae) in northern Malawi. I. Regional variation. *Economic Botany* 41:190–203.

Morales-Valderrama, C. and T. Quiñones-Vega. 2000. Social, cultural and economic data collection and analysis including gender: Methods used for increasing access, participation and decision-making. In D. Jarvis, B. Sthapit, and L. Sears, eds., *Conserving Agricultural Biodiversity In Situ: A Scientific Basis for Sustainable Agriculture*, 49–50. Rome: IPGRI.

Mulumba, W. J., C. Nkwiine, K. B. Male, A. Kalanzi, and D. Karamura. 2004. Evaluation of farmers' best practices for on-farm conservation of rare banana (*Musa*) landraces in the semi-arid region of Lwengo sub-county, Masaka district—Uganda. *Uganda Journal of Agriculture* 9(1):275–281.

Pandey, Y. R., D. K. Rijal, M. P. Upadhyay, B. R. Sthapit, and B. K. Joshi. 2003. In situ characterization of morphological traits of sponge gourd at Begnas ecosite, Kaski, Nepal. In B. R. Sthapit, M. P. Upadhyaya, B. K. Baniya, A. Subedi, and B. K. Joshi, eds., *On-Farm Management of Agricultural Biodiversity in Nepal*, 63–70. Proceedings of a national workshop, April 24–26, 2001, Lumle, Nepal. Kathmandu, Nepal: NARC/LI-BIRD/IPGRI.

Pham, J. L., S. Quilloy, L. D. Huong, T. V. Tuyen, T. V. Minh, and S. Morin. 1999. *Molecular Diversity of Rice Varieties in Central Vietnam*. Paper presented at the workshop of the participants of the project "Safeguarding and Preserving the Biodiversity of the Rice Genepool. Component II: On-Farm Conservation," May 17–22, 1999, International Rice Research Institute, Los Baños, Philippines.

Pressoir, G. and J. Berthaud. 2004. Patterns of population structure in maize landraces from the Central Valleys of Oaxaca in Mexico. *Heredity* 92:88–94.

Quiros, C. F., S. B. Brush, D. S. Douches, K. S. Zimmerer, and G. Huestis. 1990. Biochemical and folk assessment of variability of Andean cultivated potatoes. *Economic Botany* 44(2):254–266.

Rana, R. B. 2004. *Influence of Socio-Economic and Cultural Factors on Agrobiodiversity Conservation On-Farm in Nepal*. PhD thesis, International and Rural Development Department, University of Reading.

Rh'rib, K., A. Amri, and M. Sadiki. 2002. Caracterisation agro morphologique des populations locales d'orge des sutes Tanant et Taounate. In A. Birouk, M. Sadiki, F. Nassif, S. Saidi, H. Mellas, A. Bammoune, and D. Jarvis, eds., *La conservation in situ de la biodiversité agricole: Un défi pour une agriculture durable,* 286–294. Rome: IPGRI.

Rijal, D. K., B. R. Sthapit, R. B. Rana, and D. I. Jarvis. 2003. Adaptation and uses of taro diversity in agroecosystems of Nepal. In B. R. Sthapit, M. P. Upadhyaya, B. K. Baniya, A. Subedi, and B. K. Joshi, eds., *On-Farm Management of Agricultural Biodiversity in Nepal,* 29–36. Proceedings of a national workshop, April 24–26, 2001, Lumle, Nepal. Kathmandu, Nepal: NARC/LI-BIRD/IPGRI.

Sadiki, M., M. Arbaoui, L. Ghaouti, and D. Jarvis. 2005. Seed exchange and supply systems and on-farm maintenance of crop genetic diversity: A case study of faba bean in Morocco. In D. I. Jarvis, R. Sevilla-Panizo, J.-L. Chavez-Servia, and T. Hodgkin, eds., *Seed Systems and Crop Genetic Diversity On-Farm,* 81–86. Proceedings of a workshop, September 16–20, 2003, Pucallpa, Peru. Rome: IPGRI.

Sadiki, M., L. Belqadi, M. Mahdi, and D. Jarvis. 2001. Identifying units of diversity management by comparing traits used by farmers to name and distinguish faba bean (*Vicia faba* L.) cultivars with measurements of genetic distinctiveness in Morocco. In *Proceedings of the LEGUMED Symposium "Grain Legumes in the Mediterranean Agriculture,"* October 25–27, 2001, Rabat, Morocco. Paris: AEP.

Sadiki, M., A. Birouk, A. Bouizzgaren, L. Belqadi, K. Rh'rrib, M. Taghouti, S. Kerfal, M. Lahbhili, H. Bouhya, R. Douiden, S. Saidi, and D. Jarvis. 2002. La diversité génétique in situ du blé dur, de l'orge, de la luzerne et de la fève: Options de stratégie pour sa conservation. In A. Birouk, M. Sadiki, F. Nassif, S. Saidi, H. Mellas, A. Bammoune, and D. Jarvis, eds., *La conservation in situ de la biodiversité agricole: Un défi pour une agriculture durable,* 37–117. Rome: IPGRI.

Sawadogo, M., J. T. Ouedraogo, R. G. Zangre, and D. Balma. 2005. Diversité biologique agricole et les facteurs de don maintien en milieu paysan. In D. Balma, B. Dossou, M. Sawadogo, R. G. Zangre, J. T. Ouedraogo, and D. I. Jarvis, eds., *La gestion de la diversité des plantes agricoles dans les agro-ecosystemes.* Compte-rendu des travaux d'un atelier abrite par CNRST, Burkina Faso et International Plant Genetic Resources Institute, Ouagadougou, Burkina Faso, December 27–28, 2001. Rome: IPGRI.

Schneider, J. 1999. Varietal diversity and farmers' knowledge: The case of sweet potato in Irian Jaya. In G. Prain, S. Fujusaka, and M. D. Warren, eds., *Biological and Cultural Diversity,* 97–114. London: IT Publications.

Sebastian, L. S., J. S. Garcia, L. R. Hipolito, S. M. Quilloy, P. L. Sanchez, M. C. Califo, and J. L. Pham. 2001. *Assessment of Diversity and Identity of Farmers' Rice Varieties Using Molecular Markers.* Paper presented at the workshop "In Situ

Conservation of Agrobiodiversity: Scientific and Institutional Experiences and Implications for National Policies," International Potato Center (CIP), August 14–17, 2001, La Molina, Peru.

Soleri, D. and D. A. Cleveland. 2001. Farmers' genetic perceptions regarding their crop populations: An example with maize in the central valleys of Oaxaca, Mexico. *Economic Botany* 55(1):106–128.

Sthapit, B., K. Joshi, R. Rana, M. P. Upadhaya, P. Eyzaguirre, and D. Jarvis. 2000. Enhancing biodiversity and production through participatory plant breeding: Setting breeding goals. In *An Exchange of Experiences from South and South East Asia*. Proceedings of the International Symposium on Participatory Plant Breeding and Participatory Plant Genetic Resources Enhancement, May 1–5, 2000, Pokhara, Nepal. Cali, Colombia: CIAT.

Taghouti, M. and S. Saidi. 2002. Perception et désignation des entités de blé dur gérées par les agriculteurs. In A. Birouk, M. Sadiki, F. Nassif, S. Saidi, H. Mellas, A. Bammoune, and D. Jarvis, eds., *La conservation in situ de la biodiversité agricole: Un défi pour une agriculture durable*, 275–279. Rome: IPGRI.

Tanto, T. 2001. Unpublished data presented at "Strengthening the Scientific Basis of *In Situ* Conservation of Agricultural Biodiversity: Genetic Diversity and On-Farm Conservation Workshop," June 11–19, 2001, Ouagadougou, Burkina Faso.

Tesfaye, B. and P. Ludders. 2003. Diversity and distribution patterns of enset landraces in Sidama, southern Ethiopia. *Genetic Resources and Crop Evolution* 50:359–371.

Teshome, A., B. R. Baum, L. Fahring, J. K. Torrance, T. J. Arnason, and J. D. Lambert. 1997. Sorghum (*Sorghum bicolor*) landrace variation and classification in North Shewa and South Welo, Ethiopia. *Euphytica* 97:225–263.

Tuyen, T. V., N. V. Truong, and H. T. T. Hoa. 2003. *Farmers' Management of Taro Diversity as a Part of Farming Systems in a Coastal Sandy Area of Phuda*. Paper presented at the national workshop "Strengthening the Scientific Basis of *In Situ* Conservation of Agricultural Biodiversity On-Farm," December 6–8, 2002, Ban Me Thuot, Vietnam.

Voss, J. 1992. Conserving and increasing on-farm genetic diversity: Farmer management of varietal bean mixtures in central Africa. In J. L. Moock and R. E. Rhoades, eds., *Diversity, Farmer Knowledge and Sustainability*, 34–51. Ithaca, NY: Cornell University Press.

Xu, J. C., Y. P. Yang, Y. D. Pu, W. G. Ayad, and P. Eyzaguirre. 2001. Genetic diversity in taro (*Colocasia esculenta* Schott, Araceae) in China: An ethnobotanical and genetic approach. *Economic Botany* 55:14–31.

Zimmerer, K. S. 2003. Just small potatoes (and ulluco)? The use of seed-size variation in "native commercialized" agriculture and agrobiodiversity conservation
among Peruvian farmers. *Agriculture and Human Values* 20:107–123.

Zimmerer, K. S. and D. S. Douches. 1991. Geographical approaches to native crop
research and conservation: The partitioning of allelic diversity in Andean potatoes. *Economic Botany* 45:176–189.

4 Seed Systems and Crop Genetic Diversity in Agroecosystems

T. HODGKIN, R. RANA, J. TUXILL, D. BALMA, A. SUBEDI, I. MAR,
D. KARAMURA, R. VALDIVIA, L. COLLADO, L. LATOURNERIE, M. SADIKI,
M. SAWADOGO, A. H. D. BROWN, AND D. I. JARVIS

In the last century, national governments have devoted major resources to modernizing their agricultural sectors, including the development and dissemination of improved crop varieties. Despite this extensive effort, the majority of rural farming communities in developing countries continue to use traditional or informal sources of seeds or vegetative planting materials (Gaifani 1992; Hardon and de Boef 1993; Tripp 2001). Either they save their own seed or they obtain seed from friends, relatives, neighbors, or local markets. In an informal system, seeds may be acquired via cash transactions, by barter, as gifts, through exchange of one variety of seed for another, as a loan to be repaid upon harvest, or even by surreptitious expropriation from another farmer's field (Badstue et al. 2002). Even seeds of varieties developed by the formal sector often are maintained and distributed informally (Mellas 2000; Bellon and Risopoulos 2001), largely independently of government institutions.

In Nepal in 1999–2000, less than 3% of rice seed was purchased from the formal certified seed sector. Informal seed systems are extensive and substantial in Burkina Faso, where less than 5% of sorghum was purchased in 1999 (Kabore 2000), and in Mexico less than 25% of maize seed was purchased from formal sectors in 1999 (Ortega-Paczka et al. 2000). In Morocco, only 13% of durum wheat seed and 2.5% of food legume seed came from certified seeds in 1999–2000, indicating that the majority of seed sown came from local crop diversity or from seed saved from earlier purchases (Mellas 2000). Furthermore, traditional or local varieties continue to constitute much of the material that circulates in these informal systems in many parts of the world. More than 50% of the area under maize production in Mexico, more than 50% of the area under rice production in Nepal, and more than 90% of the area under millet production in Burkina Faso continues to be cultivated with traditional varieties (Upadhaya 1996; Perales 1998; Zangre 1998).

There have been a number of studies on the functioning of informal seed systems, particularly with respect to their ability to meet users' needs during emergencies and disasters such as flood, drought, or war (Almekinders et al. 1994; Richards and Ruivenkamp 1997; Sperling 2001). Other studies have been concerned with the social institutions involved in informal seed networks or with the ways in which they meet farmers' needs for appropriate varieties (Weltzien and vom Brocke 2000). Much of this work has been concerned primarily with seed system function rather than with the materials present in the system. Thus McGuire (2001) wrote in terms of the processes involved in seed provision, and Dominguez and Jones (2005) described seed systems as the ways in which farmers produce, select, save, and acquire seeds. Similarly, Almekinders et al. (1994) discussed seed systems in terms of the flows of seed and other planting materials through the production system and the roles of both formal and informal sector institutions and farmers in these flows.

Seed systems are clearly important to the maintenance of crop genetic diversity on farm. The numbers and proportions of different varieties and their availability, relationships, and movement within an area often depend significantly on the functioning of local informal seed systems (Jarvis et al. 2005), which can be quite dynamic and vary from year to year. The characteristics of the systems and the ways in which they change over time seem likely to have a substantial impact on the genetic diversity present in individual crops. Some of the most important features of seed systems that might be expected to affect genetic diversity include the availability, accessibility, and sources of different materials, the maintenance methods and selection practices used, and the extent to which these change over time.

The seed systems of specific crops are subject to substantial variation in the availability of different materials as a result of variation in production, markets, and climate and of catastrophes such as droughts and hurricanes. The units of maintenance also show great variation. In some instances separate populations are maintained by individual households. In others, populations are combined and mixed and then separated into different seed lots grown in new sites. Both natural selection and farmer selection can have substantial effects on the seed produced for future crops, and farmers may differ in their perspectives and practices in managing their seed stocks and introducing new material, which can depend on gender, wealth status, and age. The area in which specific varieties occur also varies substantially, and whereas some are maintained very locally, others may be part of extremely extensive seed systems extending over more than one region or country (Louette et al. 1997; Zimmerer 2003; Valdivia 2005).

In this chapter, work from the International Plant Genetic Resources Institute (IPGRI) global project concerned with on-farm conservation (Jarvis and Hodgkin 2000) and other relevant information on seed systems and genetic diversity are reviewed. The operation of different components of seed systems (e.g., seed source, seed flow, seed production, farmer selection, and seed storage) is explored in relation to the evolutionary forces that shape the genetic structure of crop variety populations on farm. The ways in which different features of seed systems contribute to gene flow, migration, selection, mutation, and recombination are examined. Finally we discuss how seed systems contribute to the maintenance of crop diversity and ask how they might best support the maintenance of sufficient adaptive capacity in crops as agricultural systems intensify.

Population Structure and Breeding Systems

Traditional varieties consist of a number of seed lots maintained by individual farmers. A first problem in any analysis of seed systems often is one of identity, establishing that different seed lots really belong to the same variety and determining the relationship between variety name and genetic makeup. This involves understanding the ways in which farmers in an area use names and understand identity (see chapter 3). Working with maize in Mexico, Louette et al. (1997) defined a seed lot as a physical unit of kernels associated with the farmer who sows it and a variety or cultivar as the set of farmers' seed lots that bear the same name or share the same origin and characteristics. Sadiki et al. (2005; chapter 3) showed that it is possible to identify a set of traits that farmers consistently use to identify varieties and suggested that these provide an effective basis for analysis of variety management, maintenance, and evolution.

Analyzing genetic diversity in seed systems requires a description of the metapopulation structure of the local varieties of a crop and the processes of seed production and supply. This involves analyzing the sizes and connectivity of the network of partially and variably isolated sub-populations that make up the plantings of varieties in a region. The links between components of the network arise from the seed supply system or from seed flows through the system. As individual seed lots become adapted to different locations and farmers carry out their own selection, the different seed lots will tend to diverge. This will be balanced by exchange or sale of seeds or by supply of materials from markets or other sources.

A primary factor determining genetic structure of traditional varieties is the breeding system of the crop (Brown 2000). Many crops such as rice, wheat, and barley are largely self-pollinated, whereas others such as pearl millet and maize are cross-pollinated. Still others—such as potato, cassava, banana, and many other fruit crops—are clonally propagated, and seed production is rare or absent. Self-pollinated crops are rarely completely so, and whereas cross-pollination may be rare in crops such as rice, in crops such as sorghum or faba beans it may reach significant levels (e.g., 84% in faba beans; Bond and Poulsen 1983).

Whereas maintaining particular properties and characteristics of specific varieties appears easy in self-pollinated or clonally propagated crops, maintaining varieties with specific complex sets of traits seems more problematic in cross-pollinated ones. Gene flow between adjacent fields with different varieties is common (Louette et al. 1997), suggesting that selection must occur in each generation to maintain the varieties' recognized traits. Yadav et al. (2003) showed that for sponge gourd, an open-pollinated crop, individual farmer households in Nepal grew very small populations of only one or two plants, yet at the community level, five distinct types were maintained. It appears that enough gene flow must occur between households to limit inbreeding depression, combined with farmer selection to maintain type identity.

In most clonally propagated crops, the "seed" actually is some other part of the plant (e.g., tuber in potato or yam, corm in taro, clonal bud in banana). Within-variety variation would be expected to be very limited (but see Brush et al. 1995 and Zimmerer and Douches 1991 for information on within-variety variation in potato). Karamura et al. (2005) suggested that this has implications for the sustainability of the system because while everything else around the plant, such as soil texture, nutrients, and water availability, may have been changing over centuries, the banana's genetic makeup may not have changed as much. This may be particularly significant where rapid changes occur over short periods, as in the adoption by a farmer of pesticides, herbicides, and fertilizers.

Seed Systems and the Operation of Evolutionary Forces

The properties of seed systems such as seed source, seed flow, seed production, farmer selection, and seed storage have their major impact on the extent and distribution of genetic diversity in traditional farming systems through their effects on the evolutionary forces that maintain or change the genetic makeup of plant populations. These forces are population size and bottlenecks and their effect on genetic drift; migration, which includes both seed exchange and pollen flow; recombination and mutation, which create new genes or gene combinations; and selection as a result of environmental forces or human actions.

Population Size, Bottlenecks, and Genetic Drift:
The Number and Size of the Populations
That Are Sources of Seed

The size of variety or seed lot populations varies very widely for different crop plants in different situations. As noted earlier, sponge gourd populations in Nepal are very small, and households seldom plant more than 10 individuals (Yadav et al. 2003). The same is true of many home garden crops (Watson and Eyzaguirre 2002; Mar et al. 2005). In contrast, farmers plant populations of many thousands of individuals of a single variety of a crop such as rice or barley.

As well as marked differences between crops with respect to population size, there may also be substantial changes between years for any single variety, and farmers' decisions regarding the size and placement of their fields will significantly affect overall population size and population structure. In Ban Mae Moot, Thailand, a village of about 100 families, the number of fields used for some varieties changed very significantly from year to year. Whereas the two most popular varieties remained unchanged in 2001 and 2002, a variety that was grown in only three fields in one area in 2001 became the third most popular in 2002, grown by 16 farmers in all five growing areas in the village. In this case the extra seed for the expansion was supplied by one of the farmers, and new farmers retained their own stocks for future years (K. Rerkasem, pers. comm., 2003).

Dramatic reductions in coverage (and population size) of varieties also are not uncommon. Chaudhary et al. (2004) recorded a reduction in the number of farmers maintaining a single traditional rice variety from 16 to 3 in a single year (in the same year the number of traditional varieties maintained dropped from 22 to 15). As well as changes in number of farmers (or number of subpopulations), significant changes can occur in areas of production at both the village and the individual farm levels.

Farmers' decisions on population sizes may also be controlled by government regulation. In Hungary, seed regulations limit population size of maize varieties because local maize varieties cannot be planted on large fields and thus are limited to small areas and home gardens (Mar et al. 2005).

Sharp drops in the number of farmers growing a variety followed by increases create genetic bottlenecks often associated with loss of genetic diversity. This can occur as a result of disasters such as floods or hurricanes where local seed availability is severely limited, as in the case of beans in Mexico. Longer-term maintenance of small population sizes is also likely to reduce genetic diversity. In any consideration of the effects on diversity of population size, one needs to take into account both the seed lot size (the population maintained by individual farmers) and the variety size (the combined population of different seed lots) and to consider the amount of exchange and mixing that occurs between seed lots over time.

The ways in which individual seed lots are linked to constitute a single larger population of a variety also depend on the breeding system of the crop and physical disposition of the units of production within an area. Farmers' fields may be large or small, close together or widely separated. This structuring can have a range of effects on the genetic diversity of crops, depending also on the extent of outcrossing. Qualset et al. (1997) suggested that small land holdings isolate variety populations from one another, thus reducing the generation of new genetic material by natural recombination. Drawing on biogeography theory (McArthur and Wilson 1967), Qualset et al. (1997) suggested that without human management, the genetic diversity in small patches of crops would suffer genetic drift, and the populations would show inbreeding depression. They also suggested that human inputs might offset these processes and introduce new genetic traits to isolated populations through seed exchange and farmer selection (see also Louette et al. 1997).

The effect of genetic drift depends on population size and often is regarded as being of limited importance where population sizes are large (Gillespie 1998), as in the case of most crop plants growing in agricultural systems. The likely extent of genetic drift in relation to allele frequencies and loss of alleles from the population can be explored through the concept of effective population size. This is an abstract standardizing parameter and is defined as the size of an idealized hypothetical population that would give rise to the same increase of inbreeding (or loss of heterozygosity, or variance in allele frequency) that is happening in the actual population under study.

For crop plants, data on the genetic effects of population and subpopulation size are very limited. Louette (2005) described genetic instability of local and exotic varieties of Mexican maize due to small population sizes. In Cuzalapa, field area is limited, and various varieties are sown in the same field. The size of the seed lots planted per variety is small, and more than 30% of seed lots sown during the six cultivation seasons covered by Louette's survey were constituted from less than 40 ears. A significant proportion of the seed lots surveyed therefore were subject to a regular reduction of their population size, leading to fluctuation in their diversity and, possibly, loss of rare alleles.

The dramatic change in population size (and the nature of the source population) that can follow poor production seasons is well illustrated for faba bean varieties in Morocco. A comparison of variety profiles after different seasons in Morocco shows that the same varieties are grown in each village. However, the frequency of each variety in the seed flows or movements (proportion of seed of each variety in the total amount of seed used in a village) changes according to type of season and the source of seed supply. In good years seed is maintained by farmers in the villages, whereas after poor years most farmers need to purchase seed of their preferred varieties from local markets. In good years they maintain larger numbers of different seed lots of larger numbers of varieties than they do in poor years. In good years there are more individual source populations, which are often rather small, whereas in poor years a single large source population (from the market) is used. Additionally, the frequency of varieties changes in terms of area planted (box 4.1).

Two other general points can be made about the effect of finite population size on genetic diversity. First, genetic drift and bottlenecks in population size have a more immediate impact on allelic richness than they have on evenness. Rare variants get lost first. There is little information on how serious this is at the genetic level in crop seed systems, although it is clear that rare varieties are the first to go when numbers of varieties are reduced over short time periods (Chaudhary et al. 2004). Thus, for example, the

Box 4.1 Statistic for Diversity Comparisons: Effective Number of Landraces in an Area

Suppose in a farm or village a survey reveals that six landraces and the observed frequencies $\{p_i\}$ are as follows:

$$\{0.5, 0.25, 0.1, 0.05, 0.05, 0.05\}$$

The concept of the effective number of entities (e.g., landraces, origins) in an area is the number of entities (n_e) with identical frequency ($1/n_e$) that would give the same probability of identical ancestry as when any two random genes are compared for their origin:

$$n_e = 1/(\Sigma p_i^2)$$

For this vector of frequencies, actual number of landraces here is 6; the effective number is 3.03 (box table 4.1).

BOX TABLE 4.1. Example: Landrace composition of farmers' own seed of faba bean in 9 villages of Ortzagh site, Morocco.

Average	Good Year 9 Villages	Medium Year 9 Villages	7 Villages*	Bad Year 9 Villages
Proportion of farm's own seed	0.93	0.82	0.4	0.31
Actual number of landraces	5.1	5.0	4.6	3.6
Effective number of landraces	3.49	3.53	2.54	1.97[†]

Sources: Arbaoui (2003); Ghaouti (2003).
   *These averages exclude two villages that planted only purchased seed.
   [†]In the two villages where no local landraces were planted, the effective number was defined as zero.

Conclusion: In the poor year, farmers had less of their own seed to plant, with a lower richness of landraces and lower evenness of frequencies.

---

total agromorphological diversity of *Phaseolus lunatus* maintained in 30 Cuban home gardens from three different parts of Cuba appeared to remain high (Castiñeiras et al. 2001a) despite maintenance in small and apparently isolated populations. Second, effective population sizes probably have to be very small if they are to be the sole agent of substantial genetic erosion. However, when combined with selection, small sizes might seriously erode the unselected diversity. This situation might exist for varieties maintained in home gardens (Castiñeiras et al. 2001b; Yadav et al. 2003; Mar et al. 2005), which could be used to investigate the possibility.

Migration: Seed and Pollen Exchange

Migration is the dispersal or movement of individual plants, vegetative propagating material, seeds, or pollen between populations or subpopulations that usually but not necessarily differ in their gene frequencies. Two kinds of seed migration can be distinguished: migration between populations of the same local variety between fields, farmers, or communities and migration between populations of different varieties as a result of deliberate or accidental mixing.

Seed-mediated migration seems to be a particularly important feature of traditional seed systems as far as dispersal or movement of seeds is concerned. Pollen-mediated gene flow is also likely to be important, but information on its occurrence (either between populations of the same variety or between different varieties) in traditional farming systems is very limited (but see Louette 2005). However, it is currently regarded as of particular importance as a result of the increasing spread of new varieties containing transgenes (Gepts and Papa 2003).

Traditional seed systems are dynamic, with frequent changes in numbers, identities, and distribution of local varieties. New varieties and materials are constantly becoming available from local markets and from commercial or national breeding programs, and this further complicates the analysis of the genetic consequences of migration in traditional seed systems. Migration generally is regarded as a powerful homogenizing force with respect to the extent and distribution of genetic diversity, and it may be that it acts as an important way of maintaining the identities of many local crop plant varieties.

SCALE OF MIGRATION

Most migration in traditional farming systems seems to be fairly local. Between 75% and 100% of the seed used by farmers in the Aguaytia Valley in Peru was exchanged within the community. Only beans, cassava, and maize seeds were exchanged outside the community, and even then this constituted the minority of seed used (25%, 15.2%, and 13.5% respectively; Riesco 2002). In a more detailed study, Collado-Panduro et al. (2005) found that seed exchange of maize, cassava, peanut, chili peppers, and cotton between 13 communities along the central Amazon River in Peru was much less than within communities. This seemed to reflect difficulties of access and communication between communities and the river that provided the main connecting route between them.

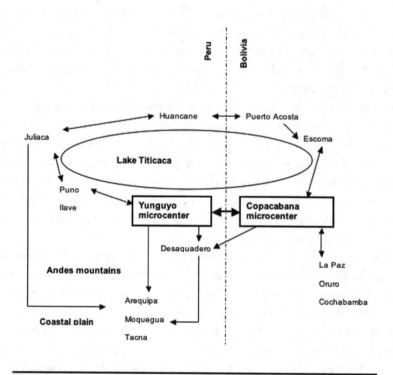

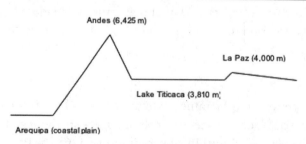

FIGURE 4.1. Distribution pattern of the oca variety *Isleño* in Bolivia and Peru.

However, the scale of migration can be much larger. Potato varieties in Peru often are transferred between different altitudes within an area as part of seed production and management practices (Zimmerer 1996). Much greater distances can be involved, as illustrated by Valdivia (2005) for some Andean root and tuber crops (figure 4.1). The Andean oca variety *Isleño,* grown in the fields of Cochabamba (in Bolivia), was sold in the local market. From Cochabamba it was taken toward Oruro and La Paz. From La Paz (El Alto), it was taken to nearby communities and to the border between Bolivia and Peru. From there it entered Peru, and some material was taken toward Yunguyo (which appears to be a conservation microcenter), where it was sold again. The destinations of this variety, as seed, were communities such as Apillani, Ollaraya, and Unicachi as well as bigger cities such as Ilave, Puno, and Juliaca, from which it was taken to nearby communities for seed and for consumption. Other destinations were coastal cities in Peru such as Tacna, Moquegua, and Arequipa, for consumption. Depending on production and climatic conditions the migration route of the seed could also go in reverse. Thus, seeds from Huancane (Peru) were moved to Puerto Acosta (Bolivia) and from there to La Paz, where they flowed to other parts of Bolivia. The distance from Cochabamba in Bolivia to Arequipa in Peru is more than 800 km.

Other examples of established movement of varieties between communities or regions include the provision of planting material of cassava in the Brazilian Amazon (Coomes 2001) or the movement of some specific varieties of barley in Nepal. However, in neither case does the movement of materials involve the complex patterns seen for oca.

SEED REPLACEMENT AND SEED SOURCES

Although most farmers prefer to save their own seeds as much as possible, over a period of years they may well have to replace them, in part or in whole, with seeds of the same variety from a different source. This source usually is a relative, a neighbor, or the local market (commonly in that order of preference). In this way, over a period of years a dynamic of movement and mixing occurs in which the progenies of individual populations are transferred between farmers, become mixed during exchange or marketing, act as sources for new exchanges, or are lost. The extent to which these kinds of movements occur varies between crops, countries, and communities, reflecting environmental factors, production problems, social relations, and socioeconomic conditions.

Current data from several systems indicate conservative strategies of seed replacement. For example, table 4.1 summarizes data from two crops in three villages in Hungary. In the case of bean, 75% (Dévaványa), 83% (Örség), and 89% (Tiszahát) of farmers replaced seed fewer than six times or had practiced no seed replacement for their varieties during the last two decades. In the case of maize the figures are slightly higher: 92% in Dévaványa, 93% in Örség, 84% in Tiszahát. However, this short-term conservatism may be misleading. From the standpoint of effective population size of the whole variety, the question is whether seed or gene flow occurs from the dwindling local populations into the new, replenishing stocks. When migration results in replacement, the effective size of the source populations determines that of the whole system. That size would be less than if some exchange takes place between old and new stocks (see Maruyama and Kimura 1980 for a model theoretical treatment).

Results from Nepal showed that the majority of local crop diversity for rice, taro, finger millet, and barley was maintained by informal seed exchange within and between communities through social networks (Baniya et al. 2003). Seed flows for finger millet were low in any one year, and about 90% of farmers saved their own seeds. However, 82% changed seeds at regular intervals, mostly on an average of three years,

Table 4.1. Farmer seed replacement practices for local bean and maize varieties in Hungary.

	Dévaványa				Örség				Tiszahát			
	Bean		Maize		Bean		Maize		Bean		Maize	
Replacement	No.	%	No.	%	No.	%	No.	%	No.	%	No.	%
No seed replacement	26	31	10	21	56	56	24	37	57	58	41	50
At least 3 but <6 times	36	44	34	71	27	27	36	56	30	31	28	34
More than 6 times	1	1	0	0	2	2	1	2	4	4	5	6
No defined replacement strategy	20	24	4	8	15	15	3	5	7	7	8	10
Total	83	100	48	100	100	100	64	100	98	100	82	100

Source: Mar et al. (2005).

with female farmers involved in such exchanges more commonly than men. Obtaining seed of local varieties from other farmers is regarded in some communities as a sign of less skill as a farmer and therefore to be avoided if possible.

The attitude and approach to seed replacement by farmers varies depending on a number of factors. In Nepal farmers are quite prepared to obtain seed of modern varieties from markets or even from formal sources. They seem to believe that seed needs to be changed frequently to provide consistent yield and that such seed is likely to be of a higher quality than their own saved seed of the same variety. However, this is not the case for seed of local varieties, which are not commonly available in local markets, and where seed maintenance is combined with careful selection.

The majority of Nepalese farmers, when replacing seeds, acquire the desired seeds from others immediately after harvest. In some cases, when seedlings fail to germinate or when they find their seedlings inferior for transplanting, farmers borrow seedlings as sources of new materials. This is a form of crisis management, and such farmers often have little choice in the variety obtained, although they may try to obtain material from a microenvironment similar to theirs.

In Yucatán, Mexico, farmers traditionally grow multicrop maize, beans, and squash together (the milpa system). The seed of maize and squash is predominantly saved by farmers themselves wherever possible. However, there is a significant dependence on farmer-to-farmer seed transactions for the local bean varieties. Farmers who are known for reliably and regularly producing a good bean crop can maintain a thriving business, at both community and regional levels, selling bean seed to other farmers who did not secure seed stock from their own plantings. Whereas beans move from farmer to farmer primarily via cash transactions, when maize and squash seed lots are transferred, it is often as gifts or as exchanges of one seed type for another. These differing seed flows may explain why beans usually are the first crop Yucatecan farmers leave out of their milpa systems when they alter their farming in response to changing agroecological and social conditions.

As shown with respect to faba beans in Morocco and crops in Mexico and Mozambique, the extent of migration can change substantially from year to year, with significant migration occurring when production is poor or as a result of major seed losses through disasters such as floods and hurricanes.

Lope (2004) has shown that in Yucatán, Mexico, varieties may exist in the village but that appropriate social ties are required to access them. In particular, Yucatecan farmers tend to rely heavily on kin networks and coparent (*compadrazgo*) or godparent relationships when searching out seed stocks to renew or replace their planting materials.

Analysis of rice seed supply networks in Nepal (Subedi et al. 2003) revealed their complexity and dependence on a range of social variables. In different communities different kinds of networks functioned. In lowland Bara, where modern varieties of rice predominate, several small nonlinked networks were found, whereas in the midhills Kaski site (still dominated by local varieties) there were fewer but larger networks. The probable reason for this is wider contacts of different individuals and choice of varieties from different farmers as well as from other seed sources. In both areas Subedi et al. identified certain individuals as nodal farmers, who were characterized by their involvement in a large number of exchanges. Nodal farmers served as recognized sources of seed for other farmers and also accumulated planting materials from within and outside the community. Interestingly, given their focal role in seed flows, there appears to be little consultation between these nodal farmers themselves. It has been suggested that these nodal farmers might act as key custodians of crop diversity in the system (Subedi et al. 2003).

Even in the larger networks, not all individuals are connected to each other at the community level. Instead, there are subnetworks, which are linked to one or the other through certain individuals. This indicates that informal flow of seed or planting materials does not necessarily occur between all the members of the community. There would be greater flow of materials through a number of spatially distributed smaller networks. In a large social network, direct contact with all the individuals may not be possible, but occasional network links may be strong in the dissemination of innovations and messages because occasional links provide opportunity to find more new information and materials (Granovetter 1973).

In the lowlands and midhills of Nepal, no separate networks for gender groups were found. Genetic materials flowed through the mixed groups of men and women in both the study areas. This is in contrast to the networks of information flow found by Subedi and Garforth (1996) in certain communities in the western hills of Nepal. In these networks men–men, men–women (men-led), women–men (women-led), and women–women networks were all found in certain communities. Similarly, the seed supply networks for rice were not based on wealth category, indicating that there was no barrier between the gender and wealth categories in the flow of genetic materials.

In Yucatán, in contrast, seed flows of maize, beans, and squash tend to be strongly gendered according to the production spaces where the crop is grown. The milpa is regarded as the center of men's sphere of influence, and it is primarily men who manage seed flows of the crops planted there, particularly maize. When the same crops are planted in home gardens and village lots, however, women often play a prominent role in seed selection, procurement, and exchange because these sites are regarded as the locus of women's influence. For a crop such as chili peppers, which is grown often in both home gardens and milpas, men and women probably play equally important roles in seed flows when viewed at a community or variety level.

In Hungary, access to local seeds and knowledge about specific production practices is limited and is realized through personal contacts. Seed sales in local markets are controlled exclusively by the National Institute for Agricultural Quality Control, so that the functioning of a local informal seed system is not legalized and markets are not part of the system. However, seeds of traditional varieties are sold as grain for food or feed in markets, and some of them may find their way back as seed for planting (figure 4.2).

MIGRATION AND SELECTION

Studies of migration by (diverging) subpopulations in model systems have shown that uneven migration rates among them reduce the effective population size of the system, particularly when the seed of one farm is replaced (Maruyama and Kimura 1980; Wang and Caballero 1999; Whitlock 2003). Thus the effects of migration on diversity depend closely on the interaction between migration and selection, on how farmers manage the continual input of diversity, and how well it is adapted to the local environment. Migrants may displace existing local varieties (or specific populations of local varieties), mix with them, hybridize and exchange genes, and ultimately fuse into one population. The genetic effect of migration is closely linked to the management and selection practices followed by farmers introducing or distributing new materials.

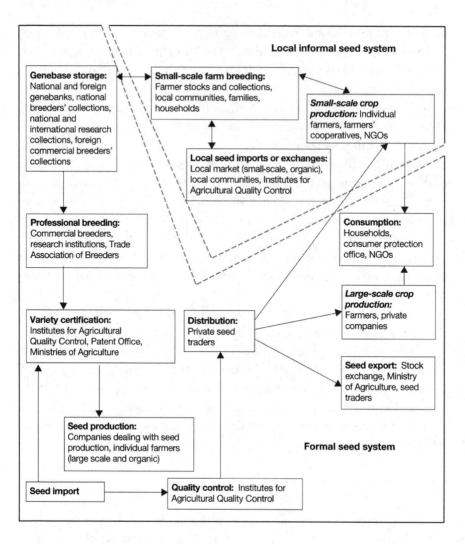

FIGURE 4.2. Seed flow through the informal and formal sectors in Hungary (Mar et al. 2005). NGO = nongovernment organization.

In Yaxcaba, Yucatán, the proportion of seed lots of improved maize planted by Mayan farmers is about equal to that of short-cycle (*xmejennal*) landraces, both of which tend to be planted in similar field microenvironments. Undoubtedly the adoption of improved maize over the past two decades has involved a certain amount of displacement of *xmejennal* populations, but the improved maize stocks are heavily creolized (*sensu* Bellon and Risopoulos 2001) in certain traits such as husk coverage, suggesting that along the way farmers have tolerated or encouraged substantial gene flow from landraces to local improved maize populations. Given the outpollinated nature of maize, one also would expect gene flow to occur from the creolized improved maize stocks to the local materials, and upon inspection, many *xmejen-nal* lots contain at least a few ears with kernel characteristics of improved maize. Louette et al. (1997) also showed that gene flow occurred in the maize varieties in Cuzalapa. Farmer selection may well minimize this effect for characteristics important to production, but it may still occur for genes or traits not under selection pressure.

It is also important to account for the genetic change that may occur within varieties as a result of the selection and use of seed on farm, depending on the breeding system of the crop. In the Cuzalapa Valley of Mexico, farmers constantly exchange small lots of maize seed, both within the region and further afield. Although small in scope, these exchanges have become an integral part of local maize cultivation because they can provide seed for planting at any time of the year, and they introduce new diversity into an existing landrace (Louette et al. 1997).

Recombination

Recombination during sexual reproduction in heterozygous plants results in the creation of new gene combinations. These may or may not survive to become part of the population, depending on natural and farmer selection. Recombination in outpollinated species such as maize and pearl millet provides a continual production of new genotypes in each generation. In self-pollinated species consisting of largely homozygous plants, recombination has a major effect only when an occasional outcross occurs. In the context of seed systems and their role in maintaining genetic diversity in crop species, the importance of recombination lies in the consequences of outcrossing between migrant and local populations after seed migration of some type or the consequences of pollen-mediated gene flow.

The fact that traditional farmers often detect and take an interest in new types occurring in their field has often been noted (Richards 1989). These may be simple contaminants (or migrants) but they may also be the progeny of outcrossing of some kind and hence a consequence of recombination. Although recombination undoubtedly is important in the maintenance of diversity, its role in or effect on seed system function seems to be slight. However, in terms of the way in which seed systems operate, it would be interesting to explore further the ways in which new materials are integrated into a crop seed system and the various ways in which they replace, compete with, or are mixed with existing components.

With the introduction of genetically modified crops, pollen flow may become increasingly significant in terms of its effect on local variety populations and their characteristics. If this occurs, it might alter seed management practices and seed system function depending on farmer selection and management practices. At present, the genes of most concern are those that confer herbicide tolerance and pest resistance. It has been reported (but not independently confirmed) that transgenic DNA has been found in local maize varieties in Oaxaca, Mexico (Quist and Chapela 2001).

Mutation

The low rates of mutation in most crops appear to rule it out as a major agent of change in the short term except in some clonally propagated species. In bananas mutation certainly leads to the occurrence of daughter suckers that differ from the mother plants. All domesticated bananas, whether diploids or triploids, are virtually seedless and are propagated clonally. Somatic mutations have been found to be more common in the different banana groups, particularly those grown on the largest scale, namely the dessert bananas, the East African Highland bananas (*Musa* AAA group), and the plantains (*Musa* AAB) (Pickersgill and Karamura 1999).

In East Africa, where the differences are slight, the daughter suckers may retain the same name and be regarded as having the same identity as the mother plant. However, where significant differences are noticed, farmers may use a new name for the propagules. Differences of this type are nearly always associated with the bunch and are significant to farmers, traders, or consumers. For example, the cooking highland bananas reputedly change into beer bananas, becoming bitter because they contain more tannin. The beer bananas may get new names, but at times they retain their original name. This is because the phenotype of beer bananas remains the same as that of the cooking banana, although they can no longer be eaten raw or cooked because of the higher tannin content. Thus we find names such as *Nakabululu-enyamuunyo* (cooking) and *Nakabululu-embiire* (beer), and *Nakabululu-embiire* also has its own name: *Enshyenyuka*. If the change is noticeable but minor, such as a change in pigmentation of pseudostem, petioles, or midrib, the daughter suckers may retain the name, with an additional part to indicate the change. For example *Nakitembe* (*Musa* AAA), an east African clone, usually has green petioles and midribs, but there is a mutant with red petioles and midribs called *Nakitembe omu-myufu* (red) (Karamura and Karamura 2004).

These mutations appear to increase the variability of genotypes in an area and, depending on the success of any mutant type, to alter the distribution of the existing diversity (box 4.2). In Uganda, *Siira*, a highland banana cooking clone with a medium-sized bunch, has a mutant (*Atwalira*) with a more cylindrical, compact, heavy bunch. The mutant has become more commercial than its original parent and has spread to areas not occupied by the original parent (D. Karamura, pers. comm., 2004). Although mutation may play a role in generating new variation in a number of clonally propagated species, the extent to which this occurs and new types enter the seed supply systems does not seem to have been investigated as such. In taro in northern Vietnam different organs are used to propagate different varieties. It would be interesting to investigate whether the different organs and methods of propagation were associated with different mutation rates and had any effect on the variability found in materials exchanged of the different varieties (table 4.2).

In some crops it is possible that farmers may have inadvertently selected mutable systems (e.g., those caused by the presence of transposable elements) because they generate new, distinctive color patterns on seeds, stalks, and flowers. One case could be the Ac/Ds system in maize, in which transposable elements induce mutations with phenotypic effects. Clegg and Durbin (2000) suggested that early human domesticators of *Ipomoea purpurea* may have seized on the flower color diversity that is a consequence of the rich variety of mobile elements residing in the morning glory genome. Such patterns can act as identifiers of varieties or provide interesting new properties. In these ways mutation may act as an identifiable if minor cause of genetic changes in varieties and in seed-propagated plants and therefore might influence variety maintenance and exchange.

Selection

FARMER SELECTION AND NATURAL SELECTION

The genetic makeup of local varieties is likely to depend on the effects of both natural selection and selection (conscious and unconscious) by farmers (box 4.3). For many characters, farmer selection may reinforce

Box 4.2 Banana Seed Flow in Uganda

In general, two banana seed pathways are recognizable—the traditional and nontraditional seed systems—although at the farm level the two pathways usually merge. By far this is the oldest and most widespread system whereby farmers deliberately select and collect seeds from friends, neighbors, or relatives far and near and plant them in their own gardens. The selection of seed follows well-defined criteria across the region (box table 4.2).

Traditional pathways are characterized by high cultivar diversity per farm, and the stand may have as many as 30 different cultivars grown in complex mixtures. Once a cultivar is selected and introduced, it is normally grown near the main house or kitchen, from which the mat is observed for several ratoon crops with respect to bunch size, food quality, response to pests and diseases, and other characteristics before it is transplanted to an appropriate site in the garden for production and conservation. Another characteristic of the system is its low-input, low-output behavior. In general, sucker seeds collected from neighbors, relatives, and friends are not cleaned and consequently carry along to the next farm a lot of soilborne pests and diseases. The system appears to have survived as a kind of barter trade whereby planting materials are exchanged without money involved. Thus any attempt to improve it would need to take into account that farmers traditionally do not buy banana planting material.

BOX TABLE 4.2. Percentages of farmers using various criteria to select planting material.

Criterion	Tanzania		Uganda	
	Chanika	Ibwera	Masaka	Bushenyi
Bunch size	29	35	32	26
Taste	22	18	17	16
Maturity period	18	10	21	12
Resistance to diseases	6	11	17	19
Plant vigor	2	6	1	0
Ratooning ability	2	8	0	2
Marketability of bunches	4	4	3	8
Drought tolerance	4	0	0	0
Large fingers	3	1	2	4
Softness of cooked food	1	2	1	0
Adaptability to soil	1	1	1	2
Longevity	1	1	1	5

Source: Karamura and Karamura (2004).

Table 4.2. Farmer methods of propagating taro varieties in northern Vietnam, 2002.

Method of Propagation	Cultivar	Agroecosystem	Distribution Pattern
Cormel and sucker	*Khoai lui doc xanh*	Lowland	Widespread
	Chat chay hau	Lowland, upland	Widespread
	Mac phuoc mong	Upland	Widespread
	Hau danh pe	Upland	Widespread
	Khoai mung tia	Upland	Widespread
	Mon tia	Lowland, home garden	Widespread
Young suckers	*Nuoc tia*	Moist area around well	Widespread
	Nuoc xanh	Moist area around well	Widespread
	Khoai ngot	Lowland, home garden	Widespread
	Bac ha	Home garden, moist area	Narrow
	Tam dao xanh	Upland, home garden	Widespread
	Hau xi	Home garden, moist area	Widespread
Stolon	*Man hua vai*	Upland	Widespread
	Hau giang	Upland	Widespread
	Khoai doi	Lowland	Widespread
Head of corm	*Kao pua*	Upland	Narrow
	Hau Danh chun	Upland	Widespread
	Mat qui	Upland	Widespread
Eyes of corm	*Hau doang*	Upland	Widespread
	Phuoc oi	Upland	Widespread
Seed and suckers	*Kay nha*	Home garden and upland	Widespread

Box 4.3 Ethnic, Social Status, Age, and Gender Differences in Relation to Variety Selection and Storage

In some regions of Burkina Faso, according to their age and social rank, women intervene equally in the selection of crop varieties in the village and neighboring areas. Women are the main actors in the processing of grain to other food and fodder products, processing more than 95% of the harvest. Retail marketing of local varieties is carried out almost entirely by women, but both men and women conduct wholesale marketing. Women, especially among the Bixa people of Médéga, were found to play an important role in choosing varieties to plant for sorghum, pearl millet, groundnut, and cowpea.

A study of farmer age and seed selection was made in Burkina Faso among women (17–90 years old) and men (23–79 years old). At all village levels women older than 50 years are involved in the breeding and conservation of seeds (box table 4.3). Any woman who breeds or conserves seeds must keep in mind some rituals in Burkina Faso; for example, on the day when seeds have to be prepared and conserved, a female farmer must not have participated in sexual activity the day before, and any woman in the family who is pregnant or menstruating must not touch seed containers, nor should she touch tamarind fruit, milk, pearl millet, or sorghum powder. To ensure good seed conservation, selection must be done during moonless periods. Similar studies in Morocco showed that for seed selection, women were as knowledgeable as their husbands or fathers about the differences between landraces and improved varieties.

BOX TABLE 4.3. Burkina Faso: Breeding systems are managed by men and women according to decision-making criteria, socioeconomic interests, and rituals.

	Decision Maker	
Crop	Male	Female
Sorghum	+++	++
Millet	+++	++
Cowpea	+++	++
Groundnut	+	+++
Frafra potato	+++	o
Okra	o	+++

+++ = high decision making (intervene always);
++ = medium decision making (intervene sometimes);
+ = low or weak decision making (intervene rarely);
o = no decision making (never intervene).

Ears of sorghum, millet, and corn are stored in granaries. Local plants are used to protect the grain in storage from pests, including *Cissus quadrangularis*, *Sansevieria senegambica*, *Hyptis spicigera*, and *Cassia migricans*. The plant is freshly ground, mixed with water, and spread in the granary before storage. Shea almond residue is also used to protect against pests. Grain is also blended with ash and stored in jars. This process is carried out in the early morning or the evening, without natural light. Pregnant and menstruating women do not participate in the process.

Source: Madibaye Djimadoum, Fédération National des Groupements Naam.

environmental selection over time, particularly selection for tolerance of adverse soil, climate, or pest environments. In Yucatán, for instance, the average time to flowering of the most widely planted landrace, *xnuuknal*, corresponds closely with the period of time between the average onset of the rainy season (when planting commences) and the peak of average monthly rainfall. The latter peak falls precisely when the male maize flowers release pollen, when the female flowers become receptive, and when the fertilized ears begin to develop—in short, when the maize plants are at a physiological stage at which water demands are most critical. This is also likely to be the case with selection for adaptation to specific abiotic or biotic stresses.

Agromorphological studies have shown that local varieties of sorghum collected in Mali in 1998 and 1999 were 7–10 days earlier in time to maturity than those collected 20 years earlier, either as the result of natural selection or as a result of farmer selection favoring materials with shorter maturation times in an environment with increasingly uncertain moisture availability (M. Grum, pers. comm., 2001).

DISASTERS AND CATASTROPHES

Extreme environmental events may trigger unusually intense selection pressures on crop genepools. In Mexico in 2002, central Yucatán State was devastated by a hurricane that flattened the maize crop at the peak of the ripening period for long-cycle landraces (these account for about 80%

of the total subsistence maize harvest in Yucatán). Although harvest losses were large, about 75% of maize farmers were able to salvage at least small amounts of seed of their *xnuuk-nal* varieties. In doing so they tended to draw on the plants in each population that were slightly more advanced in physiological maturation when the hurricane struck and thus were able to complete development of viable seed. In effect, the hurricane acted as a selection force, potentially shifting the mean ripening time forward in many Yucatán maize populations (for any given population the intensity of selection was also influenced by the planting date).

The hurricane also reconfigured local crop populations, most prominently in the case of common beans (*Phaseolus vulgaris* var. *xkolibu'ul*) and lima beans (*Phaseolus lunatus*). Before 2002, annual surveys in the community of Yaxcaba revealed common beans to be grown by 66–70% of farmers and lima beans by 45–65%. In 2002, more than 90% of common bean and 83% of lima bean seed lots were lost entirely by farmers through hurricane-related damage. Most of the remaining seed lots were reduced to a fraction of their normal size (i.e., a handful of seeds rather than several kilograms). In 2003 the proportion of farmers replanting with common bean and lima bean stocks was only 20% for each, and the majority of these farmers obtained their seed off farm, particularly via exchanges and purchases from farmers in other communities where bean supplies could be found. The net effect was a major reduction in local bean population numbers, with many local subpopulations disappearing entirely, to be replaced by new materials from alternative sources that were themselves subject to local selection forces and, presumably, some shift in their own characteristics.

During emergencies (droughts and floods) farmers travel to other villages with similar environmental conditions to exchange or buy seed. The local market is an important source of seed, especially during emergencies, but often the poorest farmers cannot afford to buy seed. Not all villages have markets, and farmers in remote areas that are far from markets have been found to be more vulnerable to seed insecurity.

Major environmental events such as hurricanes and floods can cause a dramatic change in planting materials, and chronic seed insecurity can also lead to continuing changes in the materials used by farmers. Sadiki et al. (2005) found that Moroccan farmers tended to depend on their own faba bean seed in good years, and in poor years they obtained seed from local markets. In these circumstances, the development of different local populations (whose genetic constitution reflects local selection and drift) maintained by individual farmers alternates with replacement from bulk sources obtained from nearby markets or elsewhere.

SEED SELECTION PRACTICES

In selecting for varieties based on agromorphological characteristics, farmers' practices influence the maintenance of genetic diversity on farm. Steps in seed selection that affect the genetic makeup of varieties over time include selection of plots or portions of the field before harvest and selection of plants or parts of plants (between and within fruits or inflorescences) at harvest (Wright et al. 1994).

In maize a whole series of selection practices has been identified that are almost certainly important in their effects on the maintenance of variety identity and the genetic diversity within varieties, although the precise genetic effects often are unclear. Thus, Louette and Smale (2000) found that the traditional seed selection practices of Mexican farmers conserved the integrity of the ear characteristics that defined their varieties even in the presence of significant gene flow caused by cross-pollination between varieties.

The timing and sequence of seed selection can vary from farmer to farmer within a single community. In Yaxcaba, Yucatán, as throughout Mexico, most farmers select their maize landrace seed every season based primarily on ear and grain characteristics, including ear size, ear health, uniformity of grain color, and grain size. In Yucatán, the selection process does not take place all at once but rather unfolds in several steps, beginning with the maize harvest in January and ending with the onset of planting at the end of May. In the initial stage of selection, when farmers separate out the highest-quality ears, most farmers store the ears in the husk and evaluate their potential seed stock based on ear characteristics such as size, weight, and husk coverage. From this pool of unhusked ears, some farmers immediately select ears for seed, put them aside, and destine the remainder for eventual consumption. However, other farmers simply store all their high-quality ears together, drawing on this stock for consumption as need arises and then selecting their seed ears shortly before the planting season. Grain characteristics come into play when the ears are husked and degrained, which farmers usually do a day or two before planting begins. However, a minority of farmers prefer to husk their maize before storage, raising the possibility that they take grain characteristics into account at an earlier stage in the selection process.

In some cases the effects of farmers' practices are much less clear and remain the subject of speculation. Throughout Mesoamerica the grains at the distal and basal ends of the maize cob generally are not used as seed (Johannessen et al. 1970). Although the different sections of the maize cob may not differ with respect to their genetic makeup, pollen competition and selection do seem to occur in maize and lead to differences in the genetic makeup of seed taken from different sections of the cob (Mulcahy et al. 1996).

In Nepal, rice farmers may select at the level of the ear or panicle or at the level of the seed (table 4.3; Rana 2004). The time of selection may depend on the appearance of reproductive organs or on vegetative characteristics, if these are considered important. There may also be ritual practices associated with selection of seeds or with the preparation of seed before planting (Rana 2004).

The amount of seed needed is also important. In Nepal the amount of seed needed for varieties grown in large areas was large, and farmers selected good plots, rogued the off-types, and then harvested the entire plot for seed (negative selection). However, for varieties grown in small areas, selection of the best panicles (positive selection) was more common (figure 4.3). Rana (2004) found that farmers took special care in seed selection of varieties grown in small areas for religious and cultural purposes because "impure" mixtures are not permissible for offering to God.

Although seed selection is important in many farming situations, it is by no means universal. Seed selection of rice from marginal and share-cropped plots in Nepal was not common unless farmers had no other choice for seeds, nor was seed selection always done on an annual or regular basis. In some circumstances farmers selected seeds only when the populations became too infested by off-types or when there were problems of disease and pests, sterility, and lodging (Rana 2004).

In Burkina Faso, pearl millet farmers harvest seed from the center of the field to maintain "purity." They harvest millet spikes and sorghum panicles from a range of plant parent types, taking into account uniformity of grain color and spikelet dehiscence. This practice appears to favor seed quality and seed vigor. When farmers followed this practice over five years (1997–2002) it appeared that 20–48% of households obtained better seed quality. Overall it was found that some seed selection practices were used at harvest, threshing, and drying as well as before storage and planting (Balma et al. 2005).

Table 4.3. Maintenance of seed quality by farmers during different operations, Kaski, Nepal, 2000.

Activity	Harvesting		Cleaning		Drying		Storing		Planting	
	No.	%*	No.	%*	No.	%*	No.	%*	No.	%*
Plant in the same place	—	—	—	—	—	—	—	—	2	4
Harvest at same time	1	1	—	—	—	—	—	—	—	—
Not harvest on Tuesday	4	6	—	—	—	—	—	—	—	—
Keep the seed cormel separate or remove off-types	5	8	3	4	1	2	10	17	1	2
Plant different varieties together	1	1	—	—	1	2	1	2	1	2
Remove other varieties from plot (rogueing)	—	—	—	—	—	—	—	—	—	—
Clean the corm and cormels	—	—	55	90	1	2	—	—	—	—
Dry for 2–4 days for seed	—	—	—	—	37	68	—	—	—	—
Keep in dry place or Kbol†	—	—	—	—	11	20	8	14	—	—

Table 4.3 Continues next page

Table 4.3: continued

Activity	Harvesting		Cleaning		Drying		Storing		Planting	
	No.	%*	No.	%*	No.	%*	No.	%*	No.	%*
Remove diseased seeds and plant after good soil preparation	—	—	1	2	—	—	1	2	6	11
Select better types from the heap	1	1	—	—	—	—	1	2	—	—
Save a cutting during harvest	55	82	1	2	—	—	—	—	—	—
Keep in the big basket (Doko†)	—	—	—	—	—	—	19	33	—	—
No more drying needed	—	—	—	—	1	2	—	—	—	—
Plant behind plow	—	—	—	—	—	—	—	—	34	65
Not much cleaning needed	—	—	—	—	—	—	—	—	—	—
Plant in the pit without breaking the sprout	—	—	1	2	1	2	1	2	4	8
Plant disease-free seed	—	—	—	—	—	—	—	—	1	2

Store in *Mach*† or *Khol*†	—	—	—	—	—	—	13	13	—	—
Store together	—	—	—	—	—	—	1	2	—	—
Cultivation area depends on needs	—	—	—	—	—	—	—	—	—	—
Keep separately in the pit	—	—	—	—	—	—	2	3	—	—
Plant separately	—	—	—	—	—	2	—	—	2	4
Seed tubers are not dried in the direct sun	—	—	—	—	1	2	—	—	—	—
Total	67	100	61	100	54	100	57	100	52	100

*The number indicates column percentages.

†*Doko* = big basket made from bamboo that is used to carry bulk material such as grasses and fodders; *Khol* = special structures made from small pieces of wood; *Mach* (*Machan*) = temporary structure made by locally available materials that is above the ground.

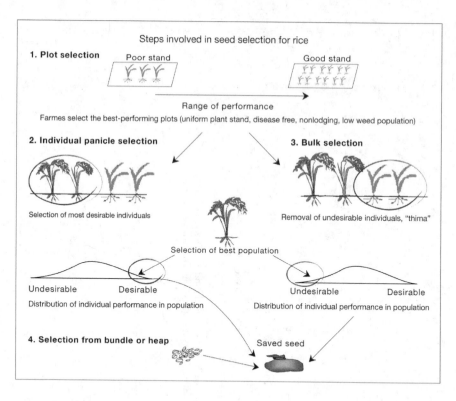

FIGURE 4.3. Rice seed selection procedures practiced by farmers in study sites (Rana 2004).

In contrast, in northern Morocco farmers do not traditionally store faba bean seeds separately from planting materials, and selection is made at the time of planting. Thus farmers appeared to make no special effort to maintain seed germination during storage.

ON-FARM SEED STORAGE AND SELECTION

Selection effects often continue after harvest. Seed storage devices and methods determine the vulnerability of seed to pests, diseases, and physiological deterioration, affecting seed quantity and quality for the next planting season (Gepts 1990). In addition to ensuring germinable clean seed, the conditions under which seed is stored may act as a selecting force on the seed lot. Seed better adapted to the conditions will be more likely to survive until the next planting season than less-adapted seed, with potential effects for the genetic diversity of the crop population over time.

In many situations the seed for next year's crop is not stored any differently from seed used for home consumption. In Yucatán, storage of maize seed under conditions different from those used for consumption takes place primarily in adverse years, when farmers have only a limited pool of high-quality seed ears from their own harvest. In such years, farmers carefully select and set aside their seed ears and store them under special conditions such as the rafters of the kitchen, where they can be bathed in smoke from the household cooking fire (Yupit-Moo 2002).

In contrast to the situation for faba bean in Morocco, the importance of finding a safe place or container suitable for seed storage is crucial to farmers to minimize damage to their grains during postharvest management. In Peru, farmers surveyed were asked to estimate the percentage of seed loss during storage for the last five years (Collado-Panduro et al. 2005). The highest loss percentage was recorded in maize, with 29.2%, 38%, and 17.6% of the Shipibo, Ashaninka, and *mestizo* households, respectively, reporting a loss of 75–100% of their seed (the highest percentage in at least one of the recent five years). Causes included weevils (*Sitophilus* spp.) and moths (*Sitotroga cerealella*), which affected mainly maize landraces with semihard grain (some hybrids between Cuban yellow and Piricinco races) and some with soft-floury grains (such as Piricinco race). Less damaged were landraces with hard grains. Beans also presented significant loss. For example, 41.2%, 19.9%, and 16% of the Ashaninka, Shipibo, and *mestizo* households, respectively, estimated a loss of more than 75% of the stored seed (the highest percentage in at least one of the recent five years). The same pests slightly affected peanut seed (3% of households). Every community faced significant losses during seed storage. Therefore, seed storage is a fragile point in the community seed supply system and postharvest management.

In Yucatán, traditional storage methods of maize and bean seed seem to be more robust than in Peru. Farmers in Yaxcaba indicated that postharvest losses normally were low, and Yupit-Moo (2002) reported that Coleopteran pests had damaged less than 20% of maize landrace ears after more than one year of storage in the husk in traditional granaries. This may reflect both the general suitability of the environment, with an extended dry season in the region, and the morphological adaptation of local landraces for storage, including an elongated, tightly fitting husk. A common complaint that Yucatán farmers voice with respect to modern varieties is their susceptibility to insect pests under traditional storage conditions.

After sowing, selection continues as a result of specific cultivation techniques. Examples include sowing many seeds at a single station, which are then thinned after germination, and the elimination of unwanted male flowering plants. However, the genetic effects of these practices have been little studied in the work reviewed here.

Both natural selection and selection by farmers on individual populations of local varieties seem likely to increase differences between the populations and their seed lots. Over time, because each farmer follows slightly different practices and material is grown in different fields, farmers' seed lots tend to diverge with respect to many selectable traits. However, this may not be the case for the key properties that characterize a variety. Where there is a common or shared understanding of some of the specific traits that a variety possesses (e.g., earliness, flavor, seed color), farmer selection maintains them as common to all populations.

Countering the tendency for local populations to diverge as a result of selection are the effects of migration caused by exchange of materials or occasional seed purchase from markets. These materials are subject to selection again, creating new local populations with improved adaptation to the particular farms and farmers in any area.

Conclusion

Farmers need healthy, viable seeds of the variety they prefer to be available at the appropriate time (Weltzien and vom Brocke 2000). Farmers look for true-to-type seeds from trusted sources. Yet it is important to remember that farmers' true-to-type criteria may differ substantially from those of crop breeders because farmers may not emphasize agromorphological uniformity but rather other traits to meet economic, environmental, or cultural needs.

As agricultural systems change through intensification, environmental variation, or stochastic events, the seed needs of farmers also change. One challenge will be to ensure that within these changing conditions the seed flow and seed production systems continue to supply material based on large enough population sizes for the adaptive capacity of the system to continue while meeting farmers' preferences. A second challenge will be to support selection practices that ensure the quality, appropriateness, and diversity of the material based on farmers' preferences. What is apparent is the dual importance of maintaining high levels of phenotypic and genetic diversity within seed systems while maintaining particular varieties to meet both present and future needs.

Although there are clearly many different and variably complex practices associated with the harvest, management, exchange, and use of seed of different crops, the genetic significance of these practices in terms of identity and the patterns of genetic diversity found in local varieties is much less clear. Given the general impression of dynamic systems in which population differentiation and exchange are characteristic, it is tempting to talk of varieties as metapopulations (Zimmerer 2003). However, we lack clear data to demonstrate that this is the case, and there is certainly a need for additional studies that analyze diversity patterns in these traditional farming systems in appropriate ways. These studies should include analyses that help us understand the role of markets as centers for mixing populations of a variety and the ways in which exchange generally supports migration of new or different genes. We also need to understand better the extent to which selection effects significant changes in the different populations of a variety.

One approach is to invoke the most plausible of available theoretical models of population genetics (e.g., "island–mainland," "stepping stone," "isolation-by-distance," "metapopulation") and compare the data with the key parameters of such models (e.g., migration rates, local population sizes, local extinction probabilities). A second approach is to build a computer model of the system that tracks the varietal composition of the standing crop in a community and simulates its behavior in time, introducing fluctuations (as suggested by the observed variability in processes) and periodic major disruptions. The processes include the fractions of seed for various varieties of different sources (from farmers' stored seed, neighbors, local markets, formal sector) and put to different uses (consumed, stored, traded, sold). This kind of modeling has been helpful in estimating the survival probabilities of endangered wild plant populations (Young et al. 2000). In this way we can aim to appraise the current trends and the resilience of the seed system and determine the critical parameters for survival of diversity. Situations could occur in which farmer-managed networks of partially isolated subpopulations are optimum under a current system but would no longer be so under greater agricultural intensification. Such results might forewarn us when interventions are needed to allow a seed system to retain its adaptive capacity.

Acknowledgments

This work is a result of the IPGRI-supported global project "Strengthening the Scientific Basis of *In Situ* Conservation of Agricultural Biodiversity On-Farm." The authors would like to thank the governments of Switzerland (Swiss Agency for Development and Cooperation), the Netherlands (Directorate-General for International Cooperation), Germany (Bundesministerium für Wirtschaftliche Zusammenarbeit/Deutsche Gesellschaft für Technische Zusammenarbeit), Canada (International Development Research Centre), Japan (Japan International Cooperation Agency), Spain, and Peru for their financial support.

References

Almekinders, C. J. M., N. P. Louwaars, and G. H de Bruijn. 1994. Local seed systems and their importance for an improved seed supply in developing countries. *Euphytica* 78:207–216.

Arbaoui, L. 2003. *Analyse des facteurs évolutifs de la diversité génétique de la fève (Vicia faba L.) in situ et leurs impacts sur son maintien et sa gestion à la ferme: Contribution au développement des bases scientifiques pour la conservation in situ de la fève au Maroc.* Mémoire de troisième cycle pour l'obtention du diplôme d'ingénieur d'état en agronomie, Option: Amélioration Génétique des Plantes. Rabat, Morocco: IAV Hassan II.

Badstue, L. B., M. Bellon, X. Juárez, I. Manuel, and A. M. Solano. 2002. *Social Relations and Seed Transactions Among Small-Scale Maize Farmers in the Central Valleys of Oaxaca, Mexico: Preliminary Findings.* CIMMYT Economics Working Paper 02–02. Mexico City: CIMMYT.

Balma, D., T. J. Ouedraogo, and M. Sawadogo. 2005. On-farm seed systems and crop genetic diversity. In D. I. Jarvis, R. Sevilla-Panizo, J.-L. Chavez-Servia, and T. Hodgkin, eds., *Seed Systems and Crop Genetic Diversity On-Farm*, 48–53. Proceedings of a workshop, September 16–20, 2003, Pucallpa, Peru. Rome: IPGRI.

Baniya, B. K., A. Subedi, R. B. Rana, R. K. Tiwari, and P. Chaudhary. 2003. Finger millet seed supply system in Kaski district of Nepal. In *On-Farm Management of*

Agricultural Biodiversity in Nepal, 171–175. Proceedings of a national workshop, April 24–26, 2001, Lumle, Nepal. Kathmandu, Nepal: NARC/LI-BIRD/IPGRI.

Bellon, M. R. and J. Risopoulos. 2001. Small-scale farmers expand the benefits of improved maize germplasm: A case study from Chiapas, Mexico. *World Development* 29(5):799–811.

Bond, D. A. and M. H. Poulsen. 1983. Pollination. In P. D. Hebblethwaite, ed., *The Faba Bean (*Vicia faba *L.),* 77–101. London: Butterworths.

Brown, A. H. D. 2000. The genetic structure of crop landraces and the challenge to conserve them in situ on farms. In S. B. Brush, ed., *Genes in the Field,* 29–48. Ottawa, Canada: IDRC/IPGRI/Lewis Publishers.

Brush, S., R. Kesseli, R. Ortega, P. Cisneros, K. Zimmerer, and C. Quiros. 1995. Potato diversity in the Andean center of crop domestication. *Conservation Biology* 9:1189–1198.

Castiñeiras, L., Z. Fundora, S. Pico, and E. Salinas. 2001a. Monitoring crop diversity in home gardens as a component in the national strategy of in situ conservation of plant genetic resources in Cuba, a pilot study. *Plant Genetic Resources Newsletter* 123:9–18.

Castiñeiras, L., Z. Fundora Mayor, T. Shagarodsky, V. Moreno, O. Barrios, L. Fernández, and R. Cristobal. 2001b. Contribution of home gardens to in situ conservation of plant genetic resources in farming systems: Cuban component. In J. W. Watson and P. B. Eyzaguirre, eds., *Contribution of Home Gardens to In Situ Conservation of Plant Genetic Resources in Farming Systems.* Proceedings of the Second International Home Gardens Workshop, July 17–19, 2001, Witzenhausen, Germany. Rome: IPGRI.

Chaudhary, P., D. Gauchan, R. B. Rana, B. R. Sthapit, and D. I. Jarvis. 2004. Potential loss of rice landraces from a Terai community in Nepal: A case study from Kachorwa, Bara. *Plant Genetic Resources Newsletter* 137:14–22.

Clegg, M. T. and M. L. Durbin. 2000. Flower color variation: A model for the experimental study of evolution. In F. J. Ayala, W. M. Fitch, and M. T. Clegg, eds., *Variation and Evolution in Plants and Microorganisms: Towards a New Synthesis 50 Years After Stebbins,* 211–234. Washington, DC: National Academy of Sciences.

Collado-Panduro, L., J. L. Chavez-Servia, A. Riesco, and R. Soto. 2005. Community systems of seed supply and storage in the central Amazon of Peru. In D. I. Jarvis, R. Sevilla-Panizo, J.-L. Chavez-Servia, and T. Hodgkin, eds., *Seed Systems and Crop Genetic Diversity On-Farm,* 103–108. Proceedings of a workshop, September 16–20, 2003, Pucallpa, Peru. Rome: IPGRI.

Coomes, O. T. 2001. Crop diversity in indigenous farming systems of Amazonia: The role and dynamics of agricultural planting stock transfers among traditional farmers. In *Abstracts. International Symposium on Managing Biodiversity in Agricultural Ecosystems,* 27, November 8–10, 2001, Montreal, Canada.

Dominguez, C. E. and R. B. Jones. 2005. The dynamics of local seed systems in Mozambique, and the roles played by women. In D. I. Jarvis, R. Sevilla-Panizo, J.-L. Chavez-Servia, and T. Hodgkin, eds., *Seed Systems and Crop Genetic Diversity On-Farm,* 141–148. Proceedings of a workshop, September 16–20, 2003, Pucallpa, Peru. Rome: IPGRI.

Gaifani, A. 1992. Developing local seed production in Mozambique. In D. Cooper, R. Vellvé, and H. Hobbelink, eds., *Growing Diversity,* 97–105. London: Intermediate Technology Publications.

Gepts, P. 1990. Genetic diversity of seed storage proteins in plants. In A. H. D. Brown, M. T. Clegg, A. L. Kakler, and B. S. Weir, eds., *Plant Population Genetics, Breeding and Genetic Resources,* 64–82. Sunderland, MA: Sinauer Associates.

Gepts, P. and R. Papa. 2003. Possible effects of (trans) gene flow from crops on the genetic diversity from landraces and wild relatives. *Environmental Biosafety Research* 2:89–103.

Ghaouti, L. 2003. *Analyse de la diversité génétique de la fève in situ et étude des mécanismes de sa maintenance à la ferme: Contribution au développement des bases scientifiques pour la conservation in situ de la fève au Maroc.* Mémoire de troisième cycle pour l'obtention du diplôme d'ingénieur d'état en agronomie, Option: Amélioration Génétique des Plantes. Rabat, Morocco: IAV Hassan II.

Gillespie, J. H. 1998. *Population Genetics: A Concise Guide.* Baltimore, MD: John Hopkins University Press.

Granovetter, M. 1973. The strength of weak ties. *American Journal of Sociology* 78:1360–1380.

Hardon, J. and W. de Boef. 1993. Linking farmers and breeders in local crop development. In W. de Boef, K. Amanor, K. Wellard, and A. Bebbington, eds., *Cultivating Knowledge: Genetic Diversity, Farmer Experimentation and Crop Research,* 64–71. London: Intermediate Technology Publications.

Jarvis, D. I. and T. Hodgkin. 2000. Farmer decision making and genetic diversity: Linking multidisciplinary research to implementation on-farm. In S. B. Brush, ed., *Genes in the Field,* 261–279. Ottawa, Canada: IDRC/IPGRI/Lewis Publishers.

Jarvis, D. I., R. Sevilla-Panizo, J.-L. Chavez-Servia, and T. Hodgkin, eds. 2005. *Seed Systems and Crop Genetic Diversity On-Farm.* Proceedings of a Workshop, September 16–20, 2003, Pucallpa, Peru. Rome: IPGRI.

Johannessen, C. L., M. R. Wilson, and W. A. Davenport. 1970. The domestication of maize: Process or event? *Geographical Review* 60(3):393–413.

Kabore, O. 2000. Burkina Faso: PPB, seed networks and grassroot strengthening. In D. I. Jarvis, B. Sthapit, and L. Sears, eds., *Conserving Agricultural Biodiversity In Situ: A Scientific Basis for Sustainable Agriculture,* 192–193. Rome: IPGRI.

Karamura, D. and E. B. Karamura. 2004. *Implications of chimerism in the East African highland bananas.* Unpublished manuscript.

Karamura, E. B., D. A. Karamura, and C. A. Eledu. 2005. Banana and plantain seed systems in the Great Lakes region of East Africa: A case for a clonal seed system. In D. I. Jarvis, R. Sevilla-Panizo, J.-L. Chavez-Servia, and T. Hodgkin, eds., *Seed Systems and Crop Genetic Diversity On-Farm,* 76–80. Proceedings of a workshop, September 16–20, 2003, Pucallpa, Peru. Rome: IPGRI.

Lope, D. 2004. *Gender Relations as a Basis for Varietal Selection in Production Spaces in Yucatan, Mexico.* MS thesis, Wageningen University.

Louette, D. 2005. Management of maize varieties in a traditional agricultural system of Mexico. In D. I. Jarvis, R. Sevilla-Panizo, J.-L. Chavez-Servia, and T. Hodgkin, eds., *Seed Systems and Crop Genetic Diversity On-Farm,* 95–102. Proceedings of a workshop, September 16–20, 2003, Pucallpa, Peru. Rome: IPGRI.

Louette, D., A. Charrier, and J. Berthaud. 1997. In situ conservation of maize in Mexico: Genetic diversity and maize seed management in a traditional community. *Economic Botany* 51:20–38.

Louette, D. and M. Smale. 2000. Farmers' seed selection practices and traditional maize varieties in Cuzalapa, Mexico. *Euphytica* 113:25–41.

Mar, I., A. Gyovai, G. Bela, and L. Holly. 2005. Multilevel seed movement across producers, consumers and key market actors: Seed marketing, exchange and seed regulatory framework in Hungary. In D. I. Jarvis, R. Sevilla-Panizo, J.-L. Chavez-Servia, and T. Hodgkin, eds., *Seed Systems and Crop Genetic Diversity On-Farm,* 54–59. Proceedings of a workshop, September 16–20, 2003, Pucallpa, Peru. Rome: IPGRI.

Maruyama, T. and M. Kimura. 1980. Genetic variability and effective population size when local extinction and recolonization of subpopulations are frequent. *Proceedings of the National Academy of Sciences USA* 77:6710–6714.

McArthur, R. H. and E. O. Wilson. 1967. *The Theory of Island Biogeography.* Princeton, NJ: Princeton University Press.

McGuire, S. 2001. Analyzing farmers' seed systems: some conceptual components. In L. Sperling, ed., *Targeting Seed Aid and Seed System Interventions: Strengthening Small Farmer Seed Systems in East and Central Africa.* Proceedings of a workshop, June 21–24, 2000, Kampala, Uganda. Kampala: CIAT.

Mellas, H. 2000. Morocco. Seed supply systems: Data collection and analysis. In D. I. Jarvis, B. Sthapit, and L. Sears, eds., *Conserving Agricultural Biodiversity In Situ: A Scientific Basis for Sustainable Agriculture,* 155–156. Rome: IPGRI.

Mulcahy, D. L., M. Sari-Gorla, and G. B. Mulcahy. 1996. Pollen-selection: Past, present and future. *Sexual Plant Reproduction* 9:353–356.

Ortega-Paczka, R., L. Dzib-Aguilar, L. Arias-Reyes, V. Cob-Vicab, J. Canul-Ku, and L. A. Burgos. 2000. Mexico. Seed supply systems: Data collection and analysis. In D. I. Jarvis, B. Sthapit, and L. Sears, eds., *Conserving Agricultural Biodiversity In Situ: A Scientific Basis for Sustainable Agriculture,* 152–154. Rome: IPGRI.

Perales, H. 1998. *Conservation and Evolution of Maize in the Valleys of Amecameca and Cuautla, Mexico.* Unpublished PhD dissertation, University of California, Davis. University Microfilms, Ann Arbor, Michigan.

Pickersgill, B. and D. Karamura. 1999. Issues and options in the classification of cultivated bananas, with particular reference to the East African Highland bananas. In S. Andrews, A. C. Leslie, and C. Alexander, eds., *Taxonomy of Cultivated Plants, Third International Symposium,* 159–167. Kew, UK: Royal Botanic Gardens.

Qualset, C. O., A. B. Damania, A. C. A. Zanatta, and S. B. Brush. 1997. Locally based crop plant conservation. In N. Maxted, B. V. Ford-Lloyd, and J. G. Hawkes, eds., *Plant Genetic Conservation: The In Situ Approach.* London: Chapman and Hall.

Quist, D. and L. Chapela. 2001. Transgenic DNA introgressed into traditional maize landraces in Oaxaca, Mexico. *Nature (London)* 414:541–543.

Rana, R. B. 2004. *Influence of Socio-Economic and Cultural Factors on Agrobiodiversity Conservation On-Farm in Nepal.* PhD thesis, Reading University.

Richards, P. 1989. Farmers also experiment: A neglected intellectual resource in African science. *Discovery and Innovation* 1(1):19–25.

Richards, P. and G. Ruivenkamp. 1997. *Seeds and Survival: Crop Genetic Resources in War and Reconstruction in Africa.* Rome: IPGRI.

Riesco, A. 2002. *Annual Report for the Project, "Strengthening the Scientific Basis of In Situ Conservation of Agricultural Biodiversity": Peru Country Component.* Rome: IPGRI.

Sadiki, M., M. Arbaoui, L. Ghaouti, and D. Jarvis. 2005. Seed exchange and supply systems and on-farm maintenance of crop genetic diversity: A case study of faba bean in Morocco. In D. I. Jarvis, R. Sevilla-Panizo, J.-L. Chavez-Servia, and T. Hodgkin, eds., *Seed Systems and Crop Genetic Diversity On-Farm,* 81–86. Proceedings of a workshop, September 16–20, 2003, Pucallpa, Peru. Rome: IPGRI.

Sperling, L., ed. 2001. *Targeting Seed Aid and Seed System Interventions: Strengthening Small Farmer Seed Systems in East and Central Africa,* 9–13. Proceedings of a workshop, June 21–24, 2000, Kampala, Uganda. Kampala: CIAT.

Subedi, A., P. Chaudhary, B. Baniya, R. Rana, R. K. Tiwari, D. Rijal, D. I. Jarvis, and B. R. Sthapit. 2003. Who maintains genetic diversity and how? Policy implications for agro-biodiversity management. In D. Gauchan, B. R. Sthapit, and

D. I. Jarvis, eds., *Agrobiodiversity Conservation On-Farm: Nepal's Contribution to a Scientific Basis for Policy Recommendations.* Rome: IPGRI.

Subedi, A. and C. Garforth. 1996. Gender information and communication networks: Implications for extension. *European Journal of Agricultural Education and Extension* 3(2):63–74.

Tripp, R. 2001. *Seed Provision and Agricultural Development.* London: Overseas Development Institute.

Upadhaya, M. P. 1996. Rice research in Nepal: Current state and future priorities. In R. E. Evenson, R. W. Herdt, and M. Hossain, eds., *Rice Research in Asia: Progress and Priorities,* 193–215. Wallingford, UK: CAB International.

Valdivia, R. F. 2005. The use and distribution of seeds in areas of traditional agriculture. In D. I. Jarvis, R. Sevilla-Panizo, J.-L. Chavez-Servia, and T. Hodgkin, eds., *Seed Systems and Crop Genetic Diversity On-Farm,* 17–21. Proceedings of a workshop, September 16–20, 2003, Pucallpa, Peru. Rome: IPGRI.

Wang, J. and A. Caballero. 1999. Developments in predicting the effective size of subdivided populations. *Heredity* 82:212–226.

Watson, J. W. and P. B. Eyzaguirre, eds. 2002. *Home Gardens and In Situ Conservation of Plant Genetic Resources in Farming Systems.* Proceedings of the Second International Home Gardens Workshop, July 17–19, 2001, Witzenhausen, Germany. Rome: IPGRI.

Weltzien, E. and K. vom Brocke. 2000. Seed systems and their potential for innovation: Conceptual framework for analysis. In L. Sperling, ed., *Targeting Seed Aid and Seed System Interventions: Strengthening Small Farmer Seed Systems in East and Central Africa,* 9–13. Proceedings of a workshop, June 21–24, 2000, Kampala, Uganda. Kampala: CIAT.

Whitlock, M. C. 2003. Fixation probabilities and time in subdivided populations. *Genetics* 164:767–779.

Wright, M., T. Donaldson, E. Cromwell, and J. New. 1994. The retention and care of seeds by small-scale farmers. *NRI Report* R2103.

Yadav, R. B., P. Chaudhary, S. P. Khatiwada, J. Bajrachara, R. K. Yadav, M. P. Upadhaya, B. R. Sthapit, A. Gautam, and B. K. Joshi. 2003. Agro-morphological diversity of sponge gourd (*Luffa cylindrica* L.) in Bara, Nepal. In *On-Farm Management of Agricultural Biodiversity in Nepal,* 42–47. Proceedings of national workshop, April 24–26, 2001, Lumle, Nepal. Kathmandu, Nepal: NARC/LI-BIRD/IPGRI.

Young, A. G., A. H. D. Brown, B. G. Murray, P. H. Thrall, and C. H. Millar. 2000. Genetic erosion, restricted mating and reduced viability in fragmented populations of the endangered grassland herb: *Rutidosis leptorrhynchoides.* In A. Young and G. Clarke, eds., *Genetics, Demography and Viability of Fragmented Populations,* 335–359. Cambridge: Cambridge University Press.

Yupit-Moo, E. 2002. *Seed Storage Systems of Milpa Crops in Yaxcaba, Yucatan.* Unpublished BS thesis, Instituto Tecnologico Agropecuario No. 2, Yucatán, Mexico.

Zangre, R. 1998. Selection by farmers of agromorphological character and genetic diversity: Methodologies for data collecting and analysis in Burkina Faso. In D. I. Jarvis and T. Hodgkin, eds., *Strengthening the Scientific Basis of In Situ Conservation of Agricultural Biodiversity on-Farm. Options for Data Collecting and Analysis.* Proceedings of a workshop to develop tools and procedures for in situ conservation on-farm, August 25–29, 1997, Rome, Italy. Rome: IPGRI.

Zimmerer, K. S. 1996. *Changing Fortunes: Biodiversity and Peasant Livelihood in the Peruvian Andes.* Los Angeles: University of California Press.

Zimmerer, K. 2003. Geographies of seed networks and approaches to agrobiodiversity conservation. *Society & Natural Resources* 16:583–601.

Zimmerer, K. S. and D. S. Douches. 1991. Geographical approaches to native crop research and conservation: The partitioning of allelic diversity in Andean potatoes. *Economic Botany* 45:176–189.

5 ❦ Measures of Diversity as Inputs for Decisions in Conservation of Livestock Genetic Resources

J. P. GIBSON, W. AYALEW, AND O. HANOTTE

Mapping Livestock Genetic Diversity

More than 6,379 documented breed populations of some 30 species of livestock have been developed in the 12,000 years since the first livestock species were domesticated (Scherf 2000). These breeds have evolved adaptations that allow livestock production in a wide range of situations, including some of the most stressful natural environments inhabited by humans. These naturally evolved genetic characteristics provide a coherent basket of sustainable options for disease resistance, survival, and efficient production that have often been ignored in the drive to find technological and management solutions to individual problems of livestock production in low-input systems. It is estimated that 35% of mammalian breeds and 63% of avian breeds are at risk of extinction and that one breed is lost every week.[1] Although it has not been clearly documented, threats to livestock genetic resources in the developing world appear to be increasing rapidly, driven primarily by rapid change in production systems and extensive use of crossbreeding. Emerging threats, such as implementation of culling policies to prevent spread of commercially important livestock diseases such as foot and mouth disease, and of zoonoses such as bovine spongiform encephalopathy and avian influenza, may risk extinction of breeds in the developed and developing worlds.

Effective conservation of livestock genetic resources, whether in situ or ex situ, entails the mobilization of substantial social and economic resources over prolonged periods of time. Such resources often are available in the developed world, where in many countries a high proportion of rare breeds of livestock is already being conserved effectively. However, the majority of livestock genetic diversity is found in the developing world, where resources for conservation are most lacking. Therefore it seems likely that hard choices will have to be made about what to conserve with the limited resources available.

Information on genetic diversity is useful in optimizing both conservation and use strategies for agricultural genetic resources. Ideally we would like to ensure that all existing genetic variation remains available for future use and to do so in the most cost-effective manner. In practice, there will often be insufficient resources to conserve the complete genetic diversity of a given species. Even where resources are adequate, we do not have full knowledge of all functional genetic variations within a species. Thus, achieving close to 100% conservation of functional variation would entail an inefficient process of conserving far more individuals or populations (e.g., races, breeds, accessions) than would be necessary if we had full information. Various measures can be used to obtain an indirect estimate of functional genetic diversity. Phenotypic characterization provides a crude estimate of the average of the functional variants of genes carried by a given individual or population. But most phenotypes of most agricultural plant, animal, and fish species have not been recorded. In the absence of reliable phenotype data, the most rapid and cost-effective measures of genetic diversity are obtained from assay of polymorphisms of anonymous molecular genetic markers. An important question is how estimates of molecular genetic diversity can be used to improve decision making in conservation and use of genetic resources.

This chapter provides an overview of different measures of phenotypic and genetic diversity and reviews how they might be used to inform conservation decisions in the developing world. Examples are provided here for livestock conservation that may have wider application in other agricultural species. Use of information on molecular genetic diversity to optimize the use of genetic diversity is not dealt with here, but one possible strategy is summarized by Gibson (2003).

Phenotypic Characterization as a Measure of Livestock Genetic Diversity

Historical Development

Enumeration and phenotypic characterization of livestock biodiversity are essential first steps for planning sustainable management programs. Widespread recording of the physical and productive characteristics of breeds of livestock began approximately 150 years ago in countries with highly developed economies and has continued, leading to an extensive scientific literature. The driving force behind such recording has been an economic interest in identifying and improving the most productive genotypes, between and within breeds. Such recording has been primarily in the most highly developed economies and has been dominated by the commercially most dominant breeds. More generally, however, awareness of the need for systematic phenotypic characterization of livestock biodiversity has been increasing, particularly since the United Nations Conference on Human Environment in Stockholm in 1972 (FAO 1984; Cunningham 1992; Swaminathan 1992) and the coming into force of the Convention on Biological Diversity (CBD) in 1993. Although it was not uppermost in the thoughts of nations when they signed the CBD,[2] such countries indirectly recognized the importance of sustainable management of livestock genetic resources through their signatures, and the CBD has greatly increased the level of debate about sustainable management of livestock genetic resources in the past decade.

Although livestock genetic resources have been part of the Food and Agriculture Organization (FAO) program since the establishment of the organization, a new approach was taken in the 1980s after an FAO Technical Consultation in Rome and an FAO/United Nations Environment Programme (UNEP) Expert Consultation in 1980, which led to the initiation of a joint FAO and UNEP global program on livestock genetic resources in 1982 and its implementation through 1990. A review of this program in 1989 laid the groundwork for the Global Strategy for the Management of Animal Genetic Resources (AnGR), which has been developed and implemented since 1993 at global and regional levels to provide a comprehensive framework for the management of farm animal genetic resources. An important component of the technical program of work of this strategy is the characterization of AnGR and the documentation of the information and dissemination of this information in a widely available and easily accessible global data and information system (FAO 1999).

Why Characterization?

Phenotypic characterization is undertaken to measure the diversity between defined breeds or populations to understand the extent, distribution, basic characteristics, comparative performance, utility value, and current status of the breeds or distinct populations within breeds. The essential activities include identification and inventory of the different breeds, a detailed description of their natural and adapted habitats, and recording of their phenotypic characteristics. The primary motivation of characterization work is to provide information for appropriate use to support human livelihoods (Cunningham 1992). Therefore the focus for application is generally on productive and adaptive attributes of the breeds. Coupled with accurate information on status and distribution, such information can provide baseline information essential to establish country, regional, and global priorities for the management of animal genetic resources (FAO 1984, 1999; Rege 1992). As discussed in this chapter, such assertions are easily stated but more difficult to achieve in practice.

Nature of Characterization

The most common descriptions of a breed's characteristics are based on phenotype. The phenotype of a given breed is determined by its underlying average genotype and the environment in which the animals are reared and recorded. Levels of performance generally are highly dependent on the environment in which the animals are reared. Although some appearance traits such as color pattern and horn size and shape are little affected by the environment and often vary little between animals within a breed, most performance traits are highly influenced by the environment and show substantial variation between animals within a breed. This requires that many animals be recorded in a well-defined environment in order to obtain an accurate and well-defined estimate of phenotype of the breed.

Recommendations on methods for comprehensive characterization as part of the global strategy for management of AnGR have been developed and documented (FAO 1984; Hodges 1987, 1992). Comprehensive lists of variables for describing the phenotypic and genetic characteristics (descriptor lists) of AnGR have been published (FAO 1986a, 1986b, 1986c). Such characterization includes description of the production environment in terms of key input and output variables, including biological, climatic, economic, social, and cultural dimensions (FAO 1984, 1986b, 1999).

In practice, the environmental factors that affect phenotype of animals are so complex that none of the descriptors proposed by various authorities describe the environment with sufficient accuracy to determine whether two or more breeds recorded at different locations and times were or were not recorded under sufficiently similar conditions to permit a valid comparison of their phenotypes. Scientifically valid comparisons between breeds can arise under limited conditions. The first is one in which two or more breeds are recorded simultaneously at the same location under identical management. The second is an indirect approach in which recording of different breeds takes place at different locations or times but different locations or times can be linked through use of common breeds. The effects of different environments can then be adjusted for through the differences in phenotype of the breeds in common across environments. Such studies rarely are undertaken deliberately, but data of this form arise naturally when many trials take place independently in different countries or at different times. Roughsedge et al. (2001) explored the possibility of making such indirect comparisons between breeds of beef cattle by analyzing data from many published experiments in the developed world. They concluded that their meta-analysis of published data yielded significantly more valuable information than the sum of the experimental datasets taken in isolation. Such meta-analysis is technically valid only where interactions between environment and genotype are negligible. This assumption is unlikely to be valid where data cover a very wide range of environments and genotypes, but otherwise, as a first approximation, it seems reasonable. The current difficulty for many breeds in the developing world is that where such data exist they are difficult to access. Overcoming this difficulty would be one valuable service of comprehensive livestock genetic resource information systems. However, it will remain a problem that many of the most important traits for lifetime productivity in the challenging environments typical of developing world livestock production systems are extremely difficult to record and generally are not recorded. Overall, although much valuable information is waiting to be extracted and used for many breeds, for the majority of breeds there is very little information on their phenotype for most traits of economic importance (see chapter 17).

Many decisions on conservation or use of appropriate germplasm involve the elimination of most alternative options, reducing candidates for action to a few breeds that are appropriate and accessible. In such cases, a substantial number of options can be eliminated safely based on gross phenotypic differences. In the absence of phenotypic data they can also be eliminated safely based on a low likelihood that they possess desirable characteristics, as assessed from their current distribution and use. For example, a breed that evolved outside of the region in which a particular disease is endemic is unlikely to possess useful resistance to that disease (the situation can be different for crops; see chapter 11). Similarly, breeds that evolved in moist temperate environments almost certainly will not be well adapted to dry and drought-prone tropical savannas. Thus even partial and inaccurate information on phenotype coupled with information on native distribution and distribution of current use provide valuable information for decision making. Such decision making would be greatly aided if current information on breed characteristics could be linked in publicly accessible databases to geographic information system (GIS) mapping showing the physical and disease challenge environments in which they evolved and are currently used.

In the pursuit of breed characterization, the knowledge of livestock keepers, both traditional and modern, is too often overlooked. Livestock keepers generally have a profound understanding of their stock. What is often lacking is a basis for comparison with other breeds. Coupled with the difficulties in converting terms that livestock keepers use into quantifiable measures, such knowledge can be difficult to capture accurately and often can be too general in nature to be of use. Thus, for example, claims by livestock keepers that their stock are generally disease resistant are of little value. In contrast, observations by livestock keepers that their stock are resistant (or susceptible) to specific endemic or epidemic diseases often have a basis in fact, particularly where they have had the opportunity to observe the performance of alternative breeds under the disease challenge.

Documentation of Global Diversity

Although information on the majority of breeds in the developing world is limited, often is of poor quality, and has not been collected in a systematic way, for many breeds a surprising amount of information has been collected over the past 100 years or so. Much of this information was published before electronic publishing became standard, or it appears in the gray literature of government and institutional publications or national or regional journals not easily accessed outside the country or region. Such information is very difficult to locate and cannot easily be found through standard literature searches. There is enormous potential value in bringing all this information together and making it accessible to the global community.

The first effort to document livestock biodiversity at the global scale was the work of Mason (1988). This book provides breed names, synonyms, and locations where breeds are found and gives a basic description of the origin, physical appearance, and main uses of each breed. It does not provide estimates of population sizes (other than occasionally indicating known breed status and population trends) or of production characteristics for the breed cited.

Another major contribution, though only on one species, is *Cattle Breeds: An Encyclopedia* (Felius 1995), which provides a brief account of more than 1,000 cattle breeds. The book provides a synthesis of origins, distribution, and development, possible relationships between the breeds, a brief description of the typical appearance and dimension of each breed, and some estimates of population size. Information on performance, adaptation, and disease resistance is lacking.

There is a substantial literature on breed comparisons and compilations of reports on breeds of given species in specific countries or regions. These studies often generated comprehensive accounts of breed-specific information as a baseline to develop realistic livestock development programs in the respective countries or regions. The utility of such information depends on how efficiently it is archived and delivered to the right stakeholders, and much of this literature is difficult to locate and access. Coupled with the fact that the volume of literature is growing rapidly, what is needed is to bring all this information together in a single location that is easily accessible. The natural solution is to develop electronic databases and information systems that can be accessed globally.

The idea for the establishment of regional ANGR databanks for developing countries emerged from the FAO/UNEP Joint Expert Panel on ANGR Conservation and Management meeting in Rome in October 1983 (FAO 1984). About 70% of global livestock biodiversity is in the hands of smallholder farmers, who do not generally share a global concern for maintenance of livestock biodiversity. Also, developing countries generally lack the capacity to respond adequately and effectively to the increasing rate of loss of genetic diversity. A lack of accurate information on the diversity and status of the existing farm ANGR is believed to contribute to current threats to livestock diversity.

Current Status and Future Needs of Globally Accessible Information Systems

At present there are several globally accessible, public domain electronic information systems on livestock biodiversity. A brief summary of the origins and content of these information systems is provided in box 5.1.

The existing information systems serve a variety of different purposes. Collectively they contain a substantial amount of information. But they also fall far short of what is necessary and possible for effective decision making for conservation and use. Thus far, in any of the information systems only a tiny proportion of the available information on most breeds of most species appears, and there is little functionality beyond simple searches by country or breed. The next generation of information systems will aim to capture a high proportion of the past and present information on the majority of livestock breeds and to classify the data in ways that allow users to make personal judgments about the value of each information item. The functionality of the information systems must be greatly increased to allow extraction and customized analysis of phenotype and molecular genetic data within and between data sources. It is to be hoped that the scope of the data acquisition can also be expanded so that breed information can be linked to GIS-based environment and production system mapping, allowing poorly documented characteristics such as disease resistance and adaptation traits to be predicted from past and current breed distributions and use. These are substantial but fully achievable functions that are urgently needed if researchers, policymakers, decision makers, and advisors to farming communities are to have the information they need to make appropriate recommendations and appropriate decisions for conservation and use of livestock genetic resources.

Box 5.1 Globally Accessible Information Systems on Livestock Genetic Resources

The FAO Global Databank on AnGR was initiated in 1987, when FAO collaborated with the European Association for Animal Production (EAAP) in developing an electronic information resource of descriptive information on all recognized livestock breeds and varieties found throughout the world. The databank is administered from two sites: The first, based in Hanover, Germany, serves the whole of Europe, and the second, based in Rome, Italy, serves the rest of the world. FAO coordinates data entry, with data provided by designated country representatives (FAO 1999).

The Global Databank is used to maintain breed inventories and monitor the conservation of livestock genetic resources as part of the global early warning system for domestic animal diversity. Currently it contains information on 14,000 breed populations from 35 mammalian and avian species. The World Watch List for Domestic Animal Diversity (WWL-DAD-3) was based on information from this global databank in 1999 (Scherf 2000).

DAD-IS

DAD-IS (www.fao.org/dad-is) is the first globally accessible database on AnGR developed by the FAO. It was initiated as a key communication and information tool for implementing the Global Strategy for the Management of Farm Animal Genetic Resources, mainly to assist countries and country networks in their respective country programs (FAO 1999). Apart from the country-level breed information, DAD-IS delivers a virtual library service on selected technical and policy documents, including tools and guidelines for research on AnGR. It also offers important Web links to relevant electronic information resources and has a facility to exchange views and address specific information requests by linking the range of stakeholders: farmers, scientists, researchers, development practitioners, and policymakers.

DAD-IS provides a summary of breed-level (or variety-level) information on the origin, population, risk status, special characteristics, morphology, and performance of breeds as provided by the FAO member countries. Currently the database lists 5,300 breeds in 35 species from 180 countries. A key feature of DAD-IS is that it provides a country-secure information storage and communication tool for use by the countries, with each country deciding when and what breed data are released through the officially designated contact person. The nature of the information means that it is of limited value to external users. A small amount of information is provided for most breeds in most countries, and

Box 5.1 continues on next page

Box 5.1 continued

the user cannot assess the origin, the context, or the accuracy of the information provided, so meaningful comparisons between breeds and countries are essentially impossible.

European Farm Animal Biodiversity Information System

This database (www.tiho-hannover.de/einricht/zucht/eaap/index.htm) was developed and is administered by the Department of Animal Breeding and Genetics, School of Veterinary Medicine, Hanover, Germany. It is part of the Global Databank but limited to 46 EAAP members and other European countries. As of October 23, 2003, the database included 1,935 European breed entries from eight livestock species (buffalo, cattle, goat, sheep, horse, ass, pig, rabbit). The database displays summary breed information on origin, development, population size, breed status, performance, and conservation activities. The database also provides important links to country-level databases in Germany, France, Switzerland, the Netherlands, and Austria. The website for Rare Breeds International of EAAP is also linked (EAAP Animal Genetic Databank 2003).

DAGRIS

DAGRIS (dagris.ilri.cgiar.org) is developed and managed by the International Livestock Research Institute. It was initiated in 1999 to compile and disseminate information on the origin, distribution, diversity, characteristics, present uses, and status of indigenous breeds. Information is taken from published research results. A unique feature of this database is that the breed information is supported by bibliographic references and abstracts of source publications. DAGRIS is designed to support research, training, public awareness, genetic improvement, and conservation. Version I of the database was released on the Web in April 2003 (DAGRIS 2003) and is also available on CD-ROM. Currently the database contains more than 16,000 trait records on 152 cattle, 96 sheep, and 62 goat breeds of Africa. Although it is limited currently to three species in Africa, there are plans to expand the scope of DAGRIS to Asia in the near future.

Plans for future developments of the database include the establishment of additional structures for remote uploading and downloading of noncurated breed information to increase the range of users participating in the development of the database, modules to incorporate decision-support tools for sustainable use and conservation of animal genetic resources in developing countries, modules for capture and analysis of molecular genetic information,

Box 5.1 continued

and facilities to link with GISS to provide overlays of various georeferenced data (Ayalew et al. 2003).

Oklahoma State University Breeds of Livestock

The Department of Animal Science of Oklahoma State University manages this database, opened in 1995 (www.ansi.okstate.edu/breeds). It provides a brief description of breeds in terms of origin, distribution, typical features, uses, and breed status, and a key reference on the breed. It presents a list of breeds from all over the world, with options to sort by region. As of October 2003, the database lists 1,074 breeds, including 289 sheep, 269 cattle, 229 horse, 106 goat, 73 swine, 8 donkey, 7 buffalo, 6 camel, 4 reindeer, 1 llama, 1 yak, 55 poultry, 10 duck, 7 turkey, 7 goose, 1 guinea fowl, and 1 black swan breed. It also provides links to useful information in their Virtual Livestock Library.

Molecular Genetic Marker–Based Estimates of Genetic Diversity

Molecular genetic markers are commonly used to estimate livestock genetic diversity parameters. Such information has been collected in a number of projects from a large number of breeds, although a comprehensive review is still lacking. Protein polymorphisms were the first markers used in livestock, and in the 1970s a large number of studies of genetic variation were conducted using blood group and allozyme systems (Baker and Manwell 1980; Manwell and Baker 1980); however, the level of polymorphism observed at these markers was often low, which greatly reduced the applicability in diversity studies. With the development of the polymerase chain reaction (PCR) technologies, DNA polymorphisms became the markers of choice for molecular-based surveys of genetic variation. Currently the two most popular classes of markers in livestock genetic characterization studies are mitochondrial DNA sequences, particularly the sequence of the hypervariable region of the *D-loop* or control region, and autosomal microsatellites loci (Sunnucks 2001).

Mitochondrial DNA is inherited as an extranuclear element, nearly exclusively through the maternal lineage. Each individual typically inherits a single haplotype from its dam. Mitochondrial genetic analysis therefore provides an incomplete picture of the diversity present in an individual or a population, in the absence of nuclear genomic diversity or the male-mediated gene flow analysis (Avise 1994). This is particularly important for livestock species, which are exclusively outbreeding, with breeding males typically producing many progeny. However, because of the lack of recombination and inheritance as a single haplotype, mitochondrial DNA studies have contributed greatly in the identification of the wild progenitors of domestic species and understanding of the complex process of domestication, which is essential information for understanding the origin and distribution of genetic diversity (see Bruford et al. 2003 for a recent review). If sequences are available for a large number of unrelated individuals, haplotype diversity may be calculated within breeds and compared between breeds. Hierarchical analysis of molecular variance (AMOVA) (Excoffier et al. 1992) allows us to compare the distribution of diversity within and between group of breeds or within and between geographic regions (Luikart et al. 2001). Mitochondrial DNA may also provide a rapid way of detecting hybridization between livestock species or subspecies (Nijman et al. 2003).

Microsatellite loci are codominant nuclear markers found at high density and randomly dispersed on all chromosomes of most (probably all) eukaryotes. They are highly polymorphic, with alleles varying in the number of tandemly repeated two to five base pair sequences. Microsatellites are of small size and can be easily amplified by PCR from DNA extracted from a variety of sources, including blood, hair, skin, or even feces. Polymorphisms can be visualized on sequencing gels, and the availability of automatic DNA sequencers allows high-throughput analysis of a large number of samples in a short period of time (Jarne and Lagoda 1996; Goldstein and Schlötterer 1999). They are now the markers of choice for diversity studies and for parentage analysis and quantitative trait locus mapping, although their current popularity might be challenged in the near future with the development of inexpensive methods for assay of single nucleotide polymorphic markers. The FAO has overseen the development of recommendations for sets of microsatellite loci to be used for diversity studies for each of the major livestock species (see dad.fao.org/en/refer/library/guidelin/marker.pdf).

Some controversy has surrounded the choice of the best mutation model applied to evolution of microsatellite loci and therefore the choice of the best population genetic models for data analysis. Microsatellite polymorphism probably is generated through the mechanism of DNA slippage (Schlötterer and Tautz 1992), and the resulting alleles are discrete. The assumption of an infinite pool of possible alleles might not hold because the size of a new mutant probably depends on the size of the allele from which it mutated, and back mutations are also likely. Therefore new measures of genetic distances and genetic differentiation based on a stepwise mutation model have been proposed (Goldstein et al. 1995). However, simulation studies have shown that use of analyses that assume an infinite allele mutation model is generally valid for within-species diversity studies using microsatellite data (Takezaki and Nei 1996).

Microsatellite data are being used to estimate within- and between-breed genetic diversity and genetic admixture between breeds. The mean number of alleles (MNA) and observed and expected heterozygosity (*Ho* and *He*) are the most commonly calculated population genetic parameters for assessing within-breed diversity. However, for MNA to be a valid comparator between breeds it is important that sample sizes be the same in all breeds. Theoretically, allelic diversity can provide information on the uniqueness of a breed, through the presence of unique (also called "private") alleles within a population. In practice the observation of private alleles must be interpreted with caution, especially if they are present at low frequency, because they might result from sampling artifacts. Similarly, the standard error of heterozygosity measurements depends on the number of animals genotyped and the level of polymorphism observed at the individual loci.

The simplest parameters for assessing diversity between breeds using microsatellite data are the genetic differentiation or fixation indices. Several estimators have been proposed (e.g., F_{ST}, G_{ST}, θ), the most widely used being F_{ST} (Weir and Basten 1990), which measures the degree of genetic differentiation between subpopulations through the calculation of the standardized variances in allele frequencies between populations. Statistical significance can be calculated for F_{ST} values between pairs of populations testing the null hypothesis of lack of genetic differentiation between populations and therefore the genetic partitioning of diversity between populations (Mburu et al. 2003). In a manner similar to that for analysis of mitochondrial DNA polymorphism, AMOVA (Excoffier et al. 1992) can be performed to assess the distribution of diversity within and between groups of breeds.

Microsatellite frequency data also are commonly used to assess the genetic relationships between populations and also between individuals, through the calculation of genetic distance measures based on microsatellite allele frequencies. The most commonly used measure of genetic distances is Nei's standard genetic distance (D_S) (Nei 1972). However, for closely related populations in which genetic drift is the main factor of genetic differentiation, as is often the case in livestock breeds particularly in the developing world, the modified Cavalli-Sforza distance (D_A) is recommended (Nei et al. 1983). Genetic relationship between breeds often is visualized through the construction of phylogenies, most often using the neighbor-joining (N-J) method (Saitou and Nei 1987), which does not assume that the evolutionary rate is the same in all lineages. Numerous articles (e.g., in the journal of the International Society of Animal Genetics, *Animal Genetics* [www.isag.org.uk]) have been published describing phylogenetic relationships between livestock breeds using genetic distances. However, a major drawback of phylogenetic tree construction is that they assume that evolution is nonreticulate (i.e., that lineages can diverge but can never result from crosses between lineages). This assumption rarely holds for livestock, where new breeds often originate from crossbreeding between two or more ancestral breeds. The visualization of evolution provided by phylogenetic reconstruction therefore must be interpreted cautiously, with the knowledge that it cannot represent the fusion of lineages.

Multivariate analysis and, more recently, Bayesian clustering approaches have been suggested for admixture analysis of microsatellite allele frequency data (Pritchard et al. 2000). Probably the most comprehensive study of this type in livestock was a continent-wide study of African cattle (Hanotte et al. 2002). Using principal component analysis, the authors were able to assess for each African breed the level of genetic admixture and infer its origins from the three currently recognized centers of domestication of cattle. Moreover, combined with archaeological information the molecular data allowed identification of the centers of origin or entry points onto the African continent of the three major genetic influences found today in African cattle. A key point here is that the use of molecular genetic data is a useful tool, in conjunction with other information such as archaeological data and written records, to understand the nature and history of origins and subsequent movements and developments of genetic diversity in livestock species.

Mapping the origins of current genetic diversity can allow inferences to be made about where functional genetic variation might be found within a species for which only limited data on phenotypic variation exist.

Use of Molecular Marker Diversity in Conservation Decisions

Although ideally we would conserve all breeds of livestock for future potential use, the necessary financial, physical, and human resources are very unlikely to be available. Therefore we will have to decide how to allocate finite resources for conservation. One goal of conservation will be to conserve the maximum amount of diversity for potential future use. There is almost no information on the distribution of potentially useful genetic polymorphisms among breeds, and very little information exists on phenotypes of developing-world breeds. In the short term, therefore, molecular marker information provides the most easily obtainable estimates of the genetic diversity within and between a given set of breeds. In the discussion that follows it is not being argued that molecular marker information is superior to phenotypic or other indirect or direct measures of functional genetic variation. On the contrary, measures of genetic diversity based on molecular markers are most valuable when all other information is lacking and become progressively less valuable as more detailed and accurate direct and indirect measures of functional genetic variation (such as accurate phenotypic assessment) become available. Ultimately, conservation decisions should be informed by optimum combinations of information on functional genetic diversity, including information based on molecular genetic markers, but current decision aids are focused primarily on use of molecular genetic marker data or on measures of diversity derived from such data. At the end of this discussion some suggestions are made about how more integrated decision aids might be developed. It should also be borne in mind that conservation decisions will include factors such as the social and cultural values of different breeds. Decision aids discussed in this chapter probably will be most valuable in helping decision makers understand the consequences of alternative courses of action, to help improve rather than drive the decision-making process.

One objective of conservation is to maximize the genetic diversity available in the future. A number of authors have suggested applying methods that maximize the total of future within- plus between-breed genetic diversity, as estimated by molecular genetic marker data (Toro et al. 1998; Eding et al. 2002). Other authors have focused on maximizing the future diversity between breeds (Thaon d'Aroldi et al. 1998; Simianer 2002), and suggestions have been made on maximizing a weighted balance of within- and between-breed diversity (Piyasatian and Kinghorn 2003).

Although high genetic variability can be found within breeds (see chapter 6), and the methods proposed for maximizing the sum of within- and between-breed genetic diversity as assessed by molecular genetic markers are elegant, we doubt that maximizing the sum of within- and between-breed diversity is an appropriate criterion for setting conservation goals for the following reasons:

• The most easily and rapidly exploitable genetic diversity lies between breeds. This is because frequencies of alleles controlling important adaptive and functional traits can be expected to be high or fixed within breeds, which is why breed substitution or grading up via crossbreeding produces much more rapid genetic change than selection within populations.

• Population genetic theory predicts that there should be a markedly nonlinear relationship between genetic distance measured by anonymous markers and functional genetic (i.e., exploitable) differences between breeds, whether evolving by genetic drift or selection. In a recent study of molecular genetic diversity among European pig breeds it was found that the diversity between breeds was much higher than expected because of drift genetic variation, consistent with the effects of selection acting across wide areas of the genome during domestication and breed evolution (L. Ollivier, pers. comm.). A further illustration supporting these first two points is the case of milk yield in cattle. Average full lactation milk yield of well-fed cattle ranges from well under 800 L for many nondairy tropical breeds to well over 6,000 L for European *Bos taurus* dairy breeds. To select a low–milk yield tropical breed for increased milk yield in a very successful breeding program could improve milk yield at about 1% per annum, so it would take 202 years to select a nondairy tropical breed to achieve yields equal to those of modern dairy breeds. Breed substitution could make that change in 5 to 10 years, and breed crossing could make half that change in 5 years and three-quarters of the change in about 10 years.

- The methods used for assessing genetic variation based on molecular marker data necessarily assume that within-breed genetic variation is functionally the same for all breeds, which may not be true.
- Conservation of a very few breeds will conserve a very high proportion of within-breed variation as assessed by molecular genetic markers.
- The metrics of diversity that are used are not monotonic or copy invariant, leading to the implausible result that adding new breeds to a set of breeds to be conserved can reduce the estimated amount of diversity conserved, and adding a breed that is already present in a set of breeds can increase the diversity conserved.

We believe that a more appropriate approach is to maximize future diversity between breeds or perhaps to put substantially greater emphasis on conserving between-breed rather than within-breed variation. Several groups have suggested application of a method proposed by Weitzman (1993, 1998) for allocation of resources to conservation of between-species diversity. This approach has been taken furthest by Simianer and colleagues (Simianer 2002; Simianer et al. 2003). The approach suggested is to first estimate genetic distances between breeds based on molecular genetic or other data. A slightly modified version of a diversity metric, D, proposed by Weitzman, was developed for assessment of genetic diversity that has the essential properties of non-negativity, monotonicity, and copy invariance. Methods were developed for estimating the extinction probability of each breed, z_i, which can then be used to calculate the expected future diversity, D_F, allowing the extinction probabilities of all breeds. The marginal contribution to diversity of each breed, m_i, can also be calculated as the difference between D_F with that breed included in the set of breeds with its probability of extinction equal to m_i and D_F if the probability of extinction $z_i = 1.0$ (i.e., the breed definitely goes extinct). The marginal contribution of a given breed is not related to its own extinction probability but it is related to the extinction probability of closely related breeds. The observation is that the breeds that are under greatest threat are not generally the breeds that are expected to have the greatest marginal contribution to diversity. This means that resources for conservation will rarely be best expended on breeds with the greatest threat of extinction.

Methods have been proposed to derive predictions of extinction probability (Reist-Marti et al. 2003), but much more research into methods of predicting extinction probabilities is needed. Methods can be developed to optimize allocation of a finite resource to conservation efforts. This requires that the relationship between extinction probability and expenditure of resources on conservation of a given breed can be defined, which should generally be possible but has not yet been attempted in any systematic way. An illustration was given by Simianer (2002) of resource optimization for conservation of breeds of African cattle based on hypothetical relationships between resource allocation and change in extinction probability. In this example, optimum allocation of resources led to an approximately 60% greater increase in future diversity than allocation of resources to all breeds equally or allocation only to the most threatened breeds. The latter is the approach most generally taken in conservation, illustrating that development and application of optimized approaches to resource allocations could have profound impacts on the effectiveness of conservation programs.

Returning to the issue of the appropriate measure of diversity one should aim to maximize in a conservation program, Barker et al. (2001) compared the use of different measures of diversity in a set of Asian goat breeds. They showed that there was essentially no correlation between the contributions to diversity of individual breeds when the goal was to maximize total (within- plus between-breed) variation and goals that maximize between-breed diversity, as measured by metrics such as Weitzman's D statistic. Although we have here taken a clear stand on which metrics we believe are most relevant to the objectives of conservation, it will be important for the international community to reach a consensus on this issue to ensure consistent and efficient use of resources for conservation.

Combining Molecular Genetic, Phenotypic, and Other Data in Decision Making

The aforementioned approaches based on anonymous measures of diversity can be extended to include direct measures of utility (e.g., disease resistance, stress resistance, productivity), such that a combination of maximum diversity and utility is conserved. Simianer (2002) proposed one possible method. Such methods need further development but in principle can deal with any situation ranging from having no information on utility, where only molecular genetic diversity is available, through having complete information on utility, where molecular genetic diversity data would carry no weight in optimization.

Deriving the optimum in such decision making is computationally intensive when more than a few breeds are involved, limiting its potential application to large problems and its use as an easily accessed tool. Genetic algorithms provide an iterative solution based on evolutionary principles that can rapidly solve these highly complex optimization problems. This was illustrated by Piyasatian and Kinghorn (2003) for a problem in maximizing a combination of within- and between-breed genetic variance, in which the user defines the relative weight to be placed on within- or between-breed variance. Such methods can be developed to generate interfaces in which users can vary many input parameters and explore the consequences of alternative conservation scenarios in real time. They could also be extended to allow more sophisticated models of the impacts of conservation decisions (e.g., the inclusion of predictions of rates of inbreeding and hence loss of within-breed genetic diversity) and, through that, the loss of future genetic improvement potential. These methods promise a new generation of tools that allow researchers and policy advisors to explore the consequences of a wide range of alternative scenarios in conservation and use of livestock genetic resources. Such tools can be supplied in conjunction with Web-based livestock genetic resource databases and information systems or as stand-alone tools that can be run on any standard desktop computer. These tools probably could be easily adapted to have important applications in decision making in conservation of other agricultural and nonagricultural species.

Note

1. These estimates come from the World Watch List for Domestic Animal Diversity, compiled from status reports on livestock genetic resources submitted by official country representatives to FAO official databases. A criticism of this approach to documenting livestock genetic diversity is that each country has the sovereign right to identify as a unique genetic resource any genetic resource present in that country. For example, many countries identify Landrace pigs as a national genetic resource, and Landrace pigs are then counted as a separate breed in each country.

In many countries there are very few Landrace pigs, and they are therefore identified as under some degree of threat, although the global population of Landrace pigs remains huge and under no threat. Although there is undoubtedly some degree of genetic differentiation between some populations of Landrace pigs, the process used to compile statistics leads to an overestimate of both the number of breeds and the proportion of breeds under threat. This problem originates almost exclusively from the reporting from developed countries. Conversely, most observers in the field think that the situation in the developing world is the opposite. In the developing world many livestock genetic resources remain unreported, and threats to genetic resources there are much higher than reported and are growing. The threats to livestock genetic resources in the developing world are driven primarily by increasing use of crossbreeding and changing agricultural practices, both of which are on the rise. There is thus an unfortunate situation in which many observers are skeptical about the threats to livestock genetic resources because of the biases induced by the reporting process in developing countries, whereas other observers believe that threats to important genetic resources are far greater than current reporting suggests. There is urgent need for an accurate, extensive documentation of the status and trends in livestock genetic resources in the developing world.

2. By May 1, 2006, there were 188 parties to the CBD, and 168 of these countries have signed (www.biodiv.org).

References

Avise, J. C. 1994. *Molecular Markers, Natural History and Evolution.* New York: Chapman and Hall.

Ayalew, R., J. E. O. Rege, E. Getahun, M. Tibbo, and Y. Mamo. 2003. Delivering systematic information on indigenous animal genetic resources: The development and prospects of DAGRIS. Proc. Deutsche Tropentag 2003: Technological and Institutional Innovations for Sustainable Rural Development, October 8–10, 2003. Goettingen, Germany.

Baker, C. M. A. and C. Manwell. 1980. Chemical classification of cattle. I. Breed groups. *Animal Blood Groups and Biochemical Genetics* 11:127–150.

Barker, J. S. F., S. G. Tan, S. S. Moore, T. K. Mukherjee, J. L. Matheson, and O. S. Silveraj. 2001. Genetic variation within and relationships among populations of Asian goats (*Capra hircus*). *Journal of Animal Breeding and Genetics* 118:213–233.

Bruford, M. W., D. G. Bradley, and G. Luikart. 2003. DNA markers reveal the complexity of livestock domestication. *Nature Reviews Genetics* 4:900–910.

Cunningham, E. P. 1992. Animal genetic resources: The perspective for developing countries. In J. O. E. Rege and M. E. Lipner, eds., *Animal Genetic Resources: Their Characterization, Conservation and Utilization.* Research Planning Workshop, ILCA, Addis Ababa, Ethiopia, February 19–21, 1992. Addis Ababa, Ethiopia: ILCA.

DAGRIS. 2003. *Domestic Animal Genetic Resources Information System (DAGRIS).* Version 1. J. E. O. Rege, W. Ayalew, and E. Getahun, eds. Addis Ababa, Ethiopia: ILRI. dagris.ilri.cgiar.org.

EAAP Animal Genetic Databank. 2003. Department of Animal Breeding and Genetics, School of Veterinary Medicine, Hanover, Germany. www.tiho-hannover.de/einricht/zucht/eaap/index.htm.

Eding, H., R. P. Crooijmans, M. A. Groenen, and T. H. Meuwissen. 2002. Assessing the contribution of breeds to genetic diversity in conservation schemes. *Genetics Selection Evolution* 34:613–633.

Excoffier, L., P. E. Smouse, and J. M. Quattro. 1992. Analysis of molecular variance inferred from metric distances among DNA haplotypes: Application of human mitochondrial DNA restriction data. *Genetics* 131:479–491.

FAO (Food and Agriculture Organization of the United Nations). 1984. *Animal Genetic Resources Conservation by Management, Data Banks and Training.* Part 1. Rome: FAO.

FAO (Food and Agriculture Organization of the United Nations). 1986a. *Animal Genetic Resources Data Banks. 1. Computer Systems Study for Regional Data Banks.* Rome: FAO.

FAO (Food and Agriculture Organization of the United Nations). 1986b. *Animal Genetic Resources Data Banks. 2. Descriptor Lists for Cattle, Buffalo, Pigs, Sheep and Goats.* Rome: FAO.

FAO (Food and Agriculture Organization of the United Nations). 1986c. *Animal Genetic Resources Data Banks. 3. Descriptor Lists for Poultry.* Rome: FAO.

FAO (Food and Agriculture Organization of the United Nations). 1999. *The Global Strategy for the Management of Farm Animal Genetic Resources.* Executive Brief. Rome: FAO.

Felius, M. 1995. *Cattle Breeds: An Encyclopedia.* Doetinchen, The Netherlands: Misset.

Gibson, J. P. 2003. Strategies for utilising molecular marker data for livestock genetic improvement in the developing world. In *Proceedings International Workshop on Marker Assisted Selection: A Fast Track to Increase Genetic Gain in Plant and Animal Breeding,* October 2003. Torino, Italy.

Goldstein, D. B. and C. Schlötterer. 1999. *Microsatellites: Evolution and Applications.* New York: Oxford University Press.

Goldstein, D. B., A. R. Linares, L. L. Cavalli-Sforza, and M. W. Feldman. 1995. An evaluation of genetic distances for use with microsatellite loci. *Genetics* 139:463–471.

Hanotte, O., D. G. Bradley, J. W. Ochieng, Y. Verjee, E. W. Hill, and J. E. O. Rege. 2002. African pastoralism: Genetic imprints of origins and migrations. *Science* 296:336–339.

Hodges, J., ed. 1987. *Animal Genetic Resources: Strategies for Improved Use and Conservation.* Proceedings of the 2nd Meeting of the FAO/UNEP Expert Panel with Proceedings of the EAAP/PSAS Symposium on Small Populations of Domestic Animals. Rome: FAO.

Hodges, J., ed. 1992. *The Management of Global Animal Genetic Resources.* Proceedings of an FAO Expert Consultation. Rome: FAO.

Jarne, P. and P. J. L. Lagoda. 1996. Microsatellites, from molecules to populations and back. *Tree* 11:424–429.

Luikart, G., L. Gielly, L. Excoffier, J. D. Vigne, J. Bouvet, and P. Taberlet. 2001. Multiple maternal origins and weak phylogeographic structure in domestic goats. *Proceedings of the National Academy of Sciences of the USA* 98:5927–5932.

Manwell, C. and C. M. A. Baker. 1980. Chemical classification of cattle. 2. Phylogenetic tree and specific status of the Zebu. *Animal Blood Groups and Biochemical Genetics* 11:151–162.

Mason, I. L. 1988. *A World Dictionary of Livestock Breeds, Types and Varieties.* Wallingford, UK: CAB International.

Mburu, D. N., J. W. Ochieng, S. G. Kuria, H. Jianlin, B. Kaufmann, J. E. O. Rege, and O. Hanotte. 2003. Genetic diversity and relationships of indigenous Kenyan camel (*Camelus dromedarius*) populations: Implications for their classification. *Animal Genetics* 34:26–32.

Nei, M. 1972. Genetic distance between populations. *The American Naturalist* 106:283–292.

Nei, M., F. Tajima, and Y. Tateno. 1983. Accuracy of estimated phylogenetic trees from molecular data. II. Gene frequency data. *Journal of Molecular Evolution* 19:153–170.

Nijman, I. J., M. Otsen, E. L. Verkaar, C. de Ruijter, E. Hanekamp, J. W. Ochieng, S. Shamshad, J. E. O. Rege, O. Hanotte, M. W. Barwegen, T. Sulawati, and J. A. Lenstra. 2003. Hybridization of banteng (*Bos javanicus*) and zebu (*Bos indicus*) revealed by mitochondrial DNA, satellite DNA, AFLP and microsatellites. *Heredity* 90:10–16.

Piyasatian, N. and B. P. Kinghorn. 2003. Balancing genetic diversity, genetic merit and population viability in conservation programmes. *Journal of Animal Breeding and Genetics* 120:1–13.

Pritchard, J. K., M. Stephens, and P. Donnelly. 2000. Inference of population structure using multilocus genotype data. *Genetics* 155:945–959.

Rege, J. E. O. 1992. Background to ILCA's AGR characterisation project, project objectives and agenda for the research planning workshop. In J. E. O. Rege and M. E. Lipner, eds., *African Animal Genetic Resources: Their Characterisation, Conservation and Utilisation,* 55–59. Addis Ababa, Ethiopia: ILCA.

Reist-Marti, S. B., D. Wakelin, H. Simianer, J. Gibson, O. Hanotte, and J. E. O. Rege. 2003. Weitzman's approach and livestock conservation: An application to African cattle breeds. *Journal of Conservation Biology* 17:1299–1311.

Roughsedge, T., R. Thompson, B. Villanueva, and G. Simm. 2001. Synthesis of direct and maternal genetic components of economically important traits from beef breed-cross evaluations. *Journal of Animal Science* 79:2307–2319.

Saitou, N. and M. Nei. 1987. The neighbor-joining method: A new method for reconstructing phylogenetic trees. *Molecular Biology and Evolution* 4:406–425.

Scherf, B., ed. 2000. *World Watch List for Domestic Animal Diversity,* 3rd ed., Part 1.9, 20, dad.fao.org/en/Home.htm, databases. Rome: FAO/UNDP.

Schlötterer, C. and D. Tautz. 1992. Slippage synthesis of simple sequence DNA. *Nucleic Acids Research* 20:211–215.

Simianer, H. 2002. Noah's dilemma: Which breeds to take aboard the ark? In *7th World Congress on Genetics Applied to Livestock Production,* 8. Montpellier, France: INRA/CIRAD.

Simianer, H., S. B. Marti, J. Gibson, O. Hanotte, and J. E. O. Rege. 2003. An approach to the optimal allocation of conservation funds to minimize loss of genetic diversity between livestock breeds. *Ecological Economics* Special Issue on AnGR 45:377–392.

Sunnucks, P. 2001. Efficient genetic markers for population biology. *Tree* 15:199–203.

Swaminathan, M. S. 1992. Biological diversity and global food security. In R. R. Lokeshwar, ed., *V International Conference on Goats. Pre-Conference Proceedings. Plenary Papers and Invited Lectures,* 1–5. New Delhi: International Goat Association and Indian Society of Sheep and Goat Production and Utilization.

Takezaki, N. and M. Nei. 1996. Genetic distances and reconstruction of phylogenetic trees from microsatellite DNA. *Genetics* 144:389–399.

Thaon d'Aroldi, C., J. L. Foulley, and L. Ollivier. 1998. An overview of the Weitzman approach to diversity. *Genetics Selection Evolution* 30:149–161.

Toro, M., L. Silió, J. Rodrigáñez, and C. Rodriguez. 1998. The use of molecular markers in conservation programmes of live animals. *Genetics Selection Evolution* 30:585–600.

Weir, B. S. and C. J. Basten. 1990. Sampling strategies for distances between DNA sequences. *Biometrics* 46:551–582.

Weitzman, M. L. 1993. What to preserve? An application of diversity theory to crane conservation. *Quarterly Journal of Economics* 108:157–183.

Weitzman, M. L. 1998. The Noah's ark problem. *Econometrica* 66:1279–1298.

6 🐦 Management of Farm Animal Genetic Resources: Change and Interaction

I. HOFFMANN

Breeds of domesticated farm animal species are the biological basis for livestock sector development and for its contribution to food security and sustainable rural development. Only 14 of the approximately 30 domesticated mammalian and bird species provide 90% of human food supply from animals, yet the value of most animal genetic resources is poorly understood. Development in the 20th century has concentrated on a very small number of breeds worldwide, often without due consideration of the effects of the local production environment on a breed's ability to survive, reproduce, and produce. The management of this biological capital has been neglected, resulting in substantial erosion, which is likely to accelerate with the global increase in demand for livestock products, often called the Livestock Revolution.

The use and development of livestock breeds and the conservation of valuable breeds of little current interest to farmers must be substantially upgraded to ensure future food security and sustainable rural development. Sustainable breed use, development, and conservation are critical and complementary elements. A range of rapidly developing molecular and reproductive biotechnologies also has important implications for animal genetic resource (AnGR) management. This chapter examines AnGR diversity management in terms of three key questions (Masinde 2001):

- To what extent do farmers want to maintain a number of species and breeds in the farming system? What are the reasons for this?

- What techniques and strategies do farmers use to maintain a number of species and breeds?
- What forces—positive and negative—help or hinder the maintenance of this diversity by farmers?

After an overview of domestication and the distribution of livestock species, the major livestock production systems, breeds, and breeding goals in such systems are reviewed. The major external forces on AnGRs are identified, and the ways farmers address them are examined. Overall, studies on valuation and management of AnGRs in small-scale and traditional agriculture are scarce, but more have been initiated in recent years. Studies on diversity within farming systems are even more scarce.

AnGR Status

Domestication and Distribution of Species

Domestication started about 12,000 years ago and followed two main lines of use for animal products. First, focus was placed on selecting animals for the supply of meat, fat, and fiber through the domestication of sheep, goats, pigs, cattle, dogs, and guinea pigs. Second, after a period of domestication that had already influenced animal behavior, animals were also used for transport and draft power. Main species selected for these purposes include cattle, buffalo, yak, donkey, horse, llama, and camel (Röhrs 1994). In most cases, humans influenced the environment in which the animals lived, but in a few production systems, such as nomadism, humans follow their animals. There are several points of domestication in time and space (Bruford et al. 2003), as shown in figure 6.1. Exchange of animals between continents and countries has always occurred to a certain extent, but this exchange increased during the time of colonialism, particularly since the nineteenth century.

Still today, there are clusters of breed diversity; for example, most of the breed diversity for buffalo and yak is located in Asia, most diversity for horses, chickens, and geese is located in Europe, and camelid diversity is concentrated in Latin America (table 6.1).

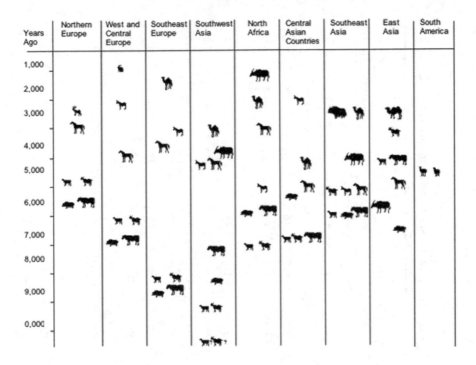

FIGURE 6.1. Place and time of domestication and distribution of domestic mammalian species (modified from Röhrs 1994).

Livestock Production Systems

Several thousand domestic animal breed populations have been developed in the 12,000 years since the first livestock species were domesticated. These breeds have evolved from adaptations that allow livestock production in a wide range of agroecological zones and production systems and under different economic regimes. Production targets also differ and include both subsistence farming with a high proportion of home consumption and commercial farming with no home consumption. Subsistence farmers do sell animals but only when obliged to do so for economic reasons, and they keep animals mainly for home consumption or for social, religious, or cultural reasons.

Livestock currently contribute between 25% and 30% of agricultural gross domestic product in developing countries, and this share is expected to rise to almost 50% in the next 20 years. Livestock provide a range of services, such as income generation, asset accumulation, insurance, buffering against cyclical

Table 6.1 Proportional share of the world's total population size and number of breeds of the major livestock species in each region.

	Africa		Asia and Pacific		Europe		Latin America and the Caribbean		Near East		North America	
	Population	Breed	Population	Breed	Population	Breed	Population	Breed	Population	Breed	Population	Breed
Buffalo	0.1	3.5	93.4	70.9	0.3	3.5	1.0	10.5	2.5	11.6	0.0	0.0
Cattle	13.2	20.5	34.9	19.3	12.3	39.4	26.9	8.7	5.4	7.0	8.4	5.1
Yak	0.0	0.0	n/a	69.2	n/a	7.7	0.0	0.0	n/a	23.1	n/a	0.0
Goat	19.4	15.6	55.4	25.6	3.7	32.8	5.8	6.0	16.2	16.5	0.2	3.5
Sheep	12.1	11.2	38.6	17.7	17.5	47.9	8.5	3.2	23.0	15.3	0.7	4.6
Pig	2.8	4.4	54.7	36.9	21.5	45.8	8.2	6.0	0.1	0.4	7.8	6.4
Ass	22.2	12.4	34.3	12.4	3.5	23.7	18.8	5.2	21.3	41.2	0.1	5.2
Horse	7.4	7.7	24.5	11.4	12.8	60.7	42.5	4.3	4.2	8.5	10.7	7.3
Camelids	17.7*	20.6*	14.8†	22.2†	0.1†	3.2†	100.0†	100.0‡	66.8†	54.0†	0.0	0.0
Chicken	5.4	7.8	45.4	17.7	14.5	64.2	16.4	5.0	7.3	3.8	13.7	1.4
Duck§	0.9	11.0	91.7	45.0	8.9	36.0	2.7	6.0	1.4	2.0	1.0	n/a
Turkey	1.0	17.6	0.8	17.6	49.5	47.1	6.3	11.8	3.1	2.9	39.6	2.9
Goose	1.4	7.6	89.8	19.7	7.0	63.6	0.2	7.6	4.8	n/a	0.1	1.5

Sources: FAO (1999) for livestock population numbers and Domestic Animal Diversity Information System for breed numbers, in Scherf (2000).
*Dromedary camels only.
†Dromedary and Bactrian camels.
‡New World camelids.
§Domestic duck and Muscovy duck.

changes, food, clothing and other goods, traction, and nutrient recycling (i.e., by using byproducts of other agricultural activities, such as crop residues). Some 70% of the world's rural poor depend on livestock as an important component of their livelihoods. Livestock provide a higher contribution to income and welfare of poor smallholders than of wealthier ones, and particularly of women and, through them, of children in such households.

The main determinants commonly used to classify livestock production systems (LPSs) are agroecological zone, mobility, on-farm integration, land assets and external inputs needed for production, and economic objectives. Sere et al. (1996) classified LPSs by the land area needed for production, into grassland-based, mixed, and landless. The grassland-based ranches or pastoral systems usually are pure livestock systems, in which livestock provide the sole source of income globally for 20 million pastoral families (Steinfeld et al. 1997). In terms of total production, grazing systems supply only 9% of global meat production. The majority of livestock is kept in mixed farming systems. Globally, mixed farming systems produce the largest share of total meat (54%) and milk (90%), and mixed farming is the main system for smallholder farmers in many developing countries. Landless systems provide more than 50% of global pork and poultry meat production and 10% of beef and mutton production. They depend on external supplies of feed, energy, and other inputs.

Breeds and Breeding

AnGRs are commonly grouped by their origin as local and traditional or exotic and modern and by their breeding history as indigenous and locally developed or commercially bred. Commercial breeds usually are derived from scientific breeding programs, which are based on animal identification and performance recording of individual animals (see the Glossary for Food and Agriculture Organization [FAO] breed definitions).

Livestock breeding starts with control of reproduction, which is difficult in some free-ranging production environments. Breeding is the most important component of use and development of animal genetic resources. Breeding goals for local breeds include adaptation to harsh environments, disease resistance, and the provision of an array of products and services, which can be tradable or nontradable. Such livestock is multifunctional in scope and provides food (meat, milk, eggs), fiber (wool, hides, and skins), draft power, manure, and fuel. In South Africa, for example, some smallholders keep geese as "watchdogs" (Bayer et al. 2003). Grazing by livestock can create socially desirable cultural landscapes and help maintain biodiversity. Indigenous societies have a deep knowledge of which types of male and female animals they need and which type best suits their environment and production conditions.

The Greeks and Romans practiced fairly well-developed animal breeding, as can be concluded from the large bone size of skeleton findings. However, such knowledge and practice disappeared in Europe in the Middle Ages, and cattle at that time had small frame sizes. The Arab horse breeders were the first to apply pedigrees in selection in the Middle Ages, and this knowledge later influenced European breeding. Modern breeding started in the 18th century, particularly in Britain, and a multitude of breeds emerged, often adapted to a specific local environment. For example, sheep or cattle breeds selected for highlands had a different phenotype than those selected for lowlands, just as thoroughbred horses were different from working horses. Breeds were also developed based on imports from other continents. British local pigs were crossed with East and South Asian and Mediterranean breeds. At that time, breeding was directed more toward phenotype than production traits, and breeds often were multipurpose, such as cattle used for milk, meat, and draft power. From these breeds (landraces), modern specialized breeds have been developed since the 1950s to produce high output in one or two major production traits such as milk, meat, eggs, or fiber. The breeds of today have been selected for at least 20 generations in a pure breeding system. Such modern breeding entails controlled mating, individual animal identification and recording, progeny and performance testing to find superior parents (particularly on the male side because of higher male reproductive capacity), and sophisticated data processing.

Today, most livestock in developed countries (and increasingly so in developing countries) is kept under controlled conditions, largely independent of the surrounding environment. Even feed ingredients are not necessarily produced locally because feed is readily available in international markets. Such environmental decoupling is most pronounced in landless LPSs such as with poultry and pigs kept in intensive conditions but can also be important for dairy and beef cattle kept in feedlots. A consequence of the uniformity of environmental conditions is the need for fewer breeds, reducing livestock diversity (Tisdell 2003).

Breeding goals include high performance in a few production traits (meat, milk, or eggs). More recently, breeding goals may also include improved animal health and metabolic stability (e.g., bone structure, integrity of vital organs in poultry), animal behavior, and product quality. Such traits have been introduced in breeding schemes to address the environmental problems created by intensive animal husbandry, increasing consumer awareness and concerns, and animal welfare movements in developed countries. In the United States there is an increased consumer interest in range- and pasture-reared poultry and eggs. External factors (e.g., waste management) and pressure from special interest groups (e.g., animal welfare) may drive breeding costs up by necessitating adaptation to unforeseen scenarios, perhaps by inclusion of fitness traits in selection programs. Possible competition with humans for food, and the high nitrogen and phosphorus pollution from high-input poultry and pig farming, must be addressed. Constant improvements of the feed conversion ratio in monogastric species are imposed on breeding companies and institutions for ethical, environmental, and economic reasons. The feed conversion ratio for egg and meat production in commercial poultry was reduced from 4.0:1 in 1950 to 2.0:1 in 2000 (Flock and Preisinger 2002) and is currently at 2.5:1 in commercial pig production.

Local breeds are kept and exchanged by a multitude of small farmers, whereas commercial breeds tend to be associated with larger scales and concentration in the sector. This concentration is independent of the legal form of the enterprises (cooperatives or companies). Farmer cooperatives can reach 100% of the market share in some breed stock markets, as do breeding companies in others (Preisinger 2004). Concentration in the animal industry depends on the reproductive rate, portability and transportation costs of breeding products, and costs associated with breeding. The reproductive rate is highest in poultry, followed by pigs (high reproduction on the female side) and cattle (high reproduction on the male side), and is much lower in small ruminants. The ease of reproductive biotechnology use is highest in cattle (deep-freezing of semen and embryos) and lower in pigs (mainly fresh semen used in commercial breeding) and poultry. In small ruminants and horses, artificial insemination is not widespread, and natural mating dominates. Because the reproductive rate is highest in poultry, and eggs and day-old hatchlings are highly portable, consolidation is highest in the poultry breeder industry. Fifty years ago there were numerous primary breeders in Western countries. In the early 1980s there were 20 breeding companies all over the world. Today, the international chicken market is dominated by three groups of primary layer breeders and four major companies in broiler breeding (Flock and Preisinger 2002). A similar trend is expected in the pig industry.

From an institutional perspective, modern breeding is highly organized and based on herdbooks with registered animals and pedigree records that support the phenotypic and performance breeding goals of the breeding organization. Breeding organizations can be breeders' associations or privately owned breeding companies. Much information on these organizations and their programs is publicly available (e.g., on the web sites of breeding associations). They usually produce only one breed, target only one or two production traits, and do not attempt to increase or maintain ANGR biodiversity but aim at managing sufficient genetic variability within a population (box 6.1). Among Holstein-Friesian cattle, for example, highly efficient reproductive technologies and the intensive use of a few sires have led to a global population of millions but an effective population size lower than 100.

Breeds express the animal genetic diversity used by humans and are defined as cultural rather than technical units (see Glossary for FAO definitions). Genetic diversity in the sense of genetic variability may be described in terms of genetic distances through molecular genetic methods, such as microsatellite markers. The more distant breeds are on a phylogenetic tree, the more genetically different they are. High genetic variability can be found within breeds with large populations and flock sizes and limited inbreeding.

Genetic diversity measured at the molecular level does not always correspond to phenotypic breed diversity because a long history of exchange, upgrading, and crossbreeding has created similar genotypes in different phenotypes or different genotypes within similar phenotypes. An example of similar genotypes in different phenotypes is the indigenous cattle breed in Namibia, which is known as the Sanga and is found in the northern and northeastern parts of the country. Four distinct ecotypes are recognized— the Ovambo, Caprivi, Kunene, and Kavango—which have evolved in their distinct environments. However, they are genetically quite similar (Nortier et al. 2002). Examples for crossings aiming at increasing genetic variability but maintaining the phenotype are the Murnau–Werdenfelser cattle, a threatened breed in Germany that has been crossed with Tarentaise; or Angler of the old type, which was crossed with Danish red. Breeders of fancy chickens are concerned mainly about the phenotype, whereas the genotype of phenotypically different breeds may be very similar. Extinct breeds may be recreated through crosses aimed at reproducing a phenotypic standard. These recreations may be desirable in phenotypic, socioculturally motivated conservation of old breeds adapted to a specific landscape and may be considered an agricultural and landscape legacy but should not be confused with maintenance of genetic variability. The two lines of argument

Box 6.1 Breeds and Genetic Variability in Poultry

The origin of all poultry seems to be the red jungle fowl from South Asia (Hillel et al. 2003). Poultry breeds in developing countries often are nondescriptive, and apart from the Fayoumi breed, developed in Egypt (Hossary and Galal 1995), there appears to be no record of a tropical adapted breed developed from indigenous chickens in Africa. The Fayoumi's genetic makeup is different from that of other chickens, and the birds are much more resistant to viral diseases than the American chicken ("Egyptian chicken plan" 1997).

High genetic variability (as high polymorphism) can be found in breeds with large populations and flock size and limited inbreeding. Based on genetically distinct local populations, pure breeds were developed that differed in many phenotypic traits, such as plumage color, plumage pattern, and comb type. The effective population size of these breeds can decrease in a short period of time if, as in the case of fancy breeds, intense selection for exhibition traits takes place. Inbreeding, genetic drift, and bottlenecks can exacerbate the situation, placing breeds at risk. The data on poultry breeds are scarce in the FAO-based Domestic Animal Diversity Information System (DAD-IS), despite its improvement over the years (Scherf 2000; Weigend and Romanov 2002). Data on 14 avian species and 1,049 breeds have been entered into the DAD-IS database, which represent only 16% of all breed entries. From the data on poultry genetic resources provided by countries to DAD-IS, it is obvious that breeds recorded in Europe and North America are mostly threatened by extinction, whereas data are insufficiently reported for other regions. About 50% of the poultry breeds registered in DAD-IS are classified as being at risk; this is the highest percentage of breeds at risk of all species contained in DAD-IS. Commercial poultry lines are not covered in DAD-IS, nor are the lines kept in reserve by the breeding companies or at universities.

Commercial poultry breeders sell various products, most of which are the outcome of three or four pure line crossings. For this to happen, the grandparent lines have to be developed continuously, and reserve lines are also kept. Commercial breeders have low levels of inbreeding and try to maintain high genetic variability (Flock and Preisinger 2002). From the genetic variability point of view, commercial breeds cover a broad range of poultry genetic diversity, which can also be found in fancy breeds. However, in recent years, breeders of commercial white egg layers have been concerned about reduced genetic variability and future response to selection because white egg layers originated with one breed, the single comb white leghorn. The genetic basis for brown egg layers or broilers has been somewhat broader, coming mainly from four breeds. The merging of breeding companies in recent years and the disposal of reserve lines for economic reasons have increased the need for conservation of genetic variation among breeds and lines (Hillel et al. 2003).

should be distinguished: the breed as a social construct with certain pheno-typic characteristics and defined by governments as custodians of biodiver-sity under the Convention on Biological Diversity, and the genetic variability at the genomic or locus level. This points at the difficulties in focusing the discussion on AnGR diversity and its characterization and management.

Indigenous AnGR Management

The main focus of this chapter is local or indigenous AnGR management. Indigenous knowledge is the actual knowledge of a population that re-flects the local experiences based on traditions and incorporates more recent experiences with modern technologies. Therefore, it is dynamic, changing through mechanisms of creativity and innovation as well as through contact with other knowledge systems, be they local or interna-tional (Richards 1985; Warren 1991; Haverkort 1993; Rajasekaran 1993; boxes 6.2 and 6.3).

From a social point of view, farmers' decisions concerning AnGRs are influenced by organizations and institutions in the community or access to and management of household and community resources (Rege 2003). Access to natural resources (land and water), type of land tenure and own-ership (private or communal) and intrahousehold (gender) issues also play a role in deciding on which species and breeds to keep. There is agreement that the concept of "breed" is a manifestation of the environment and community values and goals; hence, conservation of agricultural diversity needs to be linked to utilization in its production environment (Rege 2003).

Why Do Pastoralists and Farmers Maintain Species and Breeds in Their Farming Systems?

Indigenous livestock keepers are mainly pastoralists and mixed farmers who have a profound knowledge of their natural environment. In these systems, animals sometimes are branded with clan or group marks but are hardly individually identifiable by outsiders. Rege (2003:27) noted that "the term breed as a formal designation has little meaning outside the areas of western influence, where pedigree recording is often nonex-istent. Nonetheless, even under these circumstances, there are strains or 'types' which owe their continuing distinct identity to a combination of

Box 6.2 Agropastoralists' Trait Preferences in N'Dama Cattle: Participatory Methods to Assess Breeding Objectives

The investigation and evaluation of livestock raisers' breeding knowledge and breeding strategies remain a challenge where no formal breeding infrastructure and no written recording systems exist. This study aimed at identifying appropriate participatory methods that facilitate a better understanding of agropastoralists' interest in indigenous cattle breeds and their preference for production and functional traits that can be applied in breed improvement programs and AnGR management.

The survey was conducted among herd owners and herders of 27 villages in three districts in the Gambia. Three study sites capture differences in the level of commercialization, tsetse challenge, and herd ownership pattern. At all study sites a traditional low-input mixed crop and livestock system prevails. Cattle are used as a multipurpose breed providing milk, meat, manure, and traction. As elsewhere in tsetse-infected zones of West Africa, exploitation of ruminant livestock is possible because of their trypanotolerance and other adaptive features, and about 95% of the Gambian cattle population consists of the trypanotolerant N'Dama cattle breed (CIRDES/ILRI/ITC 2000). Nevertheless, proximity to the arid savanna climate makes immigration of zebu-type Gobra cattle easy.

Different survey techniques were used to identify and evaluate breed preferences of agropastoralists and to make them available for the definition of breeding objectives. Focus group discussions in seven villages were used to investigate agropastoralists' production objectives; breeding strategies, including breed and trait preferences; and breeding practices.

Focus group discussions revealed that although N'Dama are the preferred cattle breed, crossbreeding with bordering Gobra is also considered in traditional breeding strategies. The most frequently mentioned evaluation criteria that agropastoralists used for N'Dama bulls are size (13.1%), strength (28.3%), libido (10.6%), and good offspring (12.3%). Agropastoralists use the word *strength* to describe a combination of vigor and fitness. In N'Dama cows, milk production (25.1%), yearly calving (24.9%), and strength (16.6%) are high-priority criteria. Health status (reflecting disease resistance) is the most important parameter in bulls and very important in cows. High-priority production traits are milk and reproduction for cows and conformation (size) and production performance for bulls.

Based on frequencies of criteria and livestock production objectives, six traits were selected for the matrix rating (box table 6.2). The N'Dama received highest ratings for adaptation to dry season stress, traction utility, and disease resistance. Gobra received lowest ratings for disease resistance and highest for size and milk yield. Results differed significantly between survey sites.

Box 6.2 continues to next page

Box 6.2 continued

BOX TABLE 6.2. Agropastoralists' ratings of cattle breeds in the Gambia, ranging from 1 to 5.

Evaluation Criteria	Gobra	N'Dama–Gobra	N'Dama
Size	4.9	4.3	3.1
Milk yield	4.7	4.3	3.2
Calving frequency	2.9	3.1	4.4
Adaptation to dry season stress	2.3	2.9	4.7
Utility for traction	2.7	3.5	4.7
Disease resistance	1.8	2.6	4.6

Participatory approaches to AnGR management are vital to identify and evaluate various aspects of traditional breeding strategies and to achieve active involvement of the livestock-keeping communities. The matrix rating tool yields quantifiable data and facilitates exchange of relevant breeding information between agropastoralists and researchers. Agropastoralists expressed a clear preference for the N'Dama breed because of its disease resistance and adaptation traits. Size is an important selection criterion in N'Dama, receives highest ratings in Gobra, and is a reason for crossbreeding. This emphasizes the need to support genetic improvement of the N'Dama breed in pure-breeding programs if their genetic integrity and adaptive traits are to be maintained in the future. Nevertheless, breeding policies should consider regional planning and support site-specific improvement programs, which in high-potential areas may even support crossbreeding endeavors already practiced by agropastoralists.

Source: Steglich and Peters (2002).

traditional 'breeding objectives' and geographical and/or cultural separation by communities which own them."

Knowledge of individual animals and the degree of control over breeding animals depend on the degree to which the societies' or farmers' livelihoods depend on livestock. Therefore they tend to be more extensive in pastoral than in mixed farming systems. There is also great variation from species to species; camel pastoralists are more likely to take an interest in breeding than do sheep and goat producers (Huelsebusch and Kaufmann 2002). Livestock are inherited, received in exchange for ser-

Box 6.3 Value of Ethiopian Boran Cattle in a Changing Environment

This study illustrates changes in the value of Boran cattle for exploiting their original habitat, caused by misconceived development interventions and higher human population density. Data were collected from two districts of Ethiopia with contrasting functionality in the traditional range management and a difference in the degree of external interference. Web district represents a traditional dry season grazing area, located in the central Borana rangelands and associated with one of the nine deep well clusters. Dida Hara district is a former wet season grazing area in the peripheral rangelands, where ponds were constructed in the 1970s to relieve grazing pressure on wet season pastures and to improve overall range use efficiency. Participatory rural appraisal techniques, interviews, global position systems, and official maps were used to assess the breed preferences of the pastoralists, the status of natural resources, and indigenous land use strategies. Body weights of adult female and male cattle of the local breed types were measured during the peak dry and wet seasons.

Ethiopian Boran cattle resulted from the pastoralists' successful breeding and selection strategies under the high-risk conditions of semiarid rangelands. Ethiopian Boran cattle were once appreciated for their high productivity in semiarid rangelands (Cousins and Upton 1988; Behnke and Abel 1996). Exported for commercial ranching to countries such as Kenya, Australia, and Mexico, the improved Boran cattle in Kenya reached body weights of up to 850 kg (Rege 1999). The indigenous land use system of Borana pastoralists was based on well-planned movements between functional rangeland categories and on herd splitting to ensure availability of adequate grazing land and water. Scarcity of water was the key variable that determined utility of the pastures, which used to be considered the best rangelands in eastern Africa. Distinct indigenous institutions matched the needs of the herds with the management of available grazing and water resources during times of plenty and scarcity. Then, the artificial ponds in Dida Hara opened up the area for permanent grazing and uncontrolled settlement, which reduced mobility of the herds and caused overgrazing in the pastures formerly used only temporarily. At the same time, the imposition of a top-down formal administration has contributed to the destruction of indigenous pasture management institutions. Moreover, the demarcation of new political and administrative boundaries—part of the recent regionalization program of the Ethiopian government, including the alienation of one-third of the Borana rangeland and important wells in favor of the Somali Regional State—has intensified conflicts between Somali and Borana tribes. The annual growth rate of the human population is estimated at 2.5–3%; this exerts further pressure on the rangelands and has reduced the per capita availability of these resources. The result is a progressive deterioration of rangeland

Box 6.3 continues to next page

Box 6.3 continued

resources, as indicated by the disappearance of preferred species and encroachment of undesirable woody browse species (Coppock 1994; Kamara 2001; Homann et al. 2004).

Changes to the Borana rangelands have threatened the maintenance of the true type of Ethiopian Boran cattle (*Qorti*) and favored smaller and more sturdy types (*Ayuna*). Under favorable range conditions the pastoralists prefer the large-framed *Qorti* type. The *Qorti* is appreciated for high fertility, good growth, and milk production. However, compared with the *Ayuna*, *Qorti* was said to show lower tolerance to drought and external parasites and poor adaptation to scarcity of forage resources. The *Ayuna* type stems from genetic introgression of highland cattle. It was described as shorter, smaller, but more sturdy and able to adapt to degraded rangeland conditions. It was judged as generally poorer than the *Qorti* in fertility and beef and milk production. Measurements of average body weights showed that adult *Qorti* were significantly heavier than *Ayuna*, but the *Ayuna* gained more weight in the rainy season.

The geographic distribution of the two cattle types reflected the adaptation of pastoralists' breeding preferences to the degrading environment. Web district was identified as a more favorable habitat where the *Qorti* can be maintained, and the frequency of occurrence of *Qorti* was significantly higher in Web than in Dida Hara. In Dida Hara, where external interference was high, grazing pressure had expanded rapidly, socioeconomic heterogeneity was pronounced, and during the drought of 1999–2001 it experienced the highest losses of cattle in the entire Borana zone. Today, because of the rapid exhaustion of pasture, only a minority of wealthy herders can afford to procure *Qorti* bulls either from the market or from the government breeding ranch.

In face of these developments, the value of the original Ethiopian Boran breed to exploit this habitat has decreased. The pastoralists recognized that the large-frame *Qorti* was in danger of gradually disappearing from the Borana rangelands. The *Qorti* type was considered not competitive when grazing resources were scarce. The majority of households maintained either only the *Ayuna* type or only a small proportion of the *Qorti* at a low level of performance. The scarcity of pasture and the increasing recurrence of droughts along with an alarming impoverishment of the majority of the population were identified as the main causes for the genetic erosion. Any attempt to conserve the true Ethiopian Boran cattle would require improvements in the quantity and quality of grazing resources available to the pastoralists.

Multiple interconnected factors have made the control of natural resource use more difficult. The pastoral communities dated the beginning of the decline of their livelihoods in the 1970s, when progressive interventions from outside started. Observing the degradation of their rangelands and being aware of their

Box 6.3 continued

declining social capital, they still considered the traditional production strategies indispensable for good range management, but they also stressed the need for subsidiary services through the formal administration. Genuine structures for continuous negotiation, including conflict mediation and arbitration between different interest groups, must be created. Sustaining extended negotiation units among natural resource stakeholders could secure shared access to natural resources at a larger scale, generate cooperative community structures, foster community management of relations with outsiders, and improve coordination of outsider activities. The advantages of this form of negotiation include facilitation of adaptive innovation, rejection of harmful practices, and arbitration of conflicting claims. Dangers include abuse of power and inequality in access to information, increasing commercialization, and political alliances, exposing disadvantaged groups to even more manipulation.

Supporting this institutional approach are demands for sufficiently pragmatic policy concepts, integration of research and development programs with socioeconomic and ecological dimensions, and measures enabling different interest groups to agree on clearly assigned tenure regimes and roles. Success depends largely on political rather than technical issues, and it certainly depends on all the actors' willingness to share information and collaborate in capacity building (Grell and Kirk 2000; Thebaud and Batterbury 2001).

Source: Homann et al. (2004).

vices (e.g., herding), received as gifts from relatives, or acquired through purchase (Hassan 2000; Gondwe and Wollny 2002; Jabbar and Diedhiou 2003).

Various studies show that breeders of local breeds use a number of criteria to judge the value of a breeding animal and also differentiate between males and females. These criteria result from the multitude of functions the animals have to fulfill. Tano et al. (2003) interviewed subsistence livestock farmers, mixed crop and livestock producers, and beef and milk livestock farmers in a tsetse-affected zone in Burkina Faso. It was discovered that all farmers prefer cattle that are not selective in the type of grass or the quality of water they consume. In bulls, traction ability, large body size, high fertility, disease resistance, and rapid weight gain are considered desirable. For cows, reproductive performance, milk yield, and body size are important criteria but differ between production systems. Farmers value traction more than pastoralists, who highly value milk yield. Mixed-crop and livestock farmers are most interested in animal traction and less interested in meat and milk production and therefore are less concerned about low reproductive performance. For pastoralists, low reproductive performance is of great concern because of its impact on herd size and productive capacity, and milk and beef often rank highly. As in the case of bulls, large frame size in cows is preferred because it has an impact on the market value of the animals (Tano et al. 2003) (see box 6.2).

A case study from northwest Nigeria (Hoffmann 2003; box 6.4) demonstrates that breed distribution in the landscape varies spatially and seasonally and that different breeds are kept by different producer groups to provide them with the necessary goods and services suitable to their ecological niche and production system.

The performance, adaptation, and disease resistance of most breeds in developing countries have not been systematically recorded, and little of the existing information is easily accessible. The majority of livestock genetic diversity is found in the developing world, where documentation is most lacking and risk of extinction is highest and increasing. The value of indigenous livestock breeds is underestimated when only marketable outputs are considered and the multitude of functions and performance sustainability are neglected. Adaptation to unfavorable production conditions is a unique attribute of many indigenous breeds but difficult to record under field conditions. Because of the lack of animal identification and performance testing, quantitative valuation of local breeds in traditional environments is difficult. Qualitative assessment has been done by several groups in the recent past, based mainly on participatory assessment of priorities and preferences of livestock keepers and their communities, mainly in traditional or modified traditional livestock systems. This methodological progress is demonstrated in "Valuing AnGR," a special issue of *Ecological Economics* (2003). Besides participatory methods, various economic tools such as conjoint analysis (Tano et al. 2003) and hedonic price models assessing buyers' preferences for certain traits and breeds on livestock markets (Mohammed 2000 on camel markets; Jabbar and Diedhiou 2003 on cattle) have proven useful for AnGR valuation. The increasing interest in local breeds and community-based management of AnGRs is also reflected in the literature (Köhler-Rollefson 2000; Mhlanga 2002).

Box 6.4 Biodiversity Management in West African Pastoral and Agropastoral Systems: A Case Study from Northwest Nigeria

Cattle are the basis of livelihood for pastoralists and agropastoralists in the northern parts of West Africa. They are used for milk, manure, meat, and draft power, and they serve as savings and insurance. Composition of pastoral herds, particularly cattle, usually has been analyzed in terms of the sex and age groups of individual herds (FDLPCS 1992a, 1992b; Vabi 1993). Data on livestock numbers are hard to obtain in northern Nigeria in interviews. First, pastoralists are reluctant to give information about their livestock numbers for cultural reasons and fear of taxation. Second, risk management practices in an ecologically and economically highly variable environment augment the problems of getting accurate data. These practices include the widespread exchange of animals in a social network, the division of herds into management units that might be herded far away from the owner's homestead, and the herding of animals of different owners in one herd.

The stocking density information for this case study was obtained from monthly animal counts along transects in the Zamfara Reserve in northwest Nigeria (Schaefer 1998). Animal counts were then converted into tropical livestock units (TLUs) of 250 kg liveweight.

Over the whole year, an average of 0.84 head of cattle, 0.55 sheep, and 0.38 goats was found per hectare of range, resulting in a stocking rate of 0.81 TLU/ha. This stocking rate exceeds the recommended one. The highest density of cattle was observed in August, with 2.3 head/ha. This coincides with the peak of rainfall, vegetation growth rate, and feed supply of pastures in quantitative and qualitative terms. The continually decreasing animal density on rangeland during the dry season reflects the decrease of feed and water availability and the subsequent migration of pastoral herders out of the Zamfara Reserve. No livestock was found in the cropland areas during the rainy season, when access is forbidden by the village authorities. Cattle numbers are kept at a rate of 1.6 head/ha of cropland from December to March, rapidly declining thereafter. In contrast, the small ruminant stocking rate declined gradually from 0.3 to 0.1 animals/ha during the dry season. The high livestock density contributes nutrient input through manure (Hoffmann et al. 2001).

Asked about their livestock holding in different surveys, pastoral and agropastoral Fulani in the reserve gave figures ranging from 69 to 75 head of cattle, 33 to 43 sheep, and 34 to 36 goats (Kyiogwom et al. 1994). Similar cattle but lower small ruminant holdings were observed among settled Fulani in northern Cameroon (Vabi 1993). Herd size of Fulani is larger than that of Hausa: 77% of the Hausa farmers keep an average of 13 sheep, and 75% of the farmers keep 11 goats. They also own a few cattle, mainly adult males for draft purposes, but only 7% of the farmers keep more than 10 cattle.

Box 6.4 continues to next page

Box 6.4 continued

The number of cattle or small ruminants observed per herding unit did not differ between cropland and rangeland. The average number of cattle in 1,264 herds was 20.2 (range 1–183). Two-thirds of the herds were made up of less than 20 head of cattle. Although the observation of herds does not allow direct conclusions on the ownership pattern, this figure can be explained by a high percentage of smaller herds owned by farmers, who keep only a few cattle with a high proportion of work bulls (Hassan 2000; Hoffmann et al. 2001). The pastoralists stated in the interviews that large herds are split and the cattle grouped by category. Bulls, cows, young stock, and calves are then herded separately to better suit their feed needs and walking ability (Schaefer 1998).

The major ecotypes or breeds in the area are the Bunaji (white Fulani), the Rahaji (red Bororo), and the Sokoto Gudali. Rahaji is a dual-purpose breed producing milk and beef. Bunaji and Sokoto Gudali supply additional draft power, Bunaji is used principally for milk, and the well-muscled Sokoto Gudali is used for meat and labor (FDLPCS 1992a).

In this area, 734 herds (58%) consisted of one breed. If herds are fairly pure, reproductive animals of a certain type are selected. This implies that the livestock owners have a good idea about adaptation of cattle for particular environments and purposes and select their bulls accordingly. Season and region within the reserve significantly influenced the distribution of these breeds (box figure 6.4).

The Rahaji, better adapted to the harsher arid environment, are found more often in the northern parts of the reserve. Rahaji is the most prestigious pastoral breed and is best adapted to arid environments. They are sensitive to humidity-related diseases (Blench 1999). The Bunaji, which is the most important breed in 42% of the herds in the north, is found more often in the central and southern parts of the reserve, where it clearly dominates (62% and 90%, respectively).

Docility of the breeds increases from Rahaji over Bunaji to Sokoto Gudali (FDLPCS 1992a). Therefore, Rahaji are kept solely by pastoralists, regardless of whether they farm. Bunaji are kept by pastoralists and agropastoralists, and Sokoto Gudali are kept mainly by farmers (Hausa and Fulani). Replacement of Rahaji and Bunaji by Sokoto Gudali cattle in the herds of settled Fulani was also observed by Vabi (1993) and Blench (1994). Thus the distribution of cattle breeds reveals a deliberate choice of particular breeds for particular purposes, for use and in light of their adaptation to ecological conditions.

The case study of the Zamfara Reserve has shown that pastoralists have developed strategies to use and manage biodiversity. In general, they are based on the local knowledge of soils, wild and domestic plants, and livestock, on the temporal and spatial variations of access to natural resources including mobility

Box 6.4 continued

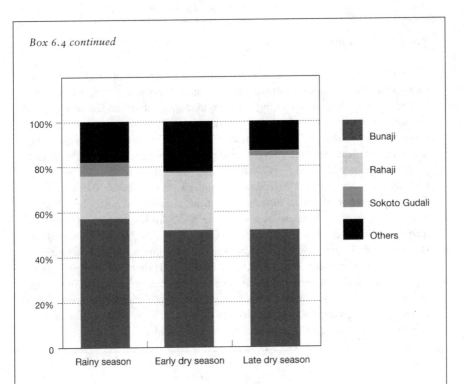

BOX FIGURE 6.4 Cattle breeds on rangeland in the Zamfara Reserve, by season (Schaefer 1998).

and flexible property rights, on the exchanges of goods and services within and between systems, and a mix of income-generating activities. This is typical for West African drylands, where both livestock and farming production systems have been maintained in the face of variable rainfall, demographic expansion, and changing market conditions. All strategies are based on high diversity, flexibility, and adaptability in order to better deal with incommensurables.

Source: Hoffmann (2003).

How Do Pastoralists and Farmers Maintain Breeds?

Even in pastoral societies across the world, strategies for controlling the breeding of livestock are extremely variable (Blench 2001). Castration of males is not acceptable in certain societies, and the separation of the sexes is difficult to manage in herding or free-roaming systems (box 6.5). Some pastoralists keep herds so large that they can select breeding

Box 6.5 Castration of Livestock

The castration of male animals is a common strategy among pastoralists in many systems. Although castrated animals may become fatter and often are less aggressive (thereby becoming easier to manage), incorrect decisions about the genetic attributes of those left intact or accidents to male animals may leave a herd breeding from poor-quality animals. One counterstrategy is the use of mechanical means to prevent reproduction; genital covers for sheep and goats are quite widespread in Western Asia. Castration probably was rarely practiced in sub-Saharan Africa in premodern times because of the risk it poses to the stock of males; however, the gradual spread of better health care has made it more widespread as a strategy. In the Andes, castration seems to have been widespread across all species. Castration also depends on social institutions for the effective circulation of males; in some ways it has structural similarities to the effects of droughts and blizzards. If there are too few potent high-quality males, when one dies it is possible to recover only by borrowing animals from beyond the household or community. This works more effectively when the community is more cohesive but can lead communities to take substantial risks in reducing the numbers of males (Blench 2001).

stock from within the herd and often favor particular breeds (Köhler-Rollefson 2003); others are better at ensuring that their herds do not mix with other herds (also for disease control reasons) than they are at controlling breeding within their own herds. Many pastoral peoples in West Africa seem indifferent to the control of breeding, even among cattle, although they are well aware of the need to introduce new breeds if their herds move into a different environmental niche (Blench 1994, 1999). Pastoralists are the owners and managers of rare and adapted livestock breeds.

In many pastoral and mixed farming systems, traditional animal exchange systems exist, which are often related to extended human families. Human ancestry is closely recorded, and there is a sense that the same should be true for livestock. Therefore livestock often are given as gifts for milestone events (birth, marriage). Mongolian herders make a clear association between human and animal bloodlines, although this process was interrupted by the collectivization of herds and the partial introduction of scientific breeding practices (Blench 2001). Beyond the exchange of genetic material, livestock exchanges are also a form of insurance and strengthen social ties. Variants of sharecropping contracts exist in which the recipient of an animal has to share the offspring with the donor (Hassan 2000; Chagunda and Wollny 2002; Gondwe and Wollny 2002).

Change and Threats to AnGRs

Genetic erosion of AnGRs continues at an increasing rate. Of the 6,300 breeds registered in DAD-IS, 1,350 are threatened by extinction or are already extinct. Globally, 35% of mammalian and 52% of avian breeds are already extinct or endangered. Europe has the highest percentage of endangered or extinct breeds (55% for mammalian and 69% for avian breeds), whereas the percentages for Africa and Asia are below average (table 6.2).

There is increasing concern that local breeds could be lost or progressively eliminated by genetic introgression or crossbreeding with exotic breeds. The loss of such breeds would lead to a corresponding loss of genetic

Table 6.2. Risk status of the world's mammalian and avian breeds as of December 1999: Absolute figures by region.

Risk Status	Africa	Asia and Pacific	Europe	Latin America and Caribbean	Near East	North America	Total
Mammalian Breeds							
Total number of breeds	632	1,031	2,512	304	562	289	5,330
Unknown	205	280	265	116	278	103	1,247
Endangered	74	99	857	43	37	69	1,179
Extinct	39	43	515	27	25	55	704
Not at risk	314	609	875	118	222	62	2,200
Avian Breeds							
Total number of breeds	106	220	611	53	34	25	1,049
Unknown	45	99	63	0	0	2	209
Endangered	21	43	391	24	7	22	508
Extinct	0	4	32	0	0	0	36
Not at risk	40	74	125	29	27	1	296

Source: Domestic Animal Diversity Information System, in Scherf (2000).

traits that have, in some cases, been built up for many centuries, such as trypanotolerance in cattle, sheep, and goats; the heat and drought resistance of Marwari goats, which can thrive with poor nutrition; or the cold resistance of Yakut cattle.

The most prominent threats to populations are as follows:

• Wars, pest and disease outbreaks (animal and human), and other natural disasters (e.g., drought, floods, earthquakes)
• Environmental changes, global warming, shifts in agroecosystems
• Social and economic changes, urbanization, market changes and intensification leading to "farmer extinction," "habitat extinction"
• Loss of traditional livelihoods and cultural diversity
• Global marketing of breeding material and the resulting breed or variety substitution or absorption, crossbreeding of local with exotic breeds
• Short-term goals, lack of recognition of current or future value of AnGRs
• Poor monitoring and management, lack of sustainable breeding programs
• Poor livestock sector development policies, lack of early warning systems, unsuitable restocking after disasters
• Land use policies that regulate common grazing grounds, displace pastoral societies, and lead to loss of animal breeds

Environmental Changes

Changes in the natural environment, such as cropland expansion into rangelands and related changes in vegetation and land use, deforestation or hunting and associated changes in habitats of vectors of animal diseases, or global warming may affect the relative advantage of livestock breeds or even species (Anderson 2004). Livestock distribution and productivity are indirectly influenced by changes in the distribution of rangelands or vector-borne livestock diseases (Tano et al. 2003; see boxes 6.2, 6.3, and 6.4).

The tsetse belt, where cattle face the threat of trypanosomiasis, has moved south because of demographic pressure and conversion of savanna into cropland. Therefore several hundred thousand square kilometers are now free of savanna tsetse in the dry subhumid zone of Africa, changing the relative value of trypanotolerant and non-trypanotolerant breeds of cattle, sheep, and goats. It is expected that demographic pressure and human impacts on the environment will result in the further retreat of tsetse and reduction in the wildlife reservoir as well. Because Zebu cattle usually are taller, farmers prefer them in non–tsetse-threatened environments. Local knowledge about tsetse pressure led to a very fine-tuned spatial distribution of trypanotolerant and susceptible breeds in the landscape. Jabbar and Diedhiou (2003) found that farmers preferred the larger zebuine, but trypano-susceptible white Fulani breed to the smaller trypanotolerant Muturu or Keteku breeds, although they acknowledged the advantages of the latter with regard to disease resistance and nonselective grazing behavior. Rege et al. (1994) described the unique social and cultural role of the Muturu in southern Nigeria. The advantage of the trypanotolerant breeds further declines in areas of lower disease challenge. Therefore Muturu and Keteku have disappeared from the savanna zones of southwestern Nigeria and are restricted to the forest zones. However, increasing resistance to drugs available for prevention and cure of trypanosomiasis may provide an incentive to continue to breed trypanotolerant breeds (Jabbar and Diedhiou 2003; Tano et al. 2003; box 6.2).

The combination of decreasing precipitation in the semiarid zones of sub-Saharan Africa and pasture degradation through overgrazing and land cultivation has resulted in a notable decline of grassland suitable for producing ruminants, bovines in particular. In the subhumid zone, new savannas are being opened through forest conversion. In cases of pasture degradation, farmers choose breeds with corresponding feeding strategies, be it with lower feed needs or different grazing behavior (see boxes 6.3 and 6.6). The ratio of small ruminants, particularly goats, to cattle increased after the droughts in the Sahel in the mid-1970s and 1980s, and the number of dromedaries is still increasing. This is because the drought tolerance and adaptation to scarce feed resources is highest in dromedaries, followed by goats and sheep and cattle. The best-suited available breeds within each species are thereby used by the farmers. In West Africa, for example, this has led to the adoption of the Sokoto Gudali cattle and in South Africa of the Nguni cattle, which are more adapted to browsing (Blench 1999; Bester et al. 2002; box 6.7).

This general movement southward of the cattle population in the Sahel has also resulted in a shift of the so-called animal traction lines, south and north of which the use of animal draft power for field preparation is not possible (Blench 1999; see box 6.6). In eastern and southern Africa, where drought has affected most of the region since 1980, farmers have had to replace cattle with the more drought-tolerant donkeys as work animals.

With regard to climate change, compensatory measures vary according to the type and extent of change. Initially, livestock producers can adapt

Box 6.6 Replacement of Cattle by Camels for Soil Fertility Maintenance in Northwest Nigeria

In northern Nigeria, a crop–livestock interaction has prevailed, based on exchange relations between segregated producer groups (McIntire et al. 1992). Mobility of pastoral livestock forms an integral part of that system, and pastoral movement to particular sites occurs regularly. Main determinants of pastoralists' choices of trekking routes are access rights to water and rangeland and the cost of access to crop residues. This crop–livestock interaction is being replaced by a more integrated system in which crop and livestock production are found on one farm. Increasing population pressure is one factor driving this transition. Livestock is more often kept in confinement rather than in free grazing on natural range. It depends increasingly on labor-intensive cut-and-carry feeding systems of crop residues, grass, and browse. Production is oriented toward manure, draft force, and milk production (Mortimore and Adams 1998). Crop residues are becoming an asset for herders and farmers.

Institutional arrangements have evolved over the years that ensure movement and help resolve conflicts between farmers and pastoralists. Farmers and herders usually exchange crop residues for manure, which are the most easily available items for the two groups involved. The farmers need their crop residues to feed their own livestock. On the other hand, they are in urgent need of fertilizer to replenish soil nutrients. They face the additional problem that cattle pastoralists leave the region and move further south because of the scarcity of grazing or water during the dry season.

Therefore, and in contrast to the traditional manuring of fields by Fulani-owned cattle herds in the region, fields are also manured by Tuareg-owned camels (*Camelus dromedarius*). As a recent development, long-distance migrating, seminomadic Tuareg camel pastoralists have taken over the role of Fulani cattle pastoralists in manuring the fields. In the villages in Dundaye district, the first camel herds were observed 25 years ago. A few farmers started contracting camel herders in 1985 to manure their fields; others have been contracting them since 1992. Of the 14 farmers who contracted herders in 1995, 8 had already contracted the same herder in 1994. Camel herds' transhumance from Niger Republic to the northern part of Nigeria starts in December and January, and the herders return to Niger Republic at the beginning of the rainy season (May and June). The seasonal migratory movement of camel herders has a substantial influence on the population of camels in northern Nigeria, particularly in the dry season. Pastoral and village camel populations in Sokoto, Kebbi, and Zamfara states were estimated at 6,800 and 36,500 head, respectively (FDLPCS 1992a, 1992b). Camel slaughter and the use of camels as draft and pack animals have increased in recent decades (Mohammed 2000).

Box 6.6 continued

Under manure contracts, pastoralists night-corral their herds on fields in the late dry season when no more crop residues are on the field and the surrounding vegetation is scarce. Animals graze and browse the nearby bush during the day. Thus corralling of livestock during the late dry season results in a net nutrient transfer from rangeland to the cropland and also contributes dung and urine to the soil. Such nutrient import from the rangeland to the harvested cropland usually is paid for by the farmers. As the next cropping season approaches, nitrogen losses through volatilization tend to be negligible as compared with early dry season manuring resulting from grazing of stubble (Hoffmann et al. 2001). The nutrient content of camel dung does not differ from that of cattle and small ruminants.

The shift from cattle to camels as manure-producing animals has allowed the use of browse as a feed stratum that still provides sufficient quantities of fodder. The camel is less dependent on herbs and grasses but prefers ligneous browse species that are still abundant in the region. The night-corralling with camels during the late dry season has three advantages. First, the crop residues are fully available for the farmers' own livestock. Second, the impact of the manure on the nutrient status of the soil is improved because it is voided right before the onset of the rainy season. Third, given the large proportion of browse in their diets, it is also likely that camel droppings contain fewer seeds of herbaceous weeds than the dung of cattle and small ruminants.

Source: Hoffmann and Mohammed (2004).

to climate change by changing management practices in controlled husbandry systems (cooling of the environment, changes in feeding and diets). In extensive systems, herds could also be moved to more hospitable locations, such as higher ground. The possible introduction of more heat-tolerant breeds may have the disadvantage of lower production potential. If maintaining productive breeds remains difficult, the shift to more resistant species could be an option, as the Sahel example demonstrates. In general, commercial and intensive livestock production systems have more potential for adaptation through the adoption of technological changes, whereas for extensive pastoral or subsistence systems the rate of technology adoption is low (Anderson 2004), and change of species is more likely to be an option.

Box 6.7 Experiences with a Large-Scale Open-Nucleus Sheep Genetic Improvement Project in the Peruvian Highlands

This box reviews experiences of development and implementation of a low-input sheep breeding program aimed at improving living standards of Andean peasant communities. Target communities are located in the Sierra Central, an isolated high mountain range environment (4,000–4,500 masl) to the east of Lima with subhumid cold weather conditions where sheep, alpacas, and cattle, in that order of importance, are run in extensive pastoral familial, communal, and multicommunal production systems. Typically, familial flocks involve 30–400 sheep, which are kept close to the owner's house and provide basic subsistence needs of the family. Communal flocks involve about 4,000 sheep, which are run as a single flock on public land in the vicinity of a peasant community. Meat and wool revenues produced by such a flock are distributed among members of the community (about 1,000 families). Multicommunal flocks (on average belonging to 6–10 communities) originated from expropriation of former privately owned land and mining companies. Such flocks consist of up to 100,000 sheep, which are run on several sites but respond to a single general breeding arrangement. The different production systems entail differences in breeding infrastructure, organizing capacity, and technological input, which in turn result in differences in productivity parameters.

Most sheep are double-purpose (meat and wool) Corriedale breed or native type sheep with different levels of upgrading. Body weights are comparable to those of commercial flocks run in more benign breeding regions of South America, although fleece weights are low and wool quality and uniformity are poor. Breeders look for improved stock in particular in view of consistent market signals for finer, high-quality wool. Rams may be homebred, bought, or exchanged, produced in independent ram-breeding flocks, or introduced from elsewhere. There is no genetic structure involving the whole Corriedale population. In addition, there is no performance recording or pedigree keeping; all selection decisions are based on visual appraisal of animals, and even in the large multicommunal populations there is no formally designed breeding program.

In the Sierra Central there is no government or private agricultural advisory service. Animal research and development activities have been hampered by terrorism and by production and marketing difficulties. A joint Peruvian–Argentinian effort to establish a sheep breeding program in the Sierra Central started in 1996 with an analysis of the traditional breeding system. After about two years of discussions, seven peasant communities and one multicommunal company agreed on the development of a breeding program aimed at improving wool production and the establishment of an appropriate extension service in order to make full use of available breeding technology.

Box 6.7 continued

The decision about the breeding strategy was to generate a collective breeding structure for the production and supply of rams for the whole Corriedale population in the region, which may be less efficient but has room for expansion and sustainability.

Each participating village establishes a multiplier flock (some already had one) and supplies the best ewes to a central nucleus. The multicommunal company participates as an additional member by supplying ewes from its top layer. An optimum open-nucleus design demands that the best ewes be concentrated in the nucleus, and culls are made for aged ewes, who are replaced with the best ewe hoggets available in the nucleus and in the participating flocks, in proportion depending on selection accuracy (Mueller 1984). However, the lack of performance records and genetic links precludes accurate between-flock selection. In addition, each participant prefers having equal access to rams. Therefore each flock contributes the same number of foundation ewes. This is clearly not efficient but is accepted for the sake of group harmony.

Considering necessary rams, effective reproductive rate, mating ratios, age structure, and inbreeding tolerance, the minimum size for the central nucleus is set at 250 ewes and 6 rams and the minimum size for multipliers at 200 ewes and 4 rams. Initially frozen semen from three Argentine rams donated by the Argentine Corriedale Breeders Association and three additional rams donated by the multicommunal company were used. Local rams have a good reputation, and foreign rams performed very well elsewhere. Performance (weaning weight, hogget weight, and fleece weight) and pedigree recording was planned in the central nucleus; best progeny tested sires would be intensively used. Performance recording was planned for the multipliers. It was expected that operational problems would limit further upward gene-flow.

Eventually, in June 1997, 432 ewes were synchronized and artificially inseminated in the central nucleus, half by laparoscopy with frozen imported semen and half with fresh local semen. Progeny testing of lambs showed that foreign sires performed better than local sires in fleece weight and fleece quality score but worse in body weight. Every year half of the ewes with their lambs returned to their village, setting the foundation of multiplier flocks. In 2001 participants increased to 15 and the central nucleus to 300 ewes. Most multiplier flocks reached the desired size. In the nucleus, ram hoggets are performance tested and visually classed. Best rams are used on nucleus and multiplier ewes. Total sheep population involved in the program is close to 160,000.

Participants wisely emphasized the need for technical support. Options were analyzed in terms of other experiences and foreseeable resources. Eventually community leaders agreed to cede land to the university for the establishment of

Box 6.7 continues to next page

Box 6.7 continued

the Research and Peasant Training Center (CICCA), which harbors the central nucleus but also serves as a demonstration farm. At the CICCA courses are held on visual selection criteria, reproductive and health inspection procedures, wool classing, and artificial insemination. Farmers judged the CICCA activities as successful and extended the agreement for another five years. This may be the single most important decision for the future of the program.

The main positive outcome of the program has been social. Cooperation and interaction between participants have fostered discussions on technical and operational aspects of the breeding program and on other aspects affecting the villages (including marketing, legal, and security matters). The establishment of the CICCA has been essential for training, confidence building, and involvement of farmers and has attracted attention for private and public, national and international cooperation and sponsorship. There were more operational difficulties than expected, mainly because many estimated husbandry skills did not correspond to reality. Selection accuracy is not increasing as quickly as planned because many difficulties in performance recording remain.

Source: Mueller et al. (2002).

Social and Economic Changes

Social and economic development affects use and survival of livestock genetic resources. Some of the most valuable and interesting animal genetic resources (e.g., fitness and behavioral traits) are kept by traditional communities, particularly pastoralists in their harsh environments. The young people from these ethnic groups often are no longer attracted to herding and prefer to migrate to the cities for employment, thereby losing indigenous knowledge (Köhler-Rollefson 2003). The increasing possibility of off-farm work in some developing countries has several consequences that may accelerate the loss of traditional local breeds used mainly for subsistence purposes or as a store of value; purchased goods tend to substitute for home-produced goods derived from animals, farm families have less time to care for livestock, and the cash economy and banks provide an alternative means to store value (Tisdell 2003). Also, the availability of motorized vehicles, stationary motors, and electricity reduces the demand for animal draft power. Thus the extension of market systems and the related changes can have important consequences for the survival of local breeds. Although the impact of such developments is readily discernible in broad terms, little is known about their impact on livestock diversity. Where livestock policy changes affect livestock genetic resources directly, the net costs and benefits of such policies usually have not been documented, and policy environments or strategies that promote conservation and appropriate use have not been defined.

There could be a conflict of interest in achieving the goals of food security and agrobiodiversity. Because of the high vertical integration and the economic efficiency of commercial poultry and pig production, a high ratio of commercial poultry or pigs in the total market supply makes it easier for countries to meet food security goals. It also may be easier to achieve food safety standards because of the ease of control of standardized production environments. Environmental concerns may also influence structural changes. Poultry production in Malaysia is expected to relocate from the present farming areas to more remote areas because of rapid urbanization and the need for large-scale operations. Poultry housing and poultry farms in general will have to become more environmentally friendly, and poultry products will have to meet sanitary and phytosanitary requirements. The recent outbreaks of infectious diseases such as avian influenza may also have policy and structural implications. One option would be to favor production systems in which biosecurity measures can be adopted easily. Another option would be to recognize the disease reservoir existent in backyard chicken flocks, composed mainly of local breeds, while encouraging vaccination and better health care. However, it seems that a focus on large-scale production, uniform genetics, and biosecurity measures are favored by exporting countries such as Thailand.

The extension of markets and economic globalization, including the global marketing of exotic breeds, have contributed significantly to the loss of local breeds through indiscriminate crossbreeding (Tisdell 2003; FAO 2001; see also chapter 17). Despite the higher milk fat content of Zebu and Criollo cattle in Latin America as compared with European breeds, they continue to be crossbred with exotic breeds, and some Criollo breeds are threatened.

In developing countries, the impact of importing exotic breeds is multifaceted and touches socioeconomic as well as genetic diversity. Importing exotic breeds into a production environment suitable to them is economically advantageous for the individual importer, as in the case of commercial poultry lines imported for industrial production systems. On the other hand, there are many examples in which upgrading of local breeds and crossbreeding programs in developing countries (e.g., Operation Coque in West Africa) have largely failed because the animals neither performed nor survived in the harsh and disease-prone environments, leading to economic losses for the small producers. Most introductions of exotic breeds in pastoral production systems have failed. Therefore loss of local AnGRs through such direct intervention may be low. Local AnGRs probably are more threatened by the indirect impact of market competition if the intensive commercial sector gains a certain market share in the country. This can even happen if markets for local and commercial livestock or their products are segmented. In such cases, the livelihoods of farmers who supply the same markets with less productive breeds may be threatened, and if they remain in production, it may no longer be economical to keep a local breed.

Conclusion

Genetic improvement programs in developing countries have failed mostly because of inappropriate strategies and lack of infrastructure and capacity. Most livestock genetic improvement efforts have been local in focus and limited in scope. Lacking essential information on the genetic resources, most livestock improvement programs to date have been unable to strategically target the most appropriate genetic resources, leading to inefficient use of scarce funds. The access of poor farmers to improved sources of AnGRs has been limited. However, the genetic characteristics of local breeds provide a basket of sustainable solutions to disease resistance, survival, and efficient production. These characteristics often have been ignored in an attempt to find technological and management solutions to individual problems of livestock production in low-input systems. Particularly in marginal environments, the cost of adjusting the production environment to the conditions needed by high-performance breeds may exceed that of improving locally adapted breeds (Wagner and Hammond 1999).

There is a need to develop breeding programs for low-input systems. A knowledge base is available to start genetic improvement programs in all livestock systems. Although opportunities to improve livestock genetic resources in low-input systems do exist, the necessary investments are substantial. Some progress in this direction has been made recently through open-nucleus breeding schemes in which the pastoralist exchanges animals with those in an improved herd, as in Uganda. Such programs work with animals that are genetically close to those in the pastoral herds and kept under conditions that are similar to those in pastoral environments (see boxes 6.7 and 6.8).

Conservation and use of livestock genetic resources in situ are critically dependent on a suitable enabling policy environment. Advanced breeding systems will support AnGR management only if equal attention is given to genetic aspects and the social and economic context. Therefore aspects such as the organizational nature of breeding initiatives, institutional demands, socioeconomic factors, and the cultural identity of the people who keep livestock must be considered. This will also lead to a better understanding of the gap between farmers' and animal breeders' goals when the two are institutionally separated. The strategic concept of community-based AnGR management is that of in situ conservation through use, ensuring that indigenous breeds remain functional parts of production systems. This promises to be a sustainable and cost-effective approach if indigenous breeds remain or become economically attractive for their owners (Rege 1999), and it calls for genetic improvement of local breeds.

Breeding for low-input production systems will continue to remain a task for the public sector and can be supported by producer cooperatives or community-based breeding programs (see boxes 6.2, 6.3, 6.7, and 6.8). However, given the choices, dynamism, and adaptation inherently embedded in indigenous knowledge and production systems on the one side and the limited availability of resources for conservation in the public sector on the other side, a certain loss of local breeds is inevitable. Most important for in situ development of AnGRs is the maintenance or creation of an environment in which livestock keepers are enabled to make informed choices about their agricultural production system and the breeds they need. It must be clearly stated that there is and will be no agrobiodiversity without active farmers, and local breeds will have the most chance of future survival if their products are consumed. Niche markets and regional products may play an important role.

National policies for livestock sector development have to take into account the AnGRs needed to achieve development goals. Therefore governments need to consider a variety of trade-offs and define their position along a spectrum of options between two extremes such as public versus private investment, large-scale versus small-scale farming, job creation versus self-employment, food security versus agrobiodiversity, and food

Box 6.8 Case Study of Nguni Cattle Breeds in South Africa

Iron Age nomads first introduced the Nguni cattle breed into South Africa in about 600 CE. These low-maintenance cattle were ideally suited to the communal farming systems of the settlers and, as far as can be established, remained unaltered during the next millennium. The advent of European colonization in the mid-19th century and the subsequent acceptance of the colonial farmer as a role model led to the introduction of exotic breeds that eventually diluted and depleted the original genepool of adapted livestock. This change was exacerbated by additional factors such as a change in the political arena, urbanization, the erosion of cultural beliefs and practices, and natural disasters.

The perception of inferiority of local breeds led to the promulgation of an act in 1934 in which populations of indigenous breeds and types were regarded as scrub (nondescript). Inspectors were empowered to inspect bulls in communal areas and to castrate them if they were regarded as inferior. Fortunately, the act was enforced only during the first few years of its existence because it proved unpopular with stockowners. A structure was developed in the country that allowed the Nguni breed to enter the growing commercial sector, and extensive recording facilitated breed improvement. Thus, while the breed was improved in the commercial sector, it was being eroded in the rural areas through crossbreeding and replacement with exotic breeds. This occurred because of the perception that the Nguni was inferior to the larger exotics, despite the fact that it was a low-maintenance breed ideally suited to the low-input farm systems of the communal farmer. Fortunately, the inherent hardiness of the breed allowed it to survive, and purebred animals are still found in limited numbers in rural communities.

The more recent realization that this hardy breed was uniquely adapted to the South African environment led to its evaluation and development in the commercial sector. In 1985, a committee was appointed to report on the desirability of having an in vitro germplasm bank for indigenous livestock and on the control of imported semen from exotic breeds.

As a selective grazer and browser, the Nguni is able to obtain optimal nutritional value from the available natural vegetation, thus enabling it to survive under conditions that would not support bulk grazers such as the European cattle breeds. Temperamentally, the Nguni is very docile. Other adaptive traits enable the Nguni to walk long distances in search of grazing and water. They are also reported to tolerate extreme temperatures and have a better ability to maintain body condition in winter than Simmental cattle.

The Nguni is now seen as a source of genetic material well suited to the management style and needs of the emergent black farmer who needs a low-maintenance and high-output animal. The initial evaluation of the Nguni showed its potential as a beef breed in both extensive and intensive farming systems. Cow

Box 6.8 continued

mass and reproductive performance of the Nguni, when compared with other breeds, showed it to be the most fertile beef breed in South Africa. It was also shown to be ideally suited as a dam line in terminal crossbreeding. In addition, its traits of heat, tick, and disease tolerance make it an ideal breed for extensive systems.

In the past, projects for the introduction of exotic cattle breeds into the communal sector invariably failed because of the introduction of complex technologies that increased production beyond the point of sustainability. Ongoing projects are designed to encourage the reintroduction of the hardy, low-maintenance Nguni breed into the communal sector in order to stem the influence of the less well-adapted exotic breeds. This reintroduction is accompanied by support technology to improve management and a marketing system to facilitate the sale of animals at market-related prices. In addition, the communities are encouraged to organize commodity groups or farmers' organizations to create an infrastructure that allows decisions to be made based on community consensus.

Source: Bester et al. (2002).

safety versus food diversity. National and international policies must identify their objectives and make decisions regarding breed maintenance, cultural diversity, and genetic variability. These decisions will have implications for breeding and conservation methods and the necessary funding. They also have implications for the research and technologies needed for characterization and valuation. Public–private partnerships are needed to achieve most objectives. Overall, there is an urgent need to raise awareness of the value of ANGRs for food and agriculture.

Chapter Glossary

BREED: A subspecific group of domestic livestock with definable and identifiable external characteristics that enable it to be separated by visual appraisal from similarly defined groups within the same species, or a group for which geographic or cultural separation from phenotypically similar groups has led to acceptance of its separate identity.

Note: Breeds have been developed according to geographic and cultural differences and to meet human food and agricultural needs. In this sense, *breed* is not a technical term. The differences, both visual and otherwise, between breeds account for much of the diversity associated with each domestic animal species. *Breed* is often accepted as a cultural rather than a technical term. **Locally adapted breeds** have been in the country for a sufficient time to be genetically adapted to one or more traditional production systems or environments in the country. **Indigenous breeds,** also called autochthonous or native breeds, originating from, adapted to, and used in a particular geographic region, form a subset of the locally adapted breeds (FAO 2001). **Exotic breeds** are maintained in an area other than the one in which they were developed and include breeds that are not locally adapted. Exotic breeds include both **recently introduced breeds,** which were imported over a short period of time within the last five or so generations, and **continually imported breeds,** whose local genepool is regularly replenished from one or more sources outside the country. Many of the breeds used in intensive production systems or marketed by international breeding companies are in this category.

BREED AT RISK: Any breed that may become extinct if the factors causing its decline in numbers are not eliminated or mitigated. Breeds may be in danger of becoming extinct for a variety of reasons. Risk of extinction may result from low population size; direct and indirect impacts of policy at the farm, country, or international levels; lack of proper breed organization; lack of adaptation to market demands; or perceived lower performance. Breeds are categorized as to their risk status on the basis of the actual numbers of male or female breeding individuals and the percentage of purebred females.

EXTINCT BREED: A breed population that can no longer be recreated. This situation becomes absolute when no breeding males or breeding females remain. In reality extinction may be realized well before the loss of the last animal, gamete, or embryo.

FARM ANIMAL GENETIC RESOURCES (AnGRs): Animal species that are used, or may be used, for the production of food and agriculture, and the populations within each of them. These populations within each species can be classified as wild and feral populations, landraces and primary populations, standardized breeds, selected lines, varieties, strains, and any conserved genetic material, all of which are currently categorized as breeds (FAO 2001).

POPULATION: A generic term that when used in a genetic sense defines an interbreeding group and may refer to all the animals within a breed. The genetic makeup of the population is concerned with the genetics of all individuals it comprises and with the transmission from generation to generation of samples of the genetic variability associated with this population (FAO 2001).

References

Anderson, S. 2004. *Environmental Effects on Animal Genetic Resources.* FAO Background Study Paper No. 28. Rome: FAO.

Bayer, W., A. von Lossau, and A. Feldmann. 2003. Smallholders and community based management of farm animal genetic resources. In *Proceedings of the Workshop on Community Based Management of Animal Genetic Resources. A Tool for Rural Development and Food Security,* 1–12. Mbabane, Swaziland, May 7–11, 2001. Rome: FAO.

Behnke, R. H. and N. Abel. 1996. Revisited: The overstocking controversy in semi-arid Africa. *World Animal Review* 12:5–27.

Bester, J., L. E. Matjuda, J. M. Rust, and H. J. Fourie. 2002. The Nguni: A Case Study. Paper presented to the Symposium on Managing Biodiversity in Agricultural Ecosystems, Montreal, Canada, November 8–10, 2001.

Blench, R. 1994. The expansion and adaptation of Fulbe pastoralism to subhumid and humid conditions in Nigeria. *Cahiers d'Études Africaines* 133–135:197–212.

Blench, R. 1999. *Traditional Livestock Breeds: Geographical Distribution and Dynamics in Relation to the Ecology of West Africa.* ODI Working Paper 122. London: ODI.

Blench, R. 2001. *Pastoralism in the New Millennium.* FAO Animal Production and Health Paper No. 150. Rome: FAO.

Bruford, M. W., D. G. Bradley, and G. Luikart. 2003. DNA markers reveal the complexity of livestock domestication. *Nature Reviews Genetics* 4:900–10.

Chagunda, M. G. G. and C. B. A. Wollny. 2002. Consequences of differences in pricing of economic values for milk yield of dairy cattle in Malawi. 7th World Congress on Genetics Applied to Livestock Production, August 19–23, 2002, Montpellier, France, Session 25: Developing Sustainable Breeding Strategies for Medium and Low-Input Systems. Communication 25–03.

CIRDES/ILRI/ITC (Centre International de Recherche–Développement sur l'Elevage en Zone Subhumide, International Livestock Research Institute, and International Trypanotolerance Centre), 2000. Collaborative research programme on trypanosomosis and trypanotolerant livestock in West Africa. In *Joint Report of Accomplishments and Results (1993–1999)*. Banjul, The Gambia: ITC.

Coppock, D. L. 1994. *The Borana Plateau of Southern Ethiopia: Synthesis of Pastoral Research, Development and Change, 1980–1991*. Addis Ababa, Ethiopia: ILCA.

Cousins, N. J. and M. Upton. 1988. Options for improvement of the Borana pastoral system. *Agricultural Systems* 27:251–278.

"Egyptian chicken plan hatches . . . 50 years later." 1997. *The Iowa Stater,* May, www.iastate.edu/IaStater/1997/may/chicken.html.

FAO (Food and Agriculture Organization of the United Nations). 1999. *The Global Strategy for the Management of Farm Animal Genetic Resources*. Executive Brief. Rome: FAO.

FAO (Food and Agriculture Organization of the United Nations). 2001. Guidelines for the development of country reports. State of the World. Annex 2, Working Definitions for Use in Developing Country Reports and Providing Supporting Data, dad.fao.org/en/Home.htm.

FDLPCS (Federal Department of Livestock and Pest Control Services). 1992a. Livestock in Sokoto State. *Nigerian Livestock Resources*. Vol. II: *National Synthesis*. Vol. III: *State Reports*. St. Helier, Jersey, UK: RIM.

FDLPCS. (Federal Department of Livestock and Pest Control Services). 1992b. *Nigerian Livestock Resources*. Vol. I: *Executive Summary and Atlas*. St. Helier, Jersey, UK: RIM.

Flock, D. K. and R. Preisinger. 2002. Breeding plans for poultry with emphasis on sustainability. 7th World Congress on Genetics Applied to Livestock Production, August 19–23, 2002, Montpellier, France, Session 24: Sustainable Breeding Plans in Developed Countries. Communication 24–02.

Gondwe, T. N. P. and C. B. A. Wollny. 2002. Traditional breeding systems in smallholder rural poultry in Malawi. 7th World Congress on Genetics Applied to Livestock Production, August 19–23, 2002, Montpellier, France, Session 25: Developing Sustainable Breeding Strategies for Medium- and Low-Input Systems. Communication 25–26.

Grell, H. and M. Kirk. 2000. The role of donors in influencing property rights over pastoral resources in Sub-Saharan Africa. In N. McCarthy, B. Swallow, M. Kirk, and P. Hazell, eds., *Property Rights, Risk, and Livestock Development in Africa,* 55–85. Washington, DC, and Nairobi, Kenya: IFPRI and ILRI.

Hassan, W. A. 2000. *Biological Productivity of Sheep and Goats Under Agro-Silvo-Pastoral Systems in Zamfara Reserve in North-Western Nigeria*. Goettingen, Germany: Cuvillier.

Haverkort, B. 1993. Agricultural development with a focus on local resources: ILEIA's view on indigenous knowledge. In D. M. Warren, D. Brokensha, and L. J. Slikkerveer, eds., *Indigenous Knowledge Systems: The Cultural Dimensions of Development*. London: Kegan Paul International.

Hillel, J., M. A. M. Groenen, M. Boichard, A. B. Korol, L. David, V. M. Kirzhner, T. Burke, A. B. Dirie, R. P. M. A. Croojimans, K. Elo, M. W. Feldman, P. J. Freidlin, A. Maki-Tanila, M. Oortwijn, P. Thomson, A. Vignal, K. Wimmers, and S. Weigend. 2003. Biodiversity of 52 chicken populations assessed by microsatellite typing of DNA pools. *Genetics Selection Evolution* 35:533–557.

Hoffmann, I. 2003. Biodiversity management in West African pastoral and agro-pastoral systems. A case study from northwest Nigeria. In *Biodiversity and the Ecosystem Approach in Agriculture, Forestry and Fisheries*, 28–49. Satellite event on the occasion of the 9th regular session of the Commission on Genetic Resources for Food and Agriculture, FAO, Rome, October 12–13, 2002. Available at www.fao.org/DOCREP/005/Y4586E/y4586e03.htm#Po_o.

Hoffmann, I., D. Gerling, U. B. Kyiogwom, and A. Mané-Bielfeldt. 2001. Farmers' management strategies to maintain soil fertility in a remote area in northwest Nigeria. *Agriculture, Ecosystems and Environment* 86(3):263–275.

Hoffmann, I. and I. Mohammed. 2004. The role of nomadic camels for manuring farmers' fields in the Sokoto Close Settled Zone, northwest Nigeria. *Nomadic Peoples* 8(1): 99–112.

Homann, S., G. Dalle, and B. Rischkowsky. 2004. Potentials and constraints of indigenous knowledge for sustainable range and water development in pastoral land use systems of Africa: A case study in the Borana Lowlands of Southern Ethiopia. Eschborn, Germany: GTZ/TÖB.

Hossary, M. A. and S. Galal. 1995. Improvement and adaptation of the Fayoumi chicken. *Animal Genetic Resources Information* 14:33–42.

Huelsebusch, C. G. and B. A. Kaufmann. 2002. *Camel Breeds and Breeding in Northern Kenya. An Account of Local Breeds of Northern Kenya and Camel Breeding Management of Turkana, Rendille, Gabra and Somali Pastoralists*. Nairobi: Kenya Agricultural Research Institute.

Jabbar, M. A. and M. L. Diedhiou. 2003. Does breed matter to cattle farmers and buyers? Evidence from West Africa. *Ecological Economics* 45(3):461–472.

Kamara, A. 2001. *Property Rights, Risk and Livestock Development in Southern Ethiopia*. PhD thesis, Wissenschaftsverlag Vauk, Kiel, Germany.

Köhler-Rollefson, I. 2000. *Managing Animal Genetic Resources at the Community Level*. Eschborn, Germany: GTZ. Available at www.gtz.de/agrobiodiv/download/koehl.pdf.

Köhler-Rollefson, I. 2003. Community based management of animal genetic resources, with special reference to pastoralists. In *Proceedings of the Workshop on Community Based Management of Animal Genetic Resources. A Tool for Rural*

Development and Food Security, 13–26. Mbabane, Swaziland, May 7–11, 2001. Rome: FAO.

Kyiogwom, U. B., I. Mohammed, H. M. Bello, S. A. Maigandi, and C. Schaefer. 1994. The economic situation of the livestock farmer in Zamfara. In *Range Development in the Endangered Sudan Savanna in Sokoto State,* 63–70. Unpublished report, Giessen.

Masinde, I. A. 2001. *Managing Biodiversity in Agricultural Ecosystems. Local Management of Agricultural Biodiversity by Communities in Kenya.* Montreal: United Nations University. Available at www.unu.edu/env/plec/cbd/Montreal/papers/Masinde.pdf.

McIntire, J., D. Bourzat, and P. Pingali. 1992. *Crop–Livestock Interaction in Sub-Saharan Africa.* Washington, DC: The World Bank.

Mhlanga, F. N. 2002. *Community-Based Management of Animal Genetic Resources: A Participatory Approaches Framework.* Eschborn, Germany: GTZ. Available at www.gtz.de/agrobiodiv/download/mhalanga.pdf.

Mohammed, I. 2000. *Study of the Integration of the Dromedary in Smallholder Crop–Livestock Production Systems in Northwestern Nigeria.* Goettingen, Germany: Cuvillier.

Mortimore, M. and W. M. Adams. 1998. Farming intensification and its implications for pastoralism in northern Nigeria. In I. Hoffmann, ed., *Prospects of Pastoralism in West Africa,* Vol. 25, 262–273. Giessener Beiträge zur Entwicklungsforschung, Reihe I. Giessen, Germany: Wissenschaftl. Zentrum Tropeninstitut.

Mueller, J. P. 1984. Single and two-stage selection on different indices in open-nucleus breeding systems. *Genetics Selection Evolution* 16:103–120.

Mueller, J. P., E. R. Flores, and G. A. Gutierrez. 2002. Experiences with a large scale sheep genetic improvement project in the Peruvian highlands. 7th World Congress on Genetics Applied to Livestock Production, August 19–23, 2002, Montpellier, France, Session 25. Developing Sustainable Breeding Strategies in Medium- to Low-Input Systems. Communication 25–12.

Nortier, C. L., J. F. Els, A. Kotze, and F. H. van der Bank. 2002. Genetic diversity of indigenous Sanga cattle in Namibia using microsatellite markers. 7th World Congress on Genetics Applied to Livestock Production, August 19–23, 2002, Montpellier, France. Session 26: Management of Genetic Diversity. Communication 26–07.

Preisinger, R. 2004. *Internationale Tendenzen der Tierzüchtung und die Rolle der Zuchtunternehmen.* Presentation, Agrobiodiversität entwickeln: Handlungsstrategien und Impulse für eine nachhaltige Tier- und Pflanzenzucht, Umweltforum Berlin, February, 3–4, www.agrobiodiversitaet.net/site/page/downloads/dateien/2.

Rajasekaran, B. 1993. A framework for incorporating indigenous knowledge systems into agricultural research, extension, and NGOs for sustainable agricultural development. *Studies in Technology and Social Change* 21. Ames: Technology and Social Change Program, Iowa State University.

Rege, J. E. O. 2003. Defining livestock breeds in the context of community based management of farm animal genetic resources. In *Proceedings of the Workshop on Community Based Management of Animal Genetic Resources. A Tool for Rural Development and Food Security,* 27–35. Mbabane, Swaziland, May 7–11, 2001. Rome: FAO.

Rege, J. E. O. 1999. The state of African genetic resources. I. Classification framework and identification of threatened and extinct breeds. *Animal Genetic Resources Information* 25:1–25.

Rege, J. E. O., G. S. Aboagye, and C. L. Tawah. 1994. Shorthorn cattle of West and Central Africa II. Ecological settings, utility, management and production systems. *World Animal Review* 78:14–21.

Richards, P. 1985. *Indigenous Agricultural Revolution: Ecology and Food Production in West Africa.* London: Hutchinson.

Röhrs, M. 1994. Entwicklung der Haustiere. In H. Kräusslich, ed., *Tierzüchtungslehre,* 4th ed., 37–55. Stuttgart, Germany: Ulmer.

Schaefer, C. 1998. *Pastorale Wiederkäuerhaltung in der Sudansavanne: Eine Untersuchung im Zamfara Forstschutzgebiet im Nordwesten Nigerias.* Göttingen, Germany: Cuveillir.

Scherf, B., ed. 2000. *World Watch List for Domestic Animal Diversity,* 3rd ed. Rome: FAO/UNDP.

Sere, C., H. Steinfeld, and J. Groenewold. 1996. World livestock production systems. Current status, issues and trends. *FAO Animal Production and Health Papers* 127. Rome: FAO.

Steglich, M. and K. J. Peters. 2002. Agro-pastoralists' trait preferences in N'dama cattle: Participatory methods to assess breeding objectives. 7th World Congress on Genetics Applied to Livestock Production, August 19–23, 2002, Montpellier, France, Session 25. Developing Sustainable Breeding Strategies in Medium- to Low-Input Systems. Communication 25-04.

Steinfeld, H., C. De Haan, and H. Blackburn. 1997. *Livestock and the Environment: Issues and Options.* Brussels: European Commission/FAO/World Bank.

Tano, K., M. Kamuanga, M. D. Faminow, and B. Swallow. 2003. Using conjoint analysis to estimate farmers' preferences for cattle traits in West Africa. *Ecological Economics* 45(3):393–408.

Tempelman, K. A. and R. A. Cardellino. In press. *Community-Based Management and Use of Animal Genetic Resources in Traditional Livestock Farming Systems.* Rome: FAO.

Thebaud, B. and S. Batterbury. 2001. Sahel pastoralists: Opportunism, struggle, conflict and negotiation. A case study from eastern Niger. *Global Environmental Change* 11:69–78.

Tisdell, C. 2003. Socioeconomic causes of loss of animal genetic diversity: Analysis and assessment: *Ecological Economics* 45(3):365–377.

Vabi, M. B. 1993. *Fulani Settlement and Modes of Adjustment in the Northwest Province of Cameroon.* ODI Pastoral Development Network Paper 35d. London: ODI.

Valuing AnGR. 2003. *Ecological Economics* Special Issue 45(3).

Wagner, H.-G. R. and K. Hammond. 1999. *The Management of Farm Animal Genetic Resources and FAO's Global Strategy.* Berlin: Deutscher Tropentag, Berlin, Humboldt University.

Warren, D. M. 1991. *Using Indigenous Knowledge in Agricultural Development.* World Bank Discussion Paper No. 127. Washington, DC: The World Bank.

Weigend, S. and M. N. Romanov. 2002. The World Watch List for Domestic Animal Diversity in the context of conservation and utilisation of poultry biodiversity. *World's Poultry Science Journal* 58(4):411–430.

M. HALWART AND D. BARTLEY

The cultivation of most rice crops in irrigated, rainfed, and deepwater systems offers a suitable environment for fish and other aquatic organisms (figure 7.1). More than 90% of the world's rice, equivalent to approximately 134 million ha (figure 7.2), is grown under flooded conditions, not only providing home to a wide range of aquatic organisms but also offering opportunities for their enhancement and culture. Aquatic production, in addition to the rice crop itself, is a critically important resource for rural livelihoods in developing countries. Local consumption and marketing are particularly important for food security because aquatic food resources are the most readily available, most reliable, and cheapest source of animal protein and fatty acids for both farming households and the landless. This chapter synthesizes recent information and highlights the important role of aquatic biodiversity from rice-based ecosystems in rural livelihoods and ecological services. This information is not commonly available but is crucial for informed policy decisions.

The Issue

Production other than rice obtained from rice-based ecosystems and its importance for rural livelihoods are generally underestimated and undervalued (e.g., FAO/MRC 2003; Halwart 2003) because local consumption or restricted marketing usually prevents this production from entering official national statistics. Additionally, the availability of this production is

FIGURE 7.1. Rice-based ecosystems often represent a dynamic and closely linked complex of rice fields, ponds, irrigation canals, and rivers (Vietnam). (Photo: FAO/M. Halwart.)

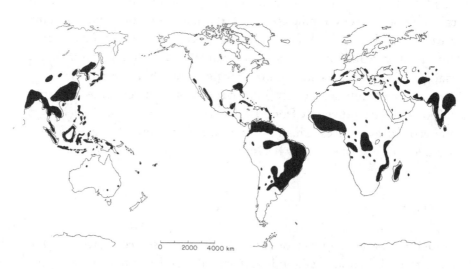

FIGURE 7.2. Rice is cultivated in approximately 151 million ha worldwide in irrigated (57%), rainfed lowland (31%), deepwater (4%), and upland (11%) environments (after Fernando 1993 and Fernando and Halwart 2001; data from IRRI World Rice Statistics and FAO database 2001 at www.irri.org/science/ricestat/index.asp).

temporally and spatially variable, and the quantity of captured, collected, or farmed organisms usually is small. Rice is generally viewed as a monoculture and considered the key commodity for local and national food security. Therefore emphasis in national crop production strategies usually is placed on enhancing rice yields, which in turn often leads to increased fertilizer and pesticide inputs. These practices and policies often ignore and threaten the other components of rice-based ecosystems.

Policymakers must base their decisions on sound information. Yet the information they need in regard to rice field fisheries and rice-based aquaculture usually is not available, and so the contribution these resources make to rural livelihoods is not recognized. Development plans that focus only on increasing yields of rice may give the people more rice to eat but may at the same time take away much of the aquatic animals and plants also harvested in and around the rice fields. Without a sound understanding of the other components of the rice field ecosystem and careful consideration of suitable extension, the aquatic animal and plant diversity can be severely reduced. Importantly, it is the poorer segments of rural society who will suffer most from the negative impacts of such development.

From Fisheries to Aquaculture: A Continuum

Rice–fish systems can be separated into capture or culture systems depending on the origin of the fish stock. All these systems are called rice-based systems or rice–fish systems because the farm economy in which aquaculture takes place usually is dominated by rice cultivation. In the capture system, wild fish enter the rice fields from adjacent water bodies and reproduce in the flooded fields. In the culture system, rice fields may be deliberately stocked with fish either simultaneously or alternately with the rice crop; this is also known as concurrent and rotational rice–fish farming. The rice fields may be used to produce fingerlings or table fish depending on the size of fish seed available for stocking, the duration of the fish culture period, and the market needs for fingerlings or table fish (Halwart 1998; Demaine and Halwart 2001).

Rice field ecosystems have rich aquatic biodiversity that is extensively used by local people. The most important group, in terms of species diversity and importance for the local community, are the fishes. Balzer et al. (2005) found 70 different fish species in Cambodia, and Luo (2005) found 52 in China (table 7.1). In addition to fish, more than 100 species of

Table 7.1a. Number of aquatic species collected from rice-based ecosystems and used by rural households.

Group	Cambodia	China
Fish	70	52
Crustaceans	6	2
Mollusks	1	4
Amphibians	2	4
Insects	2	3
Reptiles	8	—
Aquatic plants	13	19

Sources: Balzer et al. (2005); Luo (2005).

crustaceans, mollusks, amphibians, insects, reptiles, and aquatic plants were found with nutritional, medicinal, decorative, and other uses (Balzer et al. 2005) (table 7.2).

Many fish species can be harvested from rice fields, but in the culture system only a few are commercially important. Among the most common and widespread are common carp and Nile tilapia. They feed low in the food chain and therefore are preferred species in the culture systems. Other popular species are *Puntius gonionotus* and *Trichogaster* spp. Many air-breathing species such as the snakehead (*Channa striata*) and catfishes (*Clarias* spp.) are well adapted to the swamplike conditions of rice fields. They are highly appreciated wild fish in the capture system because they fetch good market prices, but they are less appreciated in the culture systems because they can decimate stocked fish populations.[1]

Traditionally a large proportion of the fish for household consumption was caught in the paddy fields. With increased fishing pressure, the conversion of many wetlands into agricultural land, and the intensification of rice production, the rice field fishery has declined in many areas, and farmers have often turned to aquaculture as an alternative source of animal protein.

Ecological Functions

Many of the aquatic organisms found in rice ecosystems play an important role as biological control agents of vectors and pests of medical and agricultural importance and are an important element of integrated pest

Table 7.1b. Fish species ($n = 70$) collected from rice fields and used by rural households in Cambodia.

Species	English Name	Species	English Name
Thynnichthys thynnoides		*Clarias batrachus*	Broadhead catfish
Mystus albolineatus		*Anabas testudineus*	Climbing perch
Osteochilus melanopleurus		*Trichogaster trichopterus*	Threespot gourami
Leptobarbus hoeveni	Mad barb	*Rasbora tornieri*	Yellowtail rasbora
Trichogaster pectoralis	Snake-skin gourami	*Rasbora trilineata*	Scissortail rasbora
Botia modesta	Red-tail botia	*Systomus partipentazona*	
Cyclocheilichthys sp.	Beardless barb	*Rasbora daniconius*	Slender rasbora
Hemibagrus splilopterus		*Rasbora borapetensis*	Blackline rasbora
Xenentodon cancila		*Cirrhinus microlepis*	
Paralaubuca typus		*Monopterus albus*	Swamp eel
Notopterus notopterus	Bronze featherback	*Trichopsis vittata*	Croaking gourami
Trichogaster pectoralis		*Botia* sp.	Sun loach
Pristolepis fasciatus	Catopra	*Pseudomystus siamensis*	Asian bumblebee catfish
Hampala macrolepidota		*Anguilla bicolor*	Shortfin eel
Oxyeleotris marmorata	Marbled sleeper	*Parambassis* sp.	Siamese glassfish
Henicorhynchus siamensis		*Ompok hypophthalmus*	
Channa micropeltes	Snakehead	*Puntius brevis*	Swamp barb
Macrognathus siamensis	Peacock eel	*Parambassis wolffi*	Duskyfin glassy perchlet
Barbodes altus		*Macrognatus taenigaster*	
Trichogaster sp.		*Osteochilus hasselti*	Silver sharkminnow

Table 7.1b. continues to next page

Table 7.1b. continued

Species	English Name	Species	English Name
Mastacembelus favur	Tire track eel		
Trichogaster sp.	Moonlight gourami	*Micronema micronema*	
Pangasius conchophilus		*Ompok bimaculatus*	Butter catfish
Puntioplites proctozysron		*Chitala ornata*	Clown featherback
Channa striata	Snakehead	*Clarias macrocephalus*	Walking catfish
Monotreta cambodgiensis		Mastacembelidae	
Acantopsis sp.		*Esomus metallicus*	Striped flying barb
Mystus mysticetus		*Paralaubuca typus*	
Labiobarbus siamensis		*Clupeichthys* sp.	Thai river sprat
Barbodes gonionotus		*Trichopsis schalleri*	Pygmy gourami
Doryichtys boaja	Long-snouted pipefish	*Macrognatus siamensis*	
Botia helodes	Tiger botia	*Parachela siamensis*	
Luciosoma bleekeri		*Trichogaster* sp.	
Nandus nandus		*Cyclocheilichthys enoplos*	
Morulius chrysophekadion	Black sharkminnow	*Channa lucius*	

Source: Balzer et al. (2005).

management. Fish that are specialized to feed on mosquito larvae or on particular snail species may control vectors of malaria and schistosomiasis. Some fish species contribute to the biological control of rice pests such as apple snails, stemborers, and caseworms (Halwart 1994, 2001; Halwart et al. 1998). Fish also feed on weeds and other insects, thereby reducing potential pest problems and maintaining the ecosystem balance. In fact, biological control has been proven to be more profitable

Table 7.2. Indicative list of uses of various aquatic organisms from rice fields.

Taxon	Scientific Name	Uses	Photo
Fish	*Cyclocheili-chthys* sp.	Fresh, fermented fish paste, fermented fish pieces, dried salted fish, fish sauce	
Reptile	*Erpeton tentaculatum*	Medicinal use	
Amphibian	*Bufo melanostictus*	Fresh, medicinal use (anthelminthic)	
Crustacean	*Somanniathel-phusa* sp.	Fresh, feed, bait	
Mollusk	*Pila* sp.	Fresh, feed, bait, sale	
Plant	*Nelumbo nucifera*	Flowers, leaves, seeds, rhizome for consumption, sale, decoration, and wrapping	
Insect	*Lethocerus* sp.	Fresh, medicinal use	

than prophylactic or threshold-based pesticide treatments (Rola and Pingali 1993). Moreover, farmers have experienced that the concurrent culture of fish with rice often increases rice yields, particularly on poorer soils and in unfertilized crops, probably because under these conditions the fertilization and nutrient cycling effects of fish are greatest. With savings on pesticides and earnings from fish sales, net income on rice–fish farms reportedly is 7–65% higher than on rice monoculture farms (Halwart 1999).

Rice fields may also harbor species that are under threat of extinction. The deepwater rice ecosystem and the adjacent flooded grass and shrub lands near the Tonle Sap, Cambodia, are habitat for many birds, among them the Bengal Florican, an endangered species of which only two populations remain worldwide (Smith 2001). The use of some endangered species, such as the Banna caecilian (*Ichthyophys bannanicus*), which has medicinal value, in the long term probably is a blessing because it is the economic value that may lead to its cultivation, ultimately ensuring the survival of the species.

Recent Activities

The high availability of wild fish usually favored development of the capture system in the rice fields associated with floodplains of large river systems. This system has been studied recently with regard to the living aquatic resources availability and use pattern of rice farmers in the Upper and Lower Mekong River floodplains in Xishuangbanna, Yunnan Province, China (Luo 2005) and in Kampong Thom Province, Cambodia (Balzer et al. 2005). A lower availability of wild fish in remote mountainous areas resulted in the emergence and evolution of the culture rice–fish system. Indigenous rice–fish systems using locally adapted strains of fish species are found in the uplands of northern Vietnam and Laos. The traditional knowledge in these rice–fish societies has been the particular focus of recent work in the Vietnamese provinces of Hoa Binh, Son La, and Lai Chau (Meusch 2005), and the Laotian provinces Xieng Khouang and Houa Phanh (Choulamany 2005).

The findings of the recent studies allow greater understanding and appreciation for the rich diversity and value of aquatic resources, the local practices related to their capture and culture, and the need to work closely with farmers to develop appropriate interventions for aquaculture production. A first step in making this rice-based aquatic biodiversity visible has been at intergovernment level, where policymakers have been urged to pay more attention to the enhancement of aquatic biodiversity and the nutritional contribution of aquatic organisms in the diet of rural people who produce or depend on rice (box 7.1). Preliminary investigations on the relationship between rice farming, living aquatic resources, and the livelihoods of the people who manage these systems revealed the value of this biodiversity to rural communities (box 7.2).

Box 7.1 Rice-Based Aquatic Biodiversity Highlighted at the 20th Session of the International Rice Commission, July 23–26, 2002

The FAO's International Rice Commission is a forum in which senior policymakers and rice specialists from rice-producing countries review their national research and development programs. Its objective is to promote national and international action in matters relating to the production, conservation, distribution, and consumption of rice. The commission meets every four years. At its twentieth session, in Bangkok in July 2002, the commission recommended the following:

- Member countries should promote the sustainable development of aquatic biodiversity in rice-based ecosystems, and policy decisions and management measures should enhance the living aquatic resource base. In areas where wild fish are depleted, rice–fish farming should be considered as a means of enhancing food security and securing sustainable rural development.
- Attention should be given to the nutritional contribution of aquatic organisms in the diet of rural people who produce or depend on rice.

Source: FAO (2002).

Productive Ecosystems Under Threat

Evidence from participatory rural appraisals of farming and fishing communities indicates that the availability of aquatic resources in rice fields is declining (Balzer et al. 2005; Luo 2005). Although the amount of aquatic organisms consumed has remained constant, a decade ago rice-based capture supplied half of this consumption, whereas today only one-fifth to one-third is derived from capture in rice-based farming, and the remainder has to be bought or farmed (Luo 2005). Farmers in Xishuangbanna claim that fish are becoming less abundant and that the amount of aquatic organisms collected in one day today is equivalent to what was collected a decade ago in one hour. Similarly, the Cambodian study (Balzer et al. 2005) points out that fish catches have decreased greatly in the past two decades. The villagers estimate that in three to five years there will not be enough fish to make a living. Human population increase and the consequent increased fishing pressure on aquatic resources is an important factor in the decline of living aquatic resources. Other related activities are

Box 7.2 Nutrition and Aquatic Resources in Quang Tri Province, Central Vietnam

Vietnam has the highest malnutrition rate among adults and children of the countries in Southeast Asia (FAO 1999). Basic nutritional needs of many children are not being met (Reinhard and Wijayaratne 2002), and according to the World Health Organization there are important public health problems: 40% of the adults have body mass index lower than 18.5, which is the threshold value for being underweight. In collaboration with a poverty alleviation program financed by the Finnish Ministry for Foreign Affairs, the Vietnamese Ministry of Planning and Investment, and Quang Tri Rural Development Programme in central Vietnam, FAO participated in a study on the nutrition status of rice-farming families and the role of aquatic resources in their daily diet in Quang Tri, one of the poorest provinces in central Vietnam. Special attention was paid to signs of malnutrition in children under five years of age.

Methods

The study used a participatory approach that concentrates mainly on the assessment of rural people's behavior and experience related to their nutrition and health status and their use of available resources. The study included three elements: a household questionnaire, anthropometric measurements of children under five years, and focus group discussions. The communes (i.e., villages) were randomly selected: five communes in the remote Dakrong district, located in a mountainous area; and two communes in Hai Lang district and one commune in Cam Lo district, located in lowland areas. In each commune 15–30% of the households were selected; the average number of people per household was five. Malnutrition in children was measured by three standard indices: underweight (weight/age), stunting (height/age), and wasting (weight/height).

Results

Fish was found to be the most frequently consumed aquatic animal, with 80% of the households in Hai Lang, 89% in Cam Lo, and 39% in Dakrong district eating fish two or more times per week (box table 7.2a). Fish is said to be a highly preferred food because of its good taste, easy availability, and health qualities. The study showed that taste is the main reason for consuming aquatic animals; most of the respondents say that they eat snakes for their health qualities and frogs, insects, and field crabs because of their easy availability. Snakes are rarely eaten in all districts, whereas insects are a basic food of almost half of the households in the mountainous Dakrong district.

Box 7.2 continued

BOX TABLE 7.2A. Consumption frequency (%) of fish and other aquatic organisms among the surveyed households.

	Never	1 ×/mo	2–3×/mo	1×/wk	2–5 ×/wk	Every Day (6–7×/wk)	Occasionally
Hai Lang (n = 70)							
Fish	2.9	—	1.4	5.7	61.4	18.3	10.3
Snakes	84.3	8.6	4.3	—	—	—	2.8
Snails	45.7	21.4	8.6	10.0	1.4	1.4	11.5
Field crabs	54.3	15.7	7.1	5.7	4.3	—	12.9
Shrimps	5.7	1.4	4.3	4.3	67.1	4.3	12.9
Insects	92.9	1.4	1.4	—	2.9	0.1	1.3
Frogs	60.0	7.1	4.3	5.7	1.4	—	21.5
Cam Lo (n=35)							
Fish	—	2.9	—	8.6	71.4	17.1	—
Snakes	94.3	—	5.7	—	—	—	—
Snails	68.6	8.6	8.6	—	2.9	—	11.3
Field crabs	51.4	2.9	8.6	11.4	11.4	—	14.3
Shrimps	20.0	8.6	5.7	25.7	37.1	—	2.9
Insects	88.6	—	—	5.7	2.9	—	2.8
Frogs	62.9	2.9	5.7	8.6	8.6	—	11.3
Dakrong (n = 169)							
Fish	1.3	21.5	27.8	10.1	33.5	5.1	0.7
Snakes	87.3	5.1	3.2	—	0.6	—	3.8
Snails	26.6	27.8	16.4	5.1	9.5	—	14.6
Field crabs	46.2	20.9	10.8	4.4	1.3	—	16.4
Shrimps	21.5	23.4	19.6	8.9	11.4	—	15.2
Insects	58.2	11.4	12.0	4.4	4.4	0.6	9
Frogs	48.1	21.5	9.5	5.7	1.3	—	13.9

Box 7.2 continues to next page

Box 7.2 continued

The study showed that households in the remote and poor Dakrong district consume on average more aquatic organisms, particularly snails but also insects and frogs, than households in the other districts. In the richer Hai Lang district, a daily average of 310 g of fish per household was consumed, whereas in Cam Lo and Dakrong the average was 260 g and 240 g, respectively (box table 7.2b)

A high percentage of women reported feeding children aged 4–12 months with fish, small shrimps, and field crabs. In Hai Lang district up to 80% of the households prepared the children's food with fish, and 64% of the households prepared food with small shrimps two to five times per week. There was a difference between the richer and poorer districts, with only 30% of the households in the poorer Dakrong district feeding their children fish two or more times per week.

Availability of Aquatic Biodiversity

Forty different fish species were reported from the wild, some of which were used for aquaculture. Participants in all districts reported that they have more difficulty collecting wild aquatic animals now than they did ten years ago. The villagers stated that possible reasons include more intensive use of pesticides and herbicides in agricultural production and a growing demand for resources from an increasing human population. Furthermore, in most of the districts unsustainable fishing methods such as electro-fishing, poisoning, and the use of mosquito nets for fishing are practiced, endangering the stocks and reproduction capacities of the aquatic animals.

Nevertheless, more than half of all households collected aquatic animals from the wild, including rice fields. In addition to being consumed by the households, such animals also generated income: 9% of the households in Cam Lo and Hai Lang districts sold fish from the wild on the market, and more than 75% of the households in the remote Dakrong district sold fish and other aquatic resources from the wild to other village members. Most of the households purchased their fish and shrimps from markets, except for those in Dakrong district, where 67% obtained these items from the wild. In all districts snakes, snails, and frogs are mainly collected from the wild, especially by the households in the remote Dakrong area. According to the survey, markets are important sources of food in light of declining biodiversity in the rice fields.

Nutritional Status

The study revealed malnutrition of children in all districts but especially in the remote, mountainous Dakrong area (box table 7.2c). Compared with the regional

Box 7.2 continued

BOX TABLE 7.2B. Consumption of aquatic organisms per household (average household size = 5 people).

District	Fish	Snakes	Snails	Field Crabs	Shrimps	Insects	Frogs	Total
				Aquatic Resources Amount (kg/day)				
Hai Lang (n = 70)	0.39	0.01	0.05	0.05	0.11	0.03	0.02	0.66
Cam Lo (n = 35)	0.26	0.02	0.04	0.15	0.21	0	0.06	0.74
Dakrong (n = 169)	0.24	0.02	0.21	0.09	0.14	0.06	0.07	0.83

Box 7.2 continues to next page

Box 7.2 continued

BOX TABLE 7.2C. Nutritional status of children under 5 years of age in the 3 study districts by severity and commune (in %).

Indicators of Malnutrition	Districts (no. of children measured) and Communes							
	Hai Lang (241)		Cam Lo (50)			Dakrong (282)		
	Hai Thuon	Hai Le	Cam Lieu	Trieu Nguyen	A Ngo	A Bung	Huc Nghi	Ta Long
Underweight Weight/age Severity (low–very high)	11.9 Medium	30.2 Very high	28 High	45.8 Very high	46.2 Very high	64.6 Very high	50.0 Very high	39.3 Very high
Stunting Height/age Severity (low–very high)	20.1 Medium	29.5 Medium	42 Very high	44.7 Very high	53.8 Very high	68.8 Very high	61.5 Very high	50.0 Very high
Wasting Weight/height Severity (low–very high)	3.0 Low	3.8 Low	2 Low	6.3 Medium	7.7 Medium	14.6 Very high	6.4 Medium	10.7 High

Box 7.2 continued

status for north central Vietnam, Hai Lang and Cam Lo had a lower or equal percentage of children suffering from underweight and stunting; the percentage of wasting is only a third of the regional average. The remote Dakrong district shows a higher rate of underweight and stunting; in some Dakrong communes the rate of wasting is lower than the regional and national averages. It is in these remote communes that wild aquatic resources are more important for households than in the richer lowland areas.

Conclusions

The study showed that aquatic biodiversity in rice-based production systems makes an important contribution to the food security of the rural population. In remote, mountainous areas the consumption of aquatic organisms is higher than in the lowlands, but diets include less fish and snakes and more snails, insects, and frogs. In general, there is insufficient availability of food from the wild or from markets to meet the demands of the entire population. However, aquatic biodiversity does provide a large portion of the population with nutritional benefits. The anthropometric data show that the nutritional status is worse in the remote areas; from other studies it is known that these communes are exposed to staple food shortages, often for several months in a year. In these areas, aquatic organisms in the diet may simply help to meet the basic energy needs of the people, and without aquatic biodiversity as a source of nutrition, malnutrition and food insecurity would be even greater. In addition, human health status may play an important role. Findings point to the need for more research to examine the nutritional role of the various aquatic organisms, particularly for resource-poor people in remote areas.

also responsible for the decline, such as pesticide use, destruction of fish breeding grounds, and illegal fishing methods such as electro-fishing and chemical poisoning. Development efforts are urgently needed to address these threats.

It is particularly the rural poor who depend on the aquatic biodiversity in rice fields. They may not have access to money, but in many areas they still have access to the biodiversity that supports them (box 7.2). Particular threats to them are the destruction of the fishery resources through overexploitation by industrial capture fisheries and the restriction of access to the fishery resources, for example when fishing grounds are leased to commercial fishery companies as fishing lots. These poor people will be hit hardest because they have no land to cultivate and depend on the capture of wild resources.

Conclusion

The diversity of aquatic species and their importance for rural livelihoods in rice has relevance for irrigated, rainfed, and deepwater rice ecosystems worldwide (figure 7.2). Effort is needed at international and national levels to assess the role of aquatic biodiversity in rice-based ecosystems and rural livelihoods. Specific studies are needed to investigate the nutritional contribution of aquatic biodiversity for rice farming households, particularly in relation to the role of fats and oils, and to raise awareness about its value for the health and well-being of people (see chapter 15 for nutritional studies). Current efforts planned by local institutions in Cambodia, China, Laos, and Vietnam include national and regional workshops during which information on the collection and use of aquatic organisms and their importance for rural livelihoods will be presented to policymakers and extension staff (FAO/NACA 2003). Similar activities are planned in other regions, particularly in West Africa and Latin America.

At the policy level, special attention must be given to aquatic resource management in rural development, food security, and poverty alleviation strategies. When targets for increased rice production are set, it must be recognized that the overall diversity and productivity of the rice field ecosystem are high. Intensification and specialization of the system to maximize rice production generally will be associated with losses in some of the other products. Therefore it is critically important to assess what those changes will be, who will benefit, and who will lose and to try to find ways to minimize the losses and maximize the gains.

Note

1. A clear distinction between the capture and culture systems is not always possible. For example, an intermediate system exists in Thailand in which the management system relies on stocked fish as prey for the wild species. These losses are accepted because of the high market value of the wild fish at local markets (Setboonsarng 1994).

References

Balzer, T., P. Balzer, and S. Pon. 2005. Traditional use and availability of aquatic biodiversity in rice-based ecosystems. I. Kampong Thom Province, Kingdom of Cambodia. In M. Halwart, D. Bartley, and H. Guttman, eds., *Aquatic Biodiversity in Rice-Based Ecosystems* (CD-ROM). Rome: FAO.

Choulamany, X. 2005. Traditional use and availability of aquatic biodiversity in rice-based ecosystems. III. Xieng Khouang and Houa Phanh provinces, Lao PDR. Northern Laos. In M. Halwart and D. Bartley, eds., *Aquatic Biodiversity in Rice-Based Ecosystems* (CD-ROM). Rome: FAO.

Demaine, H. and M. Halwart. 2001. An overview of rice-based small-scale aquaculture. In *Utilizing Different Aquatic Resources for Livelihoods in Asia: A Resource Book*, 189–197. Cavite, Philippines: International Institute of Rural Reconstruction, International Development Research Centre, Food and Agriculture Organization of the United Nations, Network of Aquaculture Centers in Asia–Pacific, and International Center for Living Aquatic Resources Management.

FAO (Food and Agriculture Organization of the United Nations). 1999. *Nutrition Country Profiles: Viet Nam*. Rome: FAO. Available at ftp.fao.org/es/esn/nutrition/ncp/viemap.pdf.

FAO (Food and Agriculture Organization of the United Nations). 2002. *Report of the 20th Session of the International Rice Commission*, Bangkok, Thailand, July 23–26, 2002. Rome: FAO.

FAO/MRC (Food and Agriculture Organization of the United Nations and Mekong River Commission). 2003. *New Approaches for the Improvement of Inland Capture Fishery Statistics in the Mekong Basin*. Report of the Ad Hoc Expert Consultation, Udon Thani, Thailand, September 2–5, 2002. Publication No. 2003/01. Bangkok: FAO/RAP.

FAO/NACA (Food and Agriculture Organization of the United Nations and Network of Aquaculture Centres in Asia–Pacific). 2003. *Traditional Use and Availability of Aquatic Biodiversity in Rice-Based Ecosystems*. Report of a Workshop, Xishuangbanna, Yunnan, P.R. China, October 21–23, 2002. Rome, Italy: FAO. Available at ftp.fao.org/fi/document/xishuangbanna/xishuangbanna.pdf.

Fernando, C. H. 1993. Rice field ecology and fish culture: An overview. *Hydrobiologia* 259:91–113.

Fernando, C. H. and M. Halwart. 2001. Fish farming in irrigation systems: Sri Lanka and global view. *Sri Lanka Journal of Aquatic Sciences* 6:1–74.

Halwart, M. 1994. *Fish as Biocontrol Agents in Rice: The Potential of Common Carp* Cyprinus carpio *and Nile Tilapia* Oreochromis niloticus. Weikersheim, Germany: Margraf Verlag.

Halwart, M. 1998. Trends in rice–fish farming. *FAO Aquaculture Newsletter* 18:3–11.

Halwart, M. 1999. Fish in rice-based farming systems: Trends and prospects. In D. van Tran, ed., *International Rice Commission: Assessment and Orientation Towards the 21st Century,* 130–141. Proceedings of the 19th Session of the International Rice Commission, Cairo, Egypt, September 7–9, 1998. Rome: FAO.

Halwart, M. 2001. Fish as biocontrol agents of vectors and pests of medical and agricultural importance. In *Utilizing Different Aquatic Resources for Livelihoods in Asia: A Resource Book,* 70–75. Cavite, Philippines: International Institute of Rural Reconstruction, International Development Research Centre, Food and Agriculture Organization of the United Nations, Network of Aquaculture Centers in Asia–Pacific, and International Center for Living Aquatic Resources Management.

Halwart, M. 2003. Recent initiatives on the availability and use of aquatic organisms in rice-based farming. In *Proceedings of the 20th Session of the International Rice Commission,* 195–206. Bangkok, Thailand, July 23–26, 2002. Rome: FAO.

Halwart, M., M. C. Viray, and G. Kaule. 1998. The potential of *Cyprinus carpio* and *Oreochromis niloticus* for the biological control of aquatic pest snails in rice fields: Effects of predator size, prey size and prey density. *Asian Fisheries Science* 10:31–42.

Luo, A. 2005. Traditional use and availability of aquatic biodiversity in rice-based ecosystems. II. Xishuangbanna, Yunnan, P.R. China. In M. Halwart, D. Bartley, and J. Margraf, eds., *Aquatic Biodiversity in Rice-Based Ecosystems* (CD-ROM). Rome: FAO.

Meusch, E. 2005. Traditional use and availability of aquatic biodiversity in rice-based ecosystems. III. Northwestern Viet Nam. In M. Halwart and D. Bartley, eds., *Aquatic Biodiversity in Rice-Based Ecosystems* (CD-ROM). Rome: FAO.

Reinhard, I. and K. B. S. Wijayaratne. 2002. *The Use of Stunting and Wasting as Indicators for Food Insecurity and Poverty.* Working Paper 27, Integrated Food Security Programme TRINCOMALEE. Available at www.sas.upenn.edu/~dludden/stunting wasting.pdf.

Rola, A. and P. Pingali. 1993. *Pesticides, Rice Productivity, and Farmers' Health: An Economic Assessment.* Manila: International Rice Research Institute and World Resources Institute.

Setboonsarng, S. 1994. Farmers' perception towards wild fish in ricefields: "Product, not predator"—An experience in rice–fish development in northeast Thailand.

In C. R. dela Cruz, ed., *Role of Fish in Enhancing Ricefield Ecology and in Integrated Pest Management*, 43–44. ICLARM Conf. Proc. 43. Manila: International Center for Living Aquatic Resources Management.

Smith, J. D., ed. 2001. *Biodiversity, the Life of Cambodia: Cambodian Biodiversity Status Report* 2001. Phnom Penh: Cambodia Biodiversity Enabling Activity.

8 🌾 Pollinator Services

P. G. KEVAN AND V. A. WOJCIK

Herbivores, predators, parasitoids, parasites, and pathogens are understood ecologically as crucial to sustaining ecosystems and their diversity. Nevertheless, just as important are the mutualistic relationships. Pollination is the hub of a multispoke productivity wheel that has all consumers—humans, livestock, and wildlife—at the rim (figure 8.1). Ecological interactions and complexity now are within the domain of conservation and sustainability. The biodiversity of the world's dominant flora (flowering plants) and dominant fauna (insects) are so intimately and coevolutionarily enmeshed through pollination that the erosion of the processes has serious environmental consequences. Indeed, pollination is now regarded as a jeopardized ecosystem service that warrants attention in all terrestrial environments, from intensive agriculture to wilderness (Buchmann and Nabhan 1996). Moreover, nonpollinating flower visitors in the web of life provide benefits, and sometimes problems, that are essential to other aspects of ecosystemic function.

Although pollination has been studied for more than 200 years, it is often overlooked and misunderstood. Therefore the roles of pollinators and flowers and the problems associated with declining biodiversity and the need for conservation deserve explanation. The importance of pollinators and other anthophiles (flower visitors) extends beyond ecosystem sustainability, plant reproduction, crop productivity, and pest management into aesthetic and ethical aspects of the quality of human life. Finally, pollinators and anthophiles may also be sensitive bioindicators of ecosystemic health.

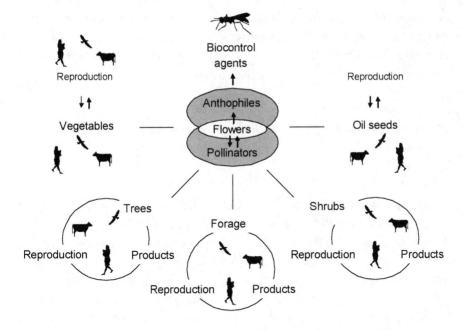

FIGURE 8.1. Pollination systems are at the center of ecosystems and ecosystem functions. Both direct and indirect interactions tie the entire biosphere to pollinator services in some manner.

Pollination, Pollinators, and Anthophiles

Pollination is simply the transfer of pollen from the anther to the stigma. After the production of reproductive structures by the plant, pollination is the next step in reproduction. Abiotic pollination occurs by wind, water, or gravity. Biotic pollination is effected by animals. There is a large and specialized vocabulary concerned with pollination and plant breeding systems, which need not be detailed in this review. For crops, Roubik (1995) and Free (1993) are encyclopedic.

Anthophiles are animals that visit flowers. They may seek pollen, nectar, oil, or floral tissue to satisfy nutritional needs (Kevan 1999). Pollinators are anthophiles that pollinate. Not all anthophiles are effective pollinators: Some are floral larcenists, removing resources sought by pollinators or eating pollen needed in abiotic pollination (Inouye 1980); others may be innocuous and merely rest in flowers or glean residual resources after pollination is over. Although pollinators are vital to plant reproduction, non-pollinating (or poorly pollinating) anthophiles may also be important in ecosystem functions. For example, many insects that are useful in biocontrol of pests need florally derived food to mate, find hosts, oviposit, and complete their life cycles (Ruppert 1993). There are also anthophiles that use floral sites for capturing prey (Kevan 1999).

Biodiversity of Anthophiles, Pollinators, and Pollination

General

Diversity among Anthophiles number in the millions of species. Almost all bees and butterflies depend on flowers. Many moths, flies, beetles, wasps, and ants visit flowers. Anthophily is common in less conspicuous insects, such as thrips and the pollen katydids. In other insect orders, anthophily is sporadic, as in lacewings and springtails. Among the vertebrates, some groups of birds (e.g., hummingbirds, flowerpeckers, honeyeaters, honeycreepers, sunbirds, and lories) and bats (fruit bats or flying foxes of the Old World Tropics and leafnosed bats of the Neotropics) are notorious as anthophiles and pollinators. There are even a few mammals specialized in gliding and climbing (the scansorial mammals) that are important pollinators, especially in Australia and Africa. Even primates are important as pollinators in some places.

In Agroecosystems

Honeybees (*Apis* spp.) are the most valuable pollinators in agriculture. They are managed in easily transportable boxes for pollination of many crops. Their biology is well known and beekeeping for pollination well established. Nevertheless, they are not the only commercially useful pollinator and do not consistently pollinate all crops (Kevan 1999). For example, bumblebees (*Bombus* spp.) are the pollinators of choice for leguminous crops that have corolla tubes too deep to allow honeybees access to nectar (Free 1993). Their behavior as buzz pollinators (pollinators that extract pollen from flowers by producing vibrations) and their capacity to forage in greenhouses set them apart from honeybees. Orchard bees (*Osmia* spp.) are more efficient and start foraging at lower temperatures than do honeybees (Kevan 1999). Recently in Malaysia, carpenter bees (*Xylocopa* spp.) have been managed by providing nesting material for pollination of passion fruit with flowers too large to be pollinated by honeybees (Mardan et al. 1991).

Pollination of some crops is effected not by bees but by other pollinators (table 8.1). In the tropics, pollination issues are especially important because the natural pollination mechanisms of many plants (crops and others) are not known (Roubik 1995; Kevan 2001). Recent research efforts have made some advances in this area. Beetles have been found to pollinate

Table 8.1. Some useful pollinators of commercial crops.

Crop	Pollinator	Reference
Red clover (*Trifolium repens*) and other legumes	Bumblebees (*Bombus* spp.)	Free 1993
Greenhouse tomatoes and other solanaceous crops	Bumblebees	Banda and Paxton 1991
Raspberries	Bumblebees (superior to honeybees)	Willmer et al. 1994
Pome fruit	Orchard bees (*Osmia* spp.)	Kevan 1999
Alfalfa and other leguminous crops	Alfalfa leafcutting bees (*Megachile* spp.)	Richards 1993
Blueberries	Blueberry bee (*Habropoda laboriosa*)	Cane and Payne 1988, 1990
Squash and pumpkin	Hoary squash bee (*Peponapis pruinosa*)	Kevan 1999
Passion fruit	Carpenter bees (*Xylocopa* spp.)	Mardan et al. 1991
Oil palm	Weevil (*Elaeidobius kamerunicus* Faust) (Curculionidae)	Kevan 1999
Various annonaceous fruit crops	Beetles	Roubik 1995
Cacao	Midges (Diptera: Ceratopogonidae)	Free 1993; Roubik 1995
Mango	Flies and other insects	Free 1993; Roubik 1995
Durian	Bats	Roubik 1995
Cashews	Honeybees (*Apis mellifera*) and a native oil-collecting bee (*Centris tarsata*)	Breno et al. 2002

annonaceous (*Annona* spp.) fruit crops such as sour sop and custard apple, but little information is available on the pollinators *most* adapted to these plants (Roubik 1995). The oil palms of Asia are now effectively pollinated by the introduced West African weevil *Elaeidobius kamerunicus* Faust (Curculionidae), which served as the pollinator in the native range of the tree (Kevan 1999). The Brazil nut tree has been found to be mellitophilous (producing male and female flowers at separate times), making self-pollination temporally impossible. Therefore it depends on the activity of the pollinator to set fruit (Maues 2002). The fact that a wide assemblage of midges and flies pollinate wild fruits indicates the need to consider alternative pollinators for many crops. Substantial efforts are still needed if a complete understanding of tropical pollination is to be achieved.

The type and amount of reward sought by pollinators influence pollination, a fact often overlooked by plant breeders. Pollen-foraging honeybees are considered better pollinators than nectar foragers, even for apples and other such crops with open bowl-shaped flowers that produce a fair amount of nectar (Free 1993). In crops with specialized floral forms (e.g., blueberry, cranberry), bees foraging for pollen are more effective than nectar foragers (Cane and Payne 1988). On these crops, honeybees often are poor pollinators because of the small proportion of pollen foragers in their colonies and their inability to shake pollen from flowers, or buzz pollinate (Buchmann 1983). Some crop plants that need or benefit from insect pollination do not produce nectar and rely on pollen-foraging insects. Examples are kiwi, tomato, and perhaps pomegranate. Lupine also produces only pollen but automatically self-pollinates.

Pollinators

Honeybees and Beekeeping

The diversity of beekeeping practices goes beyond that represented by hives of European races and hybrids of honeybees (*Apis mellifera* spp. *ligustica, caucasica,* and *carnica,* or the Italian, Caucasian, and Carniolan honeybees). Nevertheless, those bees are the best understood and easily managed for pollination and hive products (see Graham 1992; Crane 1990).

Throughout the African and Middle Eastern parts of the range of *A. mellifera,* indigenous races are kept in various types of hives and by various management techniques. Many of these races are defensive and prone to abscond from their hives, whether artificial or natural. Therefore they are difficult to keep. The so-called killer bee better known as the Africanized bee (a hybrid between European and African races), has spread from Brazil throughout the tropical and subtropical Americas since the introduction there of the African parent stock of *A. m. scutellata* from southeastern Africa in 1956. It is notoriously defensive and easily provoked to attack intruders.

In Asia, other honeybee species are kept or encouraged for human exploitation. The most important of those is the Asiatic hive bee (*Apis cerana*), which comprises as much racial biodiversity as does *A. mellifera* (Kevan 1995). Although the Asiatic hive bee has been much maligned as a manageable species, recently more attention has been paid to its potential, and the wisdom of transplanting European honeybees beyond their natural range has been seriously questioned. In tropical and subtropical Asia, other species of honeybees are used commercially. The honey of *Apis dorsata,* the giant or rock honeybee, is harvested in India, Bangladesh, Sri Lanka, Malaysia, Thailand, Vietnam, Kumpuchea, and Laos, as is honey of its races or sister species, *Apis laboriosa* in the Himalayan foothills and *A. d. binghami* in parts of the Southeast Asian Archipelago. The smallest honeybee, *Apis florea,* is also exploited commercially.

In the tropical and subtropical Americas, where there are no indigenous *Apis* species, stingless bees (Meliponinae) have been traditionally kept since pre-Columbian times. Meliponine bees occur throughout the world's tropics and present immense potential for managed pollination in agriculture. However, little attention has been paid to their biology as pollinators (Roubik 1995).

Native Pollinators

Little is known about the significance and involvement of native bees in crop pollination (Kremen et al. 2002). In all likelihood, much of the credit given to honeybees for pollination in reality belongs to other species. Castro (2002) studied some 32 fruit tree species in Bahia, Brazil, and found that though not as abundant as the other exotic species, the native stingless bees (Apidae: Meliponinae) were important as pollinators. In *agronatural* settings—agricultural settings that are situated in or retain some of the traditional natural landscape—native pollinators provide "free services" (Kremen et al. 2002). However, these services are not truly free; they depend on a healthy ecosystem that provides habitat for the native species.

Nonnative honeybees and other such exotic species have a significant influence over native pollinator systems (Kremen and Ricketts 2000).

Other Managed Pollinators

Other bees that do not produce harvestable quantities of honey and are managed—or have potential for management—as pollinators include leaf-cutting bees, alkali bees, orchard bees, squash bees, blueberry bees, and carpenter bees. Those bees were mentioned earlier with respect to particular crops. Crane (1990) lists about 50 species of bees that have been managed either commercially (very few) or experimentally for pollination. The huge economic value of some crops has sparked interest in developing proper management practices for their successful pollination. The case of cashew pollination in Brazil (box 8.1) shows just how complicated and involved the process can be.

Demise of the Pollinators

The demise of pollinators has come about through four major human activities: pesticide use, habitat destruction, spread of pathogens and parasites, and competition from introduced flower visitors.

Pesticides

The dangers of pesticides, especially insecticides, to pollinators are well documented and understood, especially with regard to the European honeybee. Less understood, and often overlooked, is the problem of sub-lethal effects that reduce longevity and adversely affect foraging, memory, and navigational abilities of some bees (MacKenzie 1993). From the few comparative studies available, it is evident that the toxicities of pesticides to honeybees are poor predictors of the hazards posed to other species (Kevan 1999).

Issues of pesticides in nonagricultural settings and agroforestry are more complex because of the importance of a wider diversity of pollinators. In eastern Canada, the use of fenitrothion in New Brunswick for spruce budworm (*Choristoneura fumiferana*) control in forests adjacent to blueberry farms caused such drastic reductions in pollinator abundance and diversity that blueberry yields fell statistically below the levels

Box 8.1 Economic Value of Cashew (*Anacardium occidentale* L.) to Brazil and Its Need for Pollination

Cashew (*Anacardium occidentale* L.) is an andromonoecious tree native to northeastern Brazil. It is of great economic importance to the region for its nut, oil, and cashew apple production. An estimate economic value of cashew to Brazil per year shows the following figures:

Total area of commercial cashew orchards	650,000 ha
Total annual yield of nuts	126,000 metric tons
Value of exports	
(nuts only)	US$135 million
(nut shell oil)	US$91 million
Value of crop (nuts, oil, and fruit) within Brazil	US$54 million

However, crop yields are disappointingly low from commercial orchards, and studies suggest that inadequate pollination could be the major cause. Cashew's flower form and presentation suggest that it is pollinated by insects, especially bees. Although numerous insects that visit cashew flowers—namely wasps, butterflies, and moths—have been cited as pollinators, they fail to set fruits in most cases, showing clearly that flower visitors and pollinators are not synonymous. Only bees visit and set cashew flowers regularly in northeastern Brazil. Two bee species are particularly efficient in pollinating cashew flowers: the solitary indigenous bee *Centris tarsata* and the exotic honeybee (*Apis mellifera*).

But there are two sides to the shortfall of adequate pollination of commercially grown cashew in northeastern Brazil. On one side, there are few or no visits to orchard-grown cashew flowers. *Apis mellifera* does not readily visit cashew flowers even when brought into orchards in large numbers because of competing weeds in bloom. The other suitable pollinator, *C. tarsata,* is rare in commercial cashew orchards because of habitat disturbance and the lack of rearing techniques to breed them in large numbers. A second side to this shortfall undoubtedly is related to horticultural practices in which a partially self-sterile clonal strain is grown over large areas without thought being given to the need for compatible sources of pollen. This problem is exacerbated as more and more cultivated areas are being planted or replanted with dwarf clones. One obvious solution is to interplant trees producing compatible pollen within main cropping strains, as is done in pome orchards. Hand-pollination experiments carried out in Australia and Brazil have identified types or strains of cashew, crossing between which produced higher yields. However, it will still be necessary to consider management of bees in commercial cashew orchards because they will be needed as the vectors of compatible pollen.

Box 8.1 continues to next page

Box 8.1 continued

One can conclude that to improve cashew crop yields in northeastern Brazil, serious consideration must be given to both conservation and management of its recognized, efficacious pollinators and also the design of orchards with appropriate mixes of compatible cashew strains.

Further Reading on Cashew Pollination in Northeastern Brazil

Freitas, B. M. 1994. Beekeeping and cashew in north-eastern Brazil: The balance of honey and nut production. *Bee World* 75(4):160–168.

Freitas, B. M. and R. J. Paxton. 1998. A comparison of two pollinators: The introduced honey bee *Apis mellifera* and an indigenous bee *Centris tarsata* on cashew *Anacardium occidentale* in its native range of NE Brazil. *Journal of Applied Ecology* 35:109–121.

Freitas, B. M., R. J. Paxton, and J. P. Holanda-Neto. 2002. Identifying pollinators among an array of flower visitors, and the case of inadequate cashew pollination in NE Brazil. In P. Kevan and V. L. Imperatriz-Fonseca, eds., *Pollinating Bees: The Conservation Link Between Agriculture and Nature*, 229–244. Brasília, Brazil: Ministry of Environment.

Holanda-Neto, J. P., B. M. Freitas, D. M. Bueno, and Z. B. Araújo. 2002. Low seed/nut productivity in cashew (*Anacardium occidentale*): Effects of self-incompatibility and honey bee (*Apis mellifera*) foraging behaviour. *Journal of Horticultural Science & Biotechnology* 77(2):226–231.

Source: Freitas et al. (2002).

expected (Kevan 1999). Finnamore and Neary (1978) note about 190 species of Canadian native bees associated with flowers of blueberry, which needs pollination by insects. Subsequent recovery seems to have taken place over periods of one or two to more than seven years, depending on the severity of damage (Kevan 1999). Today the diversity and reproductive potential of blueberry pollinators are being reduced by use of herbicides that kill alternative forage for the pollinators when blueberries are not in bloom.

Most pesticide problems stem from accidents, carelessness in application, and deliberate misuse despite label warnings and recommendations (Johansen and Mayer 1990). As pesticide applications become more regulated and applicators are required to take courses in safety and use before certification, the problem should diminish. Methods such as not spraying blooming plants or spraying when pollinators are not foraging are commonsense approaches to reducing problems associated with pesticide applications, even when regulation is poor.

Habitat Destruction

There are three ways in which habitat destruction affects pollinator populations, as with populations of any organism: destruction of food sources, destruction of nesting or oviposition sites, and destruction of resting or mating sites. The most common means of habitat destruction are through the establishment of monocultures, overgrazing, land clearing, and irrigation.

The destruction of food sources in agricultural areas is best illustrated by examples of the removal of vegetation that provides the pollinators' food when crops are not in bloom (Kevan 2001). Very often the vegetation removed is regarded as unwanted, as weeds or competition for the crop plants, yet it is invaluable to pollinators and other beneficial insects. Roadside and right-of-way sprayings of herbicides can reduce the diversity and abundance of alternative food supplies for pollinators.

The destruction of nesting and oviposition sites was documented in the 1950s in Manitoba for the demise of populations of leafcutting bees, which were left without nesting sites in stumps and logs as fields of alfalfa for seed production expanded (Stephen 1955). In Europe the decline of bumblebees occurred as the amount of undisturbed land in hedgerows and other noncultivated areas declined (Corbet et al. 1991). In the tropics the inadequate pollination of cacao by midges in plantations is related to the loss of oviposition substrates (i.e. rotting vegetation), which had been too fastidiously removed (Winder 1977). In Malaysia, additional substrate of rotting palm trunks is provided to increase pollinator populations (Kevan 1999).

Habitat manipulations associated with agriculture often adversely affect availability of both food sources and nest sites, creating a double problem for native pollinators. This is especially noticeable in populations of pollinations that are long-lived, such as colonies of bumblebees. Developing parts of the world, such as Africa, are facing the same pollination resource problems as habitats are manipulated for agricultural purposes. The Kenyan horticultural industry described in box 8.2 is just one of many examples of how wild habitats increase the vigor and persistence of natural pollinator services.

The general issue of habitat destruction for pollinators has evoked concern on a broad scale. Daniel Janzen's (1974) article "The Deflowering of

Box 8.2 Wild Habitat Provides Pollination Services to Kenyan Horticultural Crops

Many areas in Kenya are being converted from natural areas to farms and fields producing horticultural crops, largely for the export market. Even group ranches of largely pastoralist communities, such as the Masai, are starting to participate in these markets.

On ol' Kirimatian group ranch, southwest of Nairobi and beyond Lake Magadi, small rivers descending from the Nguruman escarpment have been channeled into open irrigation furrows, permitting the cultivation of horticultural crops. The land being converted to cropland is largely *Acacia tortilis* riverine forests. Horticultural crops are grown year-round for the market and are purchased by middlemen, who transport the vegetables to the airport for sale shortly afterwards in the markets of London and elsewhere in Europe.

Many horticultural crops grown on ol' Kirimatian need or benefit from pollination services, such as eggplant, okra, and bitter gourd. In the case of eggplant, the crop may be entirely dependent on native bee pollinators, particularly those that buzz pollinate—that is, bite the flower and vibrate their wing muscles at a certain frequency, such that pollen flies out of small pores in the flower and can be carried to another flower to produce fruit. Honeybees cannot buzz pollinate, but two species of solitary bees that occur naturally in the forest that is being cleared for farms are very effective pollinators. One is a carpenter bee, nesting in old wood; the other is a ground-nesting bee called *Macronomia rufipes*. The bees get only pollen from eggplant; it does not produce nectar. Therefore they cannot live exclusively on agricultural land, and they make use of different resources along farm paths and in the plots of forest that have not yet been cleared.

The natural environments around the farmland were surveyed to note the degree to which crop pollinators also use the floral resources in these wild habitats. In most months the key pollinators of eggplant used the floral resources found on farm, primarily ruderal weeds along farm paths. But in the very driest month, before the onset of rains, key pollinators of eggplant were more present than at any other time in the small remnants of *Acacia tortilis* riverine forest and made greater use of its floral resources.

One might argue that the wild habitat is not providing the majority of pollination services. But if pollinators did not have the alternative floral resources of the wild habitat to exploit—perhaps only during the most severe periods of the dry season—they might not be able to persist in this otherwise arid and marginal ecosystem. The opportunity cost of preserving this wild habitat among the fields undoubtedly will depend on its several values—including *Acacia tortilis* seed pod forage for goats—offset against its potential income if cleared and converted to crop production. Adding in its value as habitat for

Box 8.2 *continued*

crop pollinators may help to tip the balance in favor of preserving some wild habitat in a developing agricultural landscape.

Source: Barbara Gemmill, African Pollination Initiative, Nairobi, and Alfred Ochieng, University of Nairobi Department of Botany.

America" illustrates the problem. He points to a vicious cycle of reduced vegetation for pollinators' resources, reduced pollination in the vegetation, the demise of the plant's reproductive success, and reductions in seed and fruit set, resulting in the failure of revegetation with the same level of biodiversity as would have otherwise existed.

Parasites and Pathogens

Mites as parasites of honeybees have evoked major concern worldwide as tracheal mites (*Acarapis woodi*) and *Varroa jacobsoni* have spread at alarming rates (Needham et al. 1988; Connor et al. 1993). It has been suggested that many amateur and small-scale beekeepers have abandoned their activities because of the additional complexities of bee management associated with monitoring for mites and controlling them once detected. The number of feral colonies of honeybees has decreased significantly as mite infestations have become common throughout the United States. The combined effects of losses of hobby beekeepers and feral colonies are already adversely affecting pollination in rural and urban settings, as predicted some years ago (Kevan 1999). Furthermore, chemical control of mites may not be acceptable to producers of pure honey because of the potential contamination of human food and other hive products.

Diseases can cause serious losses if not properly controlled through monitoring and treatment. For beekeeping with European honeybees, American foulbrood, a bacterial disease of the larvae, is the most serious. Other brood diseases, such as European foulbrood (bacterial), chalkbrood (fungal), and sacbrood (viral), are less problematic. The only disease of adult European honeybees that is of concern is dysentery (protozoan *Nosema*). For the Asiatic hive bee the viral disease Thai sacbrood has caused widespread losses as epidemics have swept parts of Asia, to be followed by resistance and recovery of populations (Kevan 1995).

Leafcutting bees also suffer from diseases. The most important is the chalkbrood fungal disease (caused by *Ascosphaera aggregata*) of the alfalfa leafcutting bee (*Megachile rotundata*) (Vandenberg and Stephen 1982). Diagnosis facilities have been established in some areas (e.g., western Canada) where these bees are the most important pollinators of alfalfa. Control of this disease involves careful and sanitary management and fumigation of pathogen-infested nesting material (Goettel et al. 1993).

Competitive Interactions

The most studied of the competitive interactions between pollinators as they relate to pollination is the effect of the Africanized honeybee on native pollinators and European races of honeybees in South and Central America. Roubik (1978) first pointed out the apparent reductions in populations of native bees in Central America after the invasion of Africanized bees. Subsequently he placed the phenomenon in a broader context (Roubik 1989), but the whole issue of the competitive interactions of Africanized bees with native pollinators in South and Central America is complex. It appears that no indigenous species have become extinct through competitive interactions with the exotic honeybee.

In Australia there has been recent debate on the effects of the introduced European races of honeybees on the native flora and fauna of pollinators. Paton (1993) concluded that there is justification for the concern that European honeybees have reduced the pollination of some native plants, especially those that are bird pollinated, by removing the nectar sought by the birds and causing changes in their populations and foraging habits. Sugden and Pyke (1991) concluded that competition with honeybees decreased the populations of native bees (e.g., *Exoneura asimillima*). The issue of the effects of European honeybees on native pollinating insects is not so clear from the botanical side, but the same trends are evident in respect to the native bees. The sequence of events is as follows:

1. Honeybees displace native pollinators by removing floral resources.
2. Honeybees may not be able to pollinate the flowers from which they remove the resources.
3. The plants then fail to reproduce sexually or at all, and their populations dwindle.

4. Remaining and reduced populations of native pollinators dwindle further.

Commercially reared bumblebees are an important component of greenhouse tomato production (Kevan 1999). At least three regionally native species are being used: *Bombus terrestris* L. in Europe, *B. impatiens* Cresson in eastern North America, and *B. occidentalis* Greene in western North America. Planned introductions of nonnative bee species should be treated with great care, with all due attention paid to quarantine and, more importantly, to the possible ecological ramifications of escapes, which are inevitable. Already, European bumblebees have been taken to New Zealand, Chile, Tasmania, Japan, and possibly Argentina (in 1993 or 1994), mostly without appropriate consideration. Dafni and Shmida (1996) also express misgivings about the impact of *Bombus terrestris* on the anthophilous fauna and pollination of the flora of Mount Carmel in Israel. In Indonesia, the introduction of the Asiatic hive bee (*Apis cerana*) from west of Wallace's and Wegener's lines into Irian Jaya has resulted in the spread of this bee to neighboring Papua New Guinea, from which it now threatens to spread to Australia. The consequences of such introductions to the natural diversity and abundance of native pollinators, and consequently to the native flora, were not assessed.

Protection and Promotion

The protection of native pollinators is critical to sustainable global productivity. Habitat destruction, from nesting sites to forage (Janzen 1974), is a major issue. Introduced diseases are threatening the health of the native honeybee races and their pollinating activities in Africa. Although the impact of pesticides is declining in importance in North American and European agriculture as the use of insecticides declines, it remains very important elsewhere. Competitive interactions between flower visitors seem important in tropical and subtropical Americas and in Australia.

Other Anthophiles

Many anthophiles are unimportant as pollinators, but floral resources are important in their lives. Other flower-visiting insects are also invaluable, especially predators and parasitoids, which are important in controlling populations of otherwise pestiferous insects in all environments. Biocontrol agents represent a particularly valuable group of insects in agroecosystems. Leius (1967) showed that the incidence of ichneumonoid parasitism of codling moth in apple orchards was greater if floral resources, such as those of weeds, were available in the orchards. Syme (1975) noted the importance of floral resources to biocontrol agents in forests as well. In fact, long ago it was suggested that the failure to establish potentially useful biocontrol agents against Japanese beetles was caused at least partially by lack of floral resources (King and Holloway 1930). Certainly, some of the successes reported in the high incidence of natural biocontrol agents of pestiferous insects in low-input agricultural systems should be ascribed to the availability of floral resources (see Kevan 1999).

Cropping Systems, Sustainability, and Biodiversity

Agriculture is changing rapidly all over the world. In North America and Europe some lands are being retired from farming (Corbet 1995), while other land is being more intensively cultivated. In other instances, more environmentally sensitive, low-input practices—such as organic methods and low- and no-till cultivation—are being used (Johnston et al. 1971; Gess and Gess 1993). These trends, coupled with reductions in the use of pesticides, generally bode well for pollinators and pollination. However, ecologically appropriate planning for these changes in land use is not being implemented, and the crucial place of pollinators is largely ignored.

In the developing world, expanding agriculture, increasing monoculture, intensification of cropping systems, growing use of agrochemicals, and rapid deterioration of natural areas are all serious problems. The case of the apple valleys of the Hindu Kush–Himalayan region is a stark example of the need to pay attention to pollination systems for the sake of a healthy ecosystem and industry (box 8.3). Lack of adequate information about the roles and biodiversity of pollinators, and their decline in natural and agricultural systems, is alarming. Although the situation in these countries is dire and the efforts of the International Bee Research Institute—through their continuing series of Conferences in Tropical Apiculture—are making some impact (Kevan 2001), pollination continues to be a neglected area. An understanding of pollination ecology leads to a better agricultural economy through better and more sustainable yields, which undoubtedly leads to better living conditions for the people involved (figure 8.2).

Box 8.3 Warning Signals from Apple Valleys of the Hindu Kush–Himalayan Region

Apples have emerged as the leading cash crop in several areas of the Hindu Kush–Himalayan (HKH) region, assuming great importance in helping many farmers move out of poverty. They can account for 60–80% of the total household income, and studies indicate that in the areas where apples are grown there is now food security and economic well-being. The estimated total annual production in the HKH of 2.2 million metric tons of apples helped to bring in an income of more than US$500 million per year to those involved in apple farming and marketing.

However, the potential for improvement is equally large: The average yield of apples in the HKH region (2.5–12.9 tons/ha) is low and declining. Problems of inadequate pollination and poor fertilization caused by a lack of pollinating insects and inclement weather have been identified in field surveys carried out by the International Centre for Integrated Mountain Development as the most important cause of the decline in productivity.

Two main factors contribute to pollination deficiencies. Many of the commercial varieties of apples planted by the farmers in the HKH region are self-incompatible and need cross-pollination with a compatible pollinizer variety. Many farmers do not understand this or are unwilling to dedicate land to the pollinizer varieties that themselves have a low market value. Overall, more than half of the farmers in the study areas had less than the necessary minimum of 20% pollinizer varieties in their orchards; most had only 7–12%.

In recent years the diversity and abundance of natural insect pollinators have been declining for several reasons, including loss of food and nesting habitats caused by clearing of forest and grassland for agricultural purposes and indiscriminate use of pesticides. The majority of farmers identified pesticide use as the main cause of loss of natural insect pollinators. The increase in orchard area was also thought to play a role, with the natural populations of insects being too small to pollinate the newly developed large areas of apple crops.

In two of the HKH areas studied—Himachal Pradesh in India and the Maoxian Valley in China—productivity dropped so severely that farmers and institutions were compelled to search for a solution. It was in these areas that the problem of insufficient pollination was first recognized, and it is here too that farmers and institutions have started investigating and testing pollination management options.

Due to the efforts of government institutions, the majority of apple farmers in Himachal Pradesh are now well aware of the apple pollination problem and the factors responsible for it. They are trying different ways to improve pollination. Management efforts include planting different varieties of pollinizers, increasing the proportion of pollinizer trees, and increasing the

Box 8.3 continues to next page

number of pollinating insects in the orchards, including greater use of managed honeybee colonies.

A different approach is being followed in Maoxian County in China. Here, hand-pollination of apples has become common practice. Because orchards are small, families try to pollinate their orchards themselves by training all members of the family, making it a community effort. Farmers also share days or hire laborers to pollinate their apple orchards. These workers have been called "human bees" because they do the work that could otherwise be done by honeybees. The advantage is that fewer pollinizer trees need to be planted, thus using the scarce land resources to a maximum for producing commercial varieties of fruit.

Source: Uma Partap, International Centre for Integrated Mountain Development.

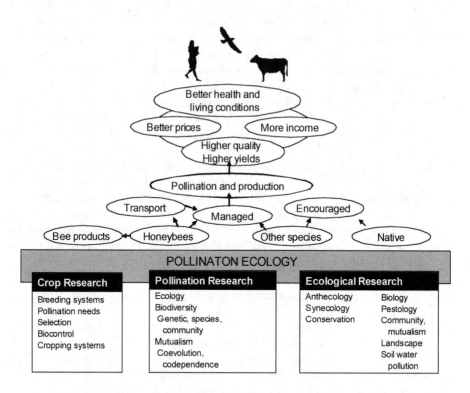

FIGURE 8.2. A strong understanding of pollination ecology is the basis for conscious and appropriate decisions regarding production systems. By association this understanding leads to better health and living for all parties involved.

Bioindicators

Bioindicators are organisms that indicate, by their presence (abundance) or absence, ongoing activities in the ecosystem. Bioindicator species usually are used to diagnose problems. There are indicators of anaerobic waters, rapid eutrophication, pollution, and pesticides. They can also indicate amelioration of problems or even suggest that the activities in an ecosystem are performing according to expectations, within normal bounds. In the latter cases, the species used may be indicators of some aspect of ecosystemic health.

Agrochemicals

Pollinators, especially honeybees, often are killed in large numbers by insecticides. They may also accumulate other pesticides in their bodies and hives. Analytical techniques for pesticide residue detection are well developed so that bees and hive products can be used to monitor for pesticides in the environment. Most often, toxic residues are assayed to determine the likely cause of bee deaths and the hazard posed by pesticides to pollinators rather than for purposes of environmental monitoring.

Pollutants

Honeybees have been investigated quite often as a way to monitor pollutants. Honey and pollen may become contaminated with various industrial pollutants. The release of arsenic and cadmium may cause mass killings of honeybees and contaminate pollen but not nectar (Krunić et al. 1989). The accumulation of radioisotopes in honey and pollen after the Chernobyl disaster in April 1986 illustrates the value of honeybee colonies as samplers of local, regional, and global environmental quality (Bunzl et al. 1988; Ford et al. 1988). They also sample fluorides (Dewey 1973), heavy metals (Stein and Umland 1987), and organic compounds (Anderson and Wojtas 1986; Morse et al. 1987) through floral nectar, pollen, and their own bodies. They have been advocated as bioindicators in natural, agricultural, industrial, and urban milieus, yet despite their proven worth, programs for their use as biomonitors do not seem to have been instituted (Kevan 1999, 2001).

Bromenshenk et al. (1991) addressed the problem of population dynamics in honeybees with respect to pollution and so expanded concern for the health of pollinators beyond pesticide hazards. Little information is available on the effects of pollutants on other pollinators. Dewey's (1973) data show that the highest levels of fluoride, associated with an aluminium reduction plant, were found in flower-visiting insects (from bumblebees to butterflies and hoverflies). Sulfur dioxide reduces activity of pollinators, including honeybees and male sweat bees (*Lasioglossum zephrum*), but may not kill them (Ginevan et al. 1980).

Ecosystem Stress and Health

The idea that concepts of health can be applied to ecosystems is not that new, but it has had difficulty gaining acceptance because there are problems as to how such a form of health might be measured (see chapter 18). One of the unifying concepts in ecology is that of competitive exclusion and niche hierarchies as arranged by degrees of overlap. Sugihara (1980) argues that in complex communities of organisms, the species occupy a hierarchy of niches with partial overlaps. The theoretical outcome of such an arrangement, given also the physical constraints of an environment, is the well-known log-normal relationship between species diversity and abundance. We accept Sugihara's (1980) argument of the biological meaning behind the log-normal relationship. Animal and plant interactions in pollination have allowed for some generalizations to be ventured concerning the structure and dynamics of ecological communities. Therefore it has been incorporated into a measure of ecosystemic health that involves pollinators.

Kevan et al. (1997) hypothesized that in the insecticide-stressed environments of blueberry fields from 1970 to 1975 in south-central New Brunswick, Canada, the log-normal relationship of species diversity and abundance of pollinating bees would be perturbed. We tested this hypothesis with data from east, central, and southwestern New Brunswick and in two periods: the years when insecticide fenitrothion was being applied in the central part of the area and the years after the cessation of its application in the vicinity of blueberry fields. Almost all our datasets were log-normal. The exceptions were those from central New Brunswick taken during the years of fenitrothion applications. We surmised that lack of log-normality in the one dataset indicated ill health.

Conclusion

Conservation of honeybees, other domesticated bees, wild bees, and other pollinators is an important issue in the global context of agricultural and natural sustainable productivity. It is a curious fact that although the major pollinators for many crops grown in the world's temperate zones are known, the quantitative relationships of pollinator populations, activities, and densities with plant and flower density and resultant seed set are largely unknown. The pollinators of many tropical crops are misidentified, unknown, or assumed to be honeybees. Furthermore, the breeding systems of many tropical crop plants are unknown or misunderstood. It is important that apiculturalists expand their horizons to embrace the culture of nonhoneybees and grasp the importance of other pollinators in agriculture. In an era of heightened concern about global environmental sustainability and conservation of biodiversity, the importance of pollination and processes that are deleterious to it embrace a wide range of interrelated issues. The need for conservation, imaginative approaches to management, and basic biological research must be fully recognized by biologists, ecologists, agriculturalists, and the general citizen in the new spirit of global environmental sustainability and conservation of biodiversity.

Acknowledgments

We thank all of our colleagues for their generosity in sharing their ideas. We are grateful for all the work that has gone into collecting and assembling the knowledge that exists on pollinators and pollinator conservation. We also thank the Convention on Biological Diversity for support.

References

Anderson, J. and M. A. Wojtas. 1986. Honey bees (Hymenoptera: Apidae) contaminated with pesticides and polychlorinated biphenyls. *Journal of Economic Entomology* 79:1200–1205.

Banda, H. J. and R. J. Paxton. 1991. Pollination of greenhouse tomatoes by bees. *Acta Horticultura* 288:194–198.

Breno, M. F., R. J. Paxton, and J. P. de Holanda-Neto. 2002. Identifying pollinators among an array of flower visitors, and the case of inadequate cashew pollination in NE Brazil. In P. G. Kevan and V. L. Imeratriz-Fonseca, eds., *Pollinating Bees: The Conservation Link Between Agriculture and Nature*, 229–244. Brasilia-DF, Brazil: Ministry of the Environment.

Bromenshenk, J. J., J. L. Gudatis, S. R. Carlson, J. M. Thomas, and M. A. Simmons. 1991. Population dynamics of honey bee nucleus colonies exposed to industrial pollutants. *Apidologie* 22:359–369.

Buchmann, S. E. 1983. Buzz pollination in angiosperms. In C. E. Jones and R. J. Little, eds., *Handbook of Experimental Pollination Biology*, 73–113. New York: Van Nostrand Reinhold.

Buchmann, S. E. and G. P. Nabhan. 1996. *The Forgotten Pollinators*. Washington, DC: Island Press.

Bunzl, K., W. Kracke, and G. Vorwohl. 1988. Transfer of Chernobyl-derived ^{134}Cs, ^{137}Cs, ^{131}I, and ^{103}Ru from flowers to honey and pollen. *Journal of Environmental Radioactivity* 6:261–269.

Cane, J. H. and J. A. Payne. 1988. Foraging ecology of *Habropoda laboriosa* (Hymenoptera: Anthophoridae), and oligolege of blueberries (Ericaceae: *Vaccinium*) in south-eastern United States. *Annals of Entomology* 81:419–427.

Cane, J. H. and J. A. Payne. 1990. Native bee pollinates rabbiteye blueberry. *Highlights in Agricultural Research, Alabama Agricultural Research Station* 37(4):4.

Castro, M. S. 2002. Bee fauna of some tropical and exotic fruits: Potential pollinators and their conservation. In P. G. Kevan and V. L. Imeratriz-Fonseca, eds., *Pollinating Bees: The Conservation Link Between Agriculture and Nature*, 275–288. Brasilia-DF, Brazil: Ministry of the Environment.

Connor, L. J., T. Rinderer, H. A. Sylvester, and S. Wongsiri, eds. 1993. Asian apiculture. In *Proceedings of the First International Conference on the Asian Honey Bees and Bee Mites*, 8. Cheshire, CT: Wicwas Press.

Corbet, S. A. 1995. Insects, plants and succession: Advantages of long-term set-aside. *Agriculture, Ecosystems and Environment* 53:201–217.

Corbet, S. A., I. H. Williams, and J. L. Osborne. 1991. Bees and the pollination of crops and wild flowers in the European Community. *Bee World* 72:47–59.

Crane, E. 1990. *Bees and Beekeeping: Science, Practice and World Resources*. Oxford, UK: Heinemann Newnes.

Dafni, A. and A. Shmida. 1996. The possible ecological implications of the invasion of *Bombus terrestris* (L.) (Apidae) at Mt Carmel, Israel. In A. S. L. Matheson, C. Buchmann, P. O'Toole, P. Westrich, and I. H. Williams, eds., *The Conservation of Bees*, 183–200. Linnean Society Symposium Series Number 18. London: Academic Press.

Dewey, J. E. 1973. Accumulation of fluorides by insects near an emission source in western Montana. *Environmental Entomology* 2:179–182.

Finnamore, B. and M. A. Neary. 1978. Blueberry pollinators of Nova Scotia, with a check list of the blueberry pollinators of eastern Canada and northeastern United States. *Annales de la Societé Entomologique de Québec* 23:161–181.

Ford, B. C., W. A. Jester, S. M. Griffith, R. A. Morse, R. R. Zall, D. M. Burgett, F. W. Bodyfelt, and D. J. Lisk. 1988. Cesium-134 and cesium-137 in honey bees and cheese samples collected in the US after the Chernobyl accident. *Chemosphere* 17:1153–1157.

Free, J. B. 1993. *Insect Pollination of Crops,* 2nd ed. London: Academic Press.

Freitas, B. M., R. J. Paxton, and J. P. Holanda-Neto. 2002. Identifying pollinators among an array of flower visitors, and the case of inadequate cashew pollination in NE Brazil. In P. Kevan and V. L. Imperatriz-Fonseca, eds., *Pollinating Bees: The Conservation Link Between Agriculture and Nature,* 229–244. Brasília, Brazil: Ministry of Environment.

Gess, F. W. and S. K. Gess. 1993. Effects of increasing land utilization on species representation and diversity of aculeate wasps and bees in the semi-arid areas of southern Africa. In J. Lasalle and I. D. Gauld, eds., *Hymenoptera and Biodiversity,* 83–114. Wallingford, UK: CAB International.

Ginevan, M. E., D. D. Lane, and L. Greenberg. 1980. Ambient air concentration of sulfur dioxide affects flight activity in bees. *Proceedings of the National Academy of Sciences of the USA* 77:5631–5633.

Goettel, M. S., K. W. Richards, and D. W. Goerzen. 1993. Decontamination of *Ascosphaera aggregata* spores from alfalfa leafcutting bee (*Megachile rotundata*) nesting material by fumigation with paraformaldehyde. *Bee Science* 3(1):22–25.

Graham, J. M. 1992. *The Hive and the Honey Bee.* Hamilton, IL: Dadant & Sons.

Inouye, D. W. 1980. The terminology of floral larceny. *Ecology* 61:1251–1253.

Janzen, D. H. 1974. The deflowering of America. *Natural History* 83:48–53.

Johansen, C. A. and D. F. Mayer. 1990. *Pollinator Protection. A Bee and Pesticide Handbook.* Cheshire, CT: Wicwas Press.

Johnston, A., J. F. Dormaar, and S. S. Smoliak. 1971. Long-term grazing effects on fescue grassland soils. *Journal of Range Management* 24:185–188.

Kevan, P. G., ed. 1995. *The Asiatic Hive Bee: Apiculture, Biology, and Role in Sustainable Development in Tropical and Subtropical Asia.* Cambridge, ON, Canada: Enviroquest Limited.

Kevan, P. G. 1999. Pollinators as bioindicators of the state of the environment: Species, activity and diversity. *Agriculture, Ecosystems and Environment* 74:373–393.

Kevan, P. G. 2001. Pollination: A plinth, pedestal and pillar for terrestrial productivity. The why, how and where of pollination protection, conservation and promotion. In C. S. Stubbs and F. A. Drummond, eds., *Bees and Crop Pollination: Crisis, Crossroads, Conservation,* 7–68. Lanham, MD: Entomological Society of America.

Kevan, P. G., C. F. Greco, and S. Belaoussoff. 1997. Log-normality of biodiversity and abundance in diagnosis and measuring of ecosystemic health: Pesticide stress on pollinators on blueberry heaths. *Journal of Applied Ecology* 34:1122–1136.

King, J. L. and J. K. Holloway. 1930. Tiphia popilliavora *Rohwer, a Parasite of the Japanese Beetle*. U.S. Department of Agriculture Circular No. 145. Washington, DC: U.S. Government Printing Office.

Kremen, C. and T. Ricketts. 2000. Global perspective on pollination disruptions. *Conservation Biology* 14:1226–1228.

Kremen, C., N. M. Williams, and R. W. Throp. 2002. Crop pollination from native bees at risk from agricultural intensification. *PNAS* 99:16812–16816.

Krunić, M. D., L. R. Terzic, and J. M. Kulincevic. 1989. Honey resistance to air contamination with arsenic from a copper processing plant. *Apidologie* 20:251–255.

Leius, K. 1967. Influence of wild flowers on parasitism of tent caterpillar and codling moth. *Canadian Entomologist* 99:444–446.

MacKenzie, K. E. 1993. Honey bees and pesticides: A complex problem. *Vector Control Bulletin of the North Central States* 1(2):123–136.

Mardan, M., I. M. Yatim, and M. R. Khalid. 1991. Nesting biology and foraging activity of carpenter bee on passion fruit. *Acta Horticultura* 288:127–132.

Maues, M. M. 2002. Reproductive phenology and pollination of the Brazil nut tree (*Bertholletia excelse* Humb. & Bonl. Lecythidaceae) in eastern Amazonia. In P. G. Kevan and V. L. Imeratriz-Fonseca, eds., *Pollinating Bees: The Conservation Link Between Agriculture and Nature*, 245–254. Brasilia-DF, Brazil: Ministry of the Environment.

Morse, R. A., T. W. Culliney, W. H. Gutenmann, C. B. Littman, and D. J. Lisk. 1987. Polychlorinated biphenyls in honey bees. *Bulletin of Environmental Contamination and Toxicology* 38:271–276.

Needham, G. R., R. E. Page, M. Delfinado-Baker, and C. E. Bowman, eds. 1988. *Africanized Honey Bees and Bee Mites*. Chichester, England: Ellis Harwood, Ltd.

Paton, D. C. 1993. Honeybees in the Australian environment. *BioScience* 43:95–103.

Richards, K. W. 1993. Non-*Apis* bees as crop pollinators. *Revue Suisse de Zoologie* 100:807–822.

Roubik, D. W. 1978. Competitive interactions between neotropical pollinators and Africanized honeybees. *Science (Washington)* 201:1030–1032.

Roubik, D. W. 1989. *Ecology and Natural History of Tropical Bees*. Cambridge, UK: Cambridge University Press.

Roubik, D. W., ed. 1995. *Pollination of Cultivated Plants in the Tropics*. FAO Agricultural Services Bulletin No. 118. Rome: FAO.

Ruppert, V. 1993. Einflusz blütenreicher Feldrandstrukturen auf die Dichte blüten-besuchender Nutzinsekten insbesondere der Syrphinae (Diptera: Syrphidae). *Agrarökologie* 8:1–149.

Stein, K. and F. Umland. 1987. Mobile und immobile Probensammlung mit Hilfe von Bienen und Birken. *Fresenius Zeitschrift für Analytlische Chemie* 327:132–141.

Stephen, W. P. 1955. Alfalfa pollination in Manitoba. *Journal of Economic Entomology* 48:543–548.

Sugden, E. A. and G. H. Pyke. 1991. Effects of honey bees on colonies of *Exoneura asimillima,* an Australian native bee. *Australian Journal of Ecology* 16:171–181.

Sugihara, G. 1980. Minimal community structure: An explanation of species abundance patterns. *American Naturalist* 116:770–787.

Syme, P. D. 1975. The effects of flowers on the longevity and fecundity of two native parasites of the European pine shoot moth in Ontario. *Environmental Entomology* 4:337–346.

Vandenberg, J. D. and W. P. Stephen. 1982. Etiology and symptomology of chalk-brood in the alfalfa leafcutting bee, *Megachile rotundata. Journal of Invertebrate Pathology* 39:133–137.

Willmer, P. G., A. A. M. Bataw, and J. P. Hughes. 1994. The superiority of bumble-bees to honeybees as pollinators: Insect visits to raspberry flowers. *Ecological Entomology* 19:271–284.

Winder, J. A. 1977. Some organic substrates which serve as insect breeding sites in Bahian cocoa plantations. *Revista Brasileira de Biologia* 37:351–356.

9 ⚡ Management of Soil Biodiversity in Agricultural Ecosystems

G. G. BROWN, M. J. SWIFT, D. E. BENNACK, S. BUNNING, A. MONTÁÑEZ, AND L. BRUSSAARD

Dimensions of Soil Biodiversity

Soil is not just an agglomeration of a little organic matter and mineral particles with ions that can be used by plants. It is a living entity and the home of countless organisms whose diversity may even surpass that of those living above ground, outside the soil.

Soil systems contain among the most diverse yet disparate assemblages of organisms on Earth (Brussaard et al. 1997; Giller et al. 1997; Wall and Moore 1999). These organisms have a broad range of body sizes, feeding strategies, and life habits, from strictly aquatic to obligatorily terrestrial (Bater 1996). They range in size from the tiniest one-celled bacteria, algae, fungi, and protozoa to the more complex nematodes and micro-arthropods and the visible earthworms, insects, small vertebrates, and plants. This community of organisms makes up the soil food web: the interactions and conversions of energy and nutrients between the primary producers (plants, lichens, moss, photosynthetic bacteria, and algae), the soil organisms that consume organic compounds derived from plants, other organisms and waste byproducts, and a few bacteria that obtain their energy from mineral compounds.

The diversity of life in the soil (soil biological diversity) exists and interacts at genetic, interspecies, and ecological levels. It is convenient to think of it as the sum of all the organisms that spend some portion of their life cycle in the soil or on its immediate surface, including the surface litter and decaying organic matter (OM). Many of the world's terrestrial insect species are soil dwellers during some stage of their life (Bater 1996). The soil biota includes familiar organisms such as termites, earthworms, and ants but also a multitude of lesser-known invertebrates and microorganisms.

Nowhere in nature are species so densely packed as they are in soil communities (Hågvar 1998). For example, a single gram of soil may contain several thousand species of bacteria and millions of individuals (Torsvik et al. 1994). A typical healthy soil might contain several species of vertebrate animals and earthworms, 20–30 species of mites, 50–100 species of insects, tens of species of nematodes, hundreds of species of fungi, and perhaps thousands of species of bacteria and actinomycetes (Ingham 1999). Soil biodiversity tends to be greater in forests and little-disturbed or undisturbed natural lands (e.g., grasslands) than in pastures and cultivated fields. However, the diversity, number, and types of organisms vary from one land use system and ecological environment to another depending on many factors, including aeration, temperature, acidity, moisture, nutrient content, and OM quality and quantity, all of which can be strongly influenced by human activities.

Soil is also a physically complex medium. Crisscrossed by an immense network of micropores, macropores, and tunnels, the soil matrix with its pore spaces and large surface area provides the habitat for a range of organisms and their biologically mediated life processes. The great spatial and temporal variability in available OM, water, and other nutrients promotes a complex niche structure in the soil. The soil structure and its food resources provide conditions for the evolution and maintenance of complex, interconnected, and sometimes even functionally redundant trophic interactions between soil organisms. Given this ecological complexity, myriad plant, animal, and microbial communities are able to coexist and provide a range of functions and services. However, this vital and dynamic subterranean soil ecosystem often is unrecognized, little understood, and therefore mismanaged.

This immense diversity, added to the technical difficulties associated with studying the soil ecosystem and the lack of taxonomists to describe it, has resulted in an appallingly poor knowledge of the world's soil biodiversity. The few currently available taxonomic inventories fall short of an accurate picture of the number of species living in soil systems. Because soil communities are so diverse yet so poorly known and described, they have been called the "other last biotic frontier" (André et al. 1994), or the "poor man's tropical rainforest" (Usher et al. 1979).

Available estimates of the number of described species for selected soil biota are given in table 9.1. However, we must emphasize that these

Table 9.1. Total number of described species of major members of the soil biota.

Size Class Organism	Number of Species Described
Microorganisms	
Bacteria and archaea	3,200
Fungi	60,000
Microfauna	
Protozoa (Protista)	36,000
Nematodes	15,000
Rotifers	2,000
Tardigrades	750
Mesofauna	
Mites (Acari)	ca. 45,000
Springtails (Collembola)	7,500
Pseudo-scorpions	3,235
Diplura	659
Symphyla	200
Pauropoda	700
Enchytraeids	800
Macrofauna	
Root herbivorous insects	>40,000
Beetles (Coleoptera)	350,000
Millipedes (Diplopoda)	10,000
Centipedes (Chilopoda)	2,500
Scorpions	1,259
Spiders	38,884
Snails (Gastropoda)	30,000
Woodlice (Isopoda)	4,250
Termites (Isoptera)	2,800
Ants (Formicidae)	11,826
Harvestmen (Opiliones)	5,500
Earthworms (Oligochaeta)	3,800
Velvet worms (Onchophora)	90

Sources: Hawksworth and Mound (1991); Brussaard et al. (1997); Wall and Moore (1999); Moreira et al. (2006); Lewinsohn and Prado (2005, 2006).

estimates are preliminary and much lower than the estimated total number of species for each group. For example, the described number of soil-dwelling, fungal species ranges from 18,000 to 35,000, but the projected number may be greater than 100,000 (Hawksworth 1991). Nematodes and mites are expected to be even more species-rich, with only 3% and 5% of their total number presently described, respectively (Hawksworth and Mound 1991). Estimates for bacteria and archaea are particularly problematic (Hawksworth and Kalin-Arroyo 1995) because scientific opinion is divided as to what criteria define a species in these groups. In addition, difficulty in isolating and culturing pure strains of these organisms complicates their identification. However, the development of molecular methods for extracting and describing the genetic composition of the soil microflora has initiated a new era of study of the bacteria and other microbes in soil and can be expected to revolutionize microbial ecology in fundamental ways (see Amman and Ludwig 2000; Torsvik and Ovreas 2002).

Ecosystem Functions, Scale Effects, and Regulatory Hierarchies

But soil organisms deserve study not just because of their great diversity and complex interrelationships but because they also perform key functions in both natural ecosystems and agroecosystems (table 9.2). Soil is the site of many key global processes mediated by soil life, notably nutrient cycling, carbon sequestration, and nitrogen fixation.

Specifically, soil biota are responsible for modifications to the soil environment, affecting its physical, chemical, and biological properties and processes. For instance, most bioturbating (soil moving or consuming) animals, plant roots, and some microbes influence the creation of soil structure, thus also affecting soil hydrological processes and water regimes (e.g., infiltration, drainage, water-holding capacity). Many microorganisms are intimately involved in symbiotic or parasitic relationships with plants and in plant protection against insect pests, microbial parasites, and diseases. Some microorganisms are asymbiotic plant growth promoters, living primarily in the rhizosphere, and other microbes are active in degrading pollutants such as pesticides and petroleum derivatives, decomposing OM, nutrient cycling, and sequestering greenhouse gases, especially methane, nitrous oxide, and carbon dioxide. Finally, many soil organisms are direct or indirect sources of food and medicines.

Therefore, the ways that certain organisms act on the soil and their contributions to ecosystem functions are highly variable. Their importance may depend on differences in body size, behavior patterns, population density and dynamics, life history strategies, living and feeding requirements, and interactions with other organisms (both synergistic and antagonistic). Spatial and temporal scales are especially critical in determining the overall functional effect of a given species in a soil environment (Anderson 2000).

Table 9.2. Ecosystem functions performed by the different members of the soil biota.

Functions	Organisms Involved
Maintenance of soil structure	Bioturbating (soil moving or consuming) invertebrates and plant roots, mycorrhizae, and some other microorganisms
Regulation of soil hydrological processes	Most bioturbating invertebrates and plant roots
Gas exchanges and carbon sequestration	Mostly microorganisms and plant roots, some carbon protected in large compact biogenic invertebrate aggregates
Soil detoxification	Mostly microorganisms
Nutrient cycling	Mostly microorganisms and plant roots, some soil- and litter-feeding invertebrates
Decomposition of organic matter	Various saprophytic and litter-feeding invertebrates (detritivores), fungi, bacteria, actinomycetes, and other microorganisms
Suppression of pests, parasites, and diseases	Plants, mycorrhizae and other fungi, nematodes, bacteria and various other microorganisms, Collembola, earthworms, various predators
Sources of food and medicines	Plant roots, various insects (crickets, beetle larvae, ants, termites), earthworms, vertebrates, microorganisms, and their byproducts
Symbiotic and asymbiotic relationships with plants and their roots	Rhizobia, mycorrhizae, actinomycetes, diazotrophic bacteria and various other rhizosphere microorganisms, ants
Plant growth control (positive and negative)	Direct effects: plant roots, rhizobia, mycorrhizae, actinomycetes, pathogens, phytoparasitic nematodes, rhizophagous insects, plant growth–promoting rhizosphere microorganisms, biocontrol agents Indirect effects: most soil biota

Many organisms and species may contribute to a specific soil process, operating at different scales of magnitude in space and time. Moreover, many organisms or species contribute to several discrete processes.

For example, nematodes feeding on bacteria and fungi at the micrometer scale may influence nitrogen mineralization (Ingham et al. 1985), and mites and Collembola feeding on nematodes and fungi at the scale of a few millimeters may affect microbial community processes over several centimeters (Anderson 1995). On the other hand, the activities of earthworms create tunnels and burrows several millimeters in diameter and centimeters in length that, in turn, may affect soil structure and hydrological processes over a scale of several meters. Finally, the wide-ranging activities associated with termite and ant colonies may affect soil physical and chemical processes over several hectares (Swift et al. 1996). These structures often are very long lasting (up to several decades), surpassing greatly the individual life span of the organisms that created them. Thus, these soil engineering (box 9.1) activities of ants, termites, and earthworms can modify the soil as a habitat for other organisms, including plants, invertebrates, and microbes.

Therefore the activities of smaller soil organisms are expressed against a background of effects attributable to larger soil organisms, such that a

---

Box 9.1 What Is an Ecosystem Engineer?

Ecosystem engineers (*sensu* Jones et al. 1994) are species that directly or indirectly modulate the availability of resources to other species (and sometimes to themselves) by causing physical changes to biotic or abiotic materials (e.g., the soil). By their activities, they alter, maintain, or create habitats.

Allogenic engineers change the environment by transforming materials (living or dead) from one physical condition to another, via mechanical or other means. Examples of allogenic soil ecosystem engineering are earthworm burrowing, casting, and feeding, which alter the physical structure of the soil and modify the availability of dead organic matter and resources to other soil organisms, including plant roots (Lavelle et al. 1997).

Autogenic engineers modify the environment by their own living or dead biomass. Examples of autogenic engineering include trees, which modulate hydrology, nutrient cycles, soil stability, temperature, humidity, wind speed and light levels, and the availability of food and other resources to organisms.

hierarchical system of top-down controls (Lavelle et al. 1997) is instituted. In such a system, the effects of biological activity at larger scales of spatiotemporal magnitude constrain biological performance at smaller scales (small box of figure 9.1). Furthermore, the activities of all soil organisms are expressed against a context of resource quality and quantity, soil properties, and climate conditions, which are also hierarchically organized.

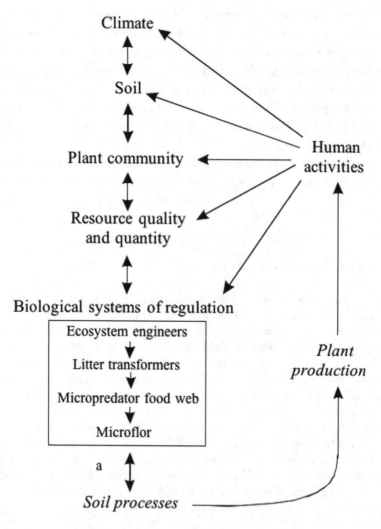

FIGURE 9.1. Hierarchical organization of soil function (after Lavelle 1996).

Bottom-up control (feedback) in the soil community occurs when a particular organism (or group of organisms) is capable of affecting other organisms at higher levels of the hierarchy (see figure 9.1). For instance, it has been suggested that earthworms may influence both the recruitment of new plant individuals and the composition of plant communities by consuming and egesting (as castings) an important proportion of the soil seed bank; selecting and consuming preferentially seeds of particular plant species, leading to preferential germination; digesting or damaging selected plant species' seeds in different manners and to various extents in their intestines, depending on the protective seed coating and the earthworm's digestive processes; dispersing particular plant species' seeds throughout the soil profile or depositing them on the soil surface; and promoting the growth of particular plant species depending on the physicochemical and biological changes induced in the soil environment by earthworm activities (Willems and Huijsmans 1994; Piearce et al. 1994; Decäens et al. 2001; Brown et al. 2004).

Functional Classifications of Soil Biota

The diversity of soil biota, coupled with its broad array of process-related roles in the environment, has led soil biologists to propose various functional group classifications for soil organisms. In such classifications, organisms are divided into groups (not necessarily taxonomically related) that perform redundant or similar functions. These groups help illustrate in a simpler manner the functions performed in the soil, the organisms that perform them, and which functions and biota may be more important in particular ecosystems. Of the different functional classifications available, perhaps the most useful ones rely on categorizations of soil organisms by body size, feeding behavior (or trophic levels), physical structures produced, and a combination of any of the former three parameters.

BODY SIZE

Body size is not always related to function but can be used as a surrogate system for ecological function of soil biota. For instance, the ability to transport, ingest, or modify greatly the soil's physical structure is generally positively related to the body size of the organism, so that larger organisms (earthworms, termites, and ants) are more able to modify soils than smaller ones (with the notable exception of the mycorrhizal fungi). On the other hand, litter decomposition and soil chemical reactions are performed mainly by the smaller biota (mites, springtails, and particularly microorganisms), although some larger biota (the litter shredders) may be particularly important in preparing the materials for and enhancing the roles of the smaller biota. Thus a classification based on body diameter produces certain correlation between taxonomy and function.

The *macrobiota* and *megabiota* (organisms generally greater than 2.0 mm in diameter and visible to the naked eye) include two large groups: the familiar vertebrates (e.g., snakes, lizards, mice, rabbits, foxes, badgers, moles) that primarily dig in the soil for food or shelter (megafauna) and the invertebrates (e.g., ants, termites, millipedes, centipedes, earthworms, pillbugs and other crustaceans, caterpillars, cicadas, ant-lions, beetle larvae and adults, fly larvae, earwigs, swordfishes, silverfishes, snails, spiders, harvestmen, scorpions, crickets, and cockroaches) that live and feed in or on the soil or surface litter and their components (macrofauna). Large insects such as bees and wasps occasionally burrow into the soil, but these are generally not considered soil organisms, even though their influence can be important. Finally, plant roots have wide-ranging, long-lasting effects on plant and animal populations above and below ground and therefore should be included among the soil biota.

The *mesobiota* (organisms generally ranging in size from 0.1 to 2.0 mm in diameter) include mainly microarthropods, such as pseudoscorpions, protura, diplura, springtails, mites, small myriapods (pauropoda and symphyla), and the wormlike enchytraeids. This group of organisms has limited burrowing ability, generally lives in soil pores, and feeds on OM, microbiota, and other invertebrates.

The *microbiota* are the smallest organisms (less than 0.1 mm in diameter) and include the extremely abundant, ubiquitous, and diverse *microflora* (algae, bacteria, archaea, cyanobacteria, fungi, yeasts, myxomycetes, and actinomycetes) that are able to decompose almost any existing natural material and include both plant pathogenic and growth-promoting species; and the *microfauna* (nematodes, protozoa, turbellarians, tardigrades, and rotifers) that generally live in soil water films and feed on microflora, plant roots, other microfauna, and sometimes larger organisms.

FEEDING BEHAVIOR

Feeding behavior can also serve as a surrogate system for ecological function of soil biota because the use of particular food resources by soil organisms can lead to cascading effects down the trophic food chain, therefore ultimately affecting soil function. These interactions between organisms and trophic levels are commonly displayed in complex soil food webs, where some organisms subsist on living plants and animals, whereas others do so on plant debris, fungi, or bacteria, and still others live off of their hosts in a parasitic or symbiotic fashion, weakening but not killing their host or helping it to grow.

PHYSICAL STRUCTURES PRODUCED AND FUNCTIONAL DOMAINS

An additional classification scheme (Lavelle 2000) groups soil organisms according to biogenic structures produced (such as pores, aggregates, and fabrics), that serve as hotspots (highly active sites) for various soil functions and processes (see table 9.2). Functional domains represent spheres of influence, or the physical location where a basic process making up part of a soil function operates at specific spatial and temporal scales (Lavelle 2002). These locations and structures usually can be physically separated from the soil matrix. Some examples of biological domains (spheres) include the drilosphere (earthworms), termitosphere (termites), myrmecosphere (ants), rhizosphere (roots), and detritusphere (plant litter).

Every structure in the soil is part of a functional domain, although some structures may be incorporated into more than one domain, the boundaries between domains are not always clear, and there may be interaction between domains (Brown et al. 2000). Functional domains can have important positive or negative effects on plant production.

Economic Benefits of Soil Biodiversity

Traditionally, soil has been viewed as a substrate for plants, which may be the most crucial role of soil for humankind. However, soil is also the place of countless interactions that control a host of services of direct and indirect use to humanity and to the natural environment—recycling of organic wastes, soil formation, nitrogen fixation, bioremediation of chemical pollution, and biological pest control—as well as a source of food and biotechnology products.

It has been estimated that the value of ecosystem services provided each year by the soil biota worldwide may exceed US\$1.5 trillion (Pimentel et al. 1997; see chapter 18), and recycling of organic wastes alone is estimated to provide some 50% of the total benefits of soil biotic activity worldwide.

Were it not for the decomposing and recycling activity of soil organisms, much of the world's land surface would be covered with organic debris.

The external benefits of soil biodiversity and other environmental goods are not commonly priced on the market. Therefore, a major and important step toward effective conservation includes adequately assessing the value of and paying for the ecosystem services derived from soil biodiversity while recognizing that many soil organisms are also detrimental to plant production and human societies.

Land Use Trends and Threats to the World's Soil Biodiversity

Soil biodiversity is threatened on a global scale by human activities that are responsible for the permanent loss of both species and habitats. The present biodiversity crisis (Wilson 1985), unlike those experienced in the past, is rooted in the patterns of human social organization, global trade and consumption of natural resources, the growth of human populations, the widespread adoption of economic systems and policies that fail to value the environment and its resources, and the inequity in ownership, management, and flow of benefits derived from the use and conservation of biological resources (McNeely et al. 1995).

Agricultural Intensification and Biodiversity

The imbalance between the short-term (socioeconomic) and long-term (ecological) human perspective when deciding how to manage the landscape (e.g., for agricultural production) may have disastrous consequences, considering the immense scale at which agricultural activities are undertaken worldwide: 11% of the total land surface is used for crop production in developing countries, 25% for livestock grazing, and 30% for forestry (FAO 2002). In general, agricultural intensification is associated with increasing specialization toward marketed commodities (e.g., soybean in developing countries) with improved technology and increased use of inputs. Increased use of pesticides and herbicides with agricultural intensification tends to be associated with high–external input agriculture (HEIA) in order to sustain high harvests and rapid returns. But it is also caused by neglect or ignorance (at policy, technical, and farmer levels) of the risks they pose to the environment and to ecosystem functioning. Thus a homogenization of cropping systems is occurring, resulting in loss of agricultural and associated biodiversity at genetic, species, and landscape levels.

Assessment of these losses in developing countries is severely limited by lack of data on quantitative and qualitative changes in pesticide use, livestock densities and wildlife populations, and land use and management practices. Knowledge of soil biodiversity is especially limited because of its complexity and the fact that it is largely invisible. It is expected that these risks will persist because in many cases the socioeconomic conditions and market forces will not favor the adaptation by small- or large-scale farmers of diverse systems and agroecological approaches that conserve biological diversity, protect the land and water resources, and ensure adequate and balanced use of organic and mineral fertilizers to compensate for soil nutrient removal by crops and grazing animals.

Catastrophic events, both past and present, serve as stark warnings against the abuse or misuse of our land. Entire ancient civilizations disappeared because of the degradation of soils under intense and unsustainable agricultural uses (Lowdermilk 1978; Hillel 1991). There is an urgent need to improve land use and management practices in order to halt soil degradation, restore already damaged lands, and enhance soil fertility and agricultural productivity.

Agricultural Practices and Soil Biota

Efforts to curb the loss of biodiversity have intensified in recent years, but they remain modest and have not kept pace with the rate of human-induced change. Furthermore, their application has been focused primarily on preserving a small number of species, especially large plants and animals for tourism and aesthetic reasons and the harvested species for food, fiber, and other products. There has been a general neglect of small organisms, particularly soil biota that dominate the structure of food webs and basic functions of natural ecosystems. Some strategies and the means by which soil biodiversity can be conserved and managed in agroecosystems were discussed in a recent workshop, as part of the activities being undertaken by the United Nations Food and Agriculture Organization (FAO) and partner organizations in the International Initiative for the Conservation and Sustainable Use of Soil Biodiversity of the Convention on Biological Diversity (box 9.2).

Nonetheless, there are some positive trends: the expansion of conservation agriculture principles and practices (no tillage or minimum tillage),

Box 9.2 The International Initiative for the Conservation and Sustainable Use of Soil Biodiversity of the Convention on Biological Diversity

In Decision VI/5 (CBD 2002:78), the Conference of the Parties to the Convention on Biological Diversity (CBD) decided "to establish the International Initiative for the Conservation and Sustainable Use of Soil Biodiversity as a cross-cutting initiative within the Programme of work on Agricultural Biological diversity" and invited the "FAO and other relevant organizations, to facilitate and coordinate this initiative" (see further information and activities of FAO and partners at www.fao.org/ag/AGL/agll/soilbiod/).

As an initial collaborative activity, an international technical workshop on the biological management of soil ecosystems for sustainable agriculture was jointly organized by FAO and Embrapa-Soybean in Londrina, Brazil, in June 2002 in order to discuss the concepts and practices of integrated soil management, share successful experiences of soil biological management, and identify priorities for action under the Soil Biodiversity Initiative. The full report of this workshop was published by FAO (2003) (www.fao.org/ag/AGL/agll/soilbiod/docs.stm), and additional documents from the workshop can be found in Brown et al. (2002a).

At the 8th Conference of the Parties to the CBD in Curitiba in March 2006, the proposed framework for action and implementation of the initiative, as presented in FAO (2003), was endorsed by the member parties of the convention, and other governments, international organizations, nongovernment organizations, and interested stakeholders were invited to support and implement the initiative and to supply additional case studies on soil biodiversity in order to strengthen the initiative.

In the framework, three strategic areas for action were identified:

- Increasing recognition of the essential services provided by soil biodiversity across all production systems and its relationship to sustainable land management
- Capacity building to promote integrated approaches and coordinated activities for the sustainable use of soil biodiversity and enhancement of agroecosystem functions, including assessment and monitoring, adaptive management, and targeted research and development
- Developing partnerships and cooperative processes through mainstreaming and coordinated actions among partners to actively promote the conservation, restoration, and sustainable use of soil biodiversity and enhanced contribution of beneficial soil organisms to the sustained productivity of agroecosystems

Progress in this initiative will depend on mobilizing policy support and investment in soil biological management and ecosystem approaches, which will also entail economic assessment of the loss of soil biodiversity, its beneficial functions, and the ecosystem services provided under specific farming systems.

especially in the Americas but increasingly in other regions, and increasing consumer support for organic agriculture. Both of these systems recognize the importance of soil protection, soil health and biological activity, and crop rotations and the risks and costs of agrochemicals.

Conservation and Management of Soil Biodiversity

Principles of Biological Management of Soil Fertility

The Green Revolution, so called because of the large increases in plant production that were obtained through its techniques, relied on overcoming soil constraints by applying external inputs such as inorganic fertilizers and other amendments in order to meet plant needs (Sanchez 1994, 1997). However, most of the world's farmers do not have access to or are unable to afford the external inputs (agrochemicals, improved crop varieties, hybrid seeds, ready access to cash and credit) needed to apply the principles and practices of HEIA (Vandermeer et al. 1998).

The optimum window for use of soil biological management techniques probably will occur in systems with intermediate disturbance and low to intermediate use of external resources and human labor (figure 9.2). Thus the potential of soil biological management may be greater in agroecosystems of intermediate complexity (agroforestry and rotational systems), in marginal lands to prevent degradation, in degraded lands in need of bioreclamation, and in regions where the availability, access to, or use of external inputs is limited, thus leading to a predominance of biological processes in the maintenance of soil fertility (Anderson 1994; Mando et al. 1997; Sanchez 1997; Senapati et al. 1999; Swift 1999).

Underlying the principle of integrated biological management of soil is the acknowledgment that

- Soil organisms and biological processes have a major role in creating and regulating soil fertility.
- A diversity of organisms creates and modifies a diversity of soil functions and processes.
- A diversity of functions and processes is essential for maintaining soil fertility and productivity (i.e., the sustainability of the agroecosystem).
- Soil organisms can be manipulated in agroecosystems through both direct and indirect interventions.

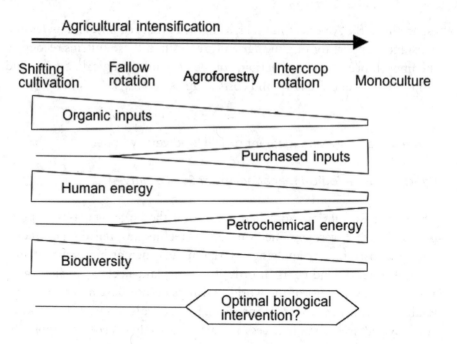

FIGURE 9.2. The relationships between agricultural intensification and the amounts of various inputs (organic, purchased, and petrochemical inputs) to the agroecosystem, biodiversity, and human energy expended. The proposed optimum window for biological soil management (intervention) is chosen by balancing each of these different factors and will depend on various human, socioeconomic, and environmental factors found at each location (drawing by M. J. Swift).

Direct and Indirect Interventions for Soil Biological Management

Regarding the different possibilities for biological soil management, Swift (1999) proposed a series of potential entry points at which management practices could be implemented. These include both direct and indirect interventions (box 9.3) such as

- Agricultural system design and management: choosing plants and their spatiotemporal organization and livestock management practices (indirect)
- Genetic control of soil function by manipulating plant resistance to disease, residue, and rhizosphere quality (root exudates) (indirect)
- Inoculation of disease antagonists, microsymbionts, rhizobacteria, and earthworms for disease control and soil fertility improvement (direct)
- Manipulation of soil biota through modification of OM quantity and quality (indirect)
- Biological control of pests and diseases (direct)

Direct methods of intervening in the production system attempt to alter the abundance or activity of specific groups of organisms (Hendrix et al. 1990).

Examples of direct interventions include inoculation of seeds or roots with rhizobia, mycorrhizae, fungi, and rhizobacteria for enhanced seedling growth and the inoculation of soil or the environment with biocontrol agents (for pest or disease control) or beneficial fauna (e.g., earthworms).

Indirect interventions are means of managing soil biotic processes by manipulating the factors that control biotic activity (habitat structure, microclimate, nutrients, and energy resources) rather than the organisms themselves (Hendrix et al. 1990).

Examples of indirect interventions include most agricultural practices (e.g., application of organic materials to soil, tillage, fertilization, irrigation, green manuring, and liming), cropping system design, and management. More recent techniques include genetic control of soil function by manipulation of plant residue and rhizosphere quality (root exudates) and resistance to diseases or pests.

Some of these interventions, particularly direct ones, such as selection of nitrogen-fixing plant species and varieties, rhizobial inoculation in grain legumes, mycorrhizal inoculation for tree establishment, and biocontrol agents for disease and pest control, are already well-developed techniques, used by many farmers and land managers in developed and developing countries. Nevertheless, they continue to be underused in many less developed countries, particularly by resource-poor farmers. The potential for the use of these direct techniques is important and should be promoted by the relevant institutions and governments responsible for agricultural development.

But the greatest benefits, particularly over the long term, are likely to come from indirect interventions such as the choice of crops and their spatiotemporal distribution, the enhancement of their natural ability to resist disease, improvements in the quality of the residues they produce, and management of OM and other external inputs such as fertilizers into the system (TSBF 1999). In a wider agricultural context, the management of mixed crop, livestock, and agroforestry systems has been shown to improve resource use efficiency and management of spatial (e.g., associations and landscape considerations) and temporal (e.g., perennials and rotations) dimensions (see also chapters 13 and 14). Furthermore, these interventions have important consequences for soil biological activity and biodiversity.

Over the last 15 years scientists have focused substantially on the manipulation of OM decomposition, in an attempt to attain optimum synchrony between the decomposition, immobilization, and mineralization processes and the nutrient demands of growing plants (Myers et al. 1994; Palm et al. 2000, 2001). Where cultivation is minimized and crop residues are retained on the soil surface (e.g., in no-till or minimum-tillage systems), it has been shown that there is much greater spatial and temporal differentiation of belowground food webs and processes compared with conventionally cultivated soils (House and Parmelee 1985; Brown et al. 2002b). In conventional tillage, bacteria-based food webs play a greater role, especially in the tilled layer, and as a result, flushes of mineralization related to tillage events may lead to greater OM loss and lower nutrient retention. In no-tillage systems, fungal-based food webs are more important, influencing nutrient availability and soil aggregate stability, tending to increase nitrogen retention and reduce leaching (Hendrix et al. 1986).

Putting Integrated Soil Biological Management into Practice

RECOGNIZING THE IMPORTANCE OF SOIL BIOTA

Integrated management of soil biota, biodiversity, and agricultural ecosystems is a holistic process that relies largely on locally available resources, climate, socioeconomic conditions, and, above all, direct involvement of farmers and other stakeholders in identifying and adapting management practices to their specific context. A seven-step process in which all stakeholders are involved in a process moving from problem diagnosis through testing and adaptation to technology adoption is demonstrated in figure 9.3 (adapted from Chambers 1991; Swift et al. 1994; Swift 1997).

Recognizing that soil biota play a key role in sustaining agricultural production is the first step toward proper management and conservation (Step 1). Farmers and agricultural practitioners of many cultures, both traditional and modern, still do not adequately recognize the roles and importance of soil biota in agricultural production (Kevan 1985; Puentes and Swift 2000). Many societies continue to fear insects and disregard earthworms, and this may explain why aggressive practices against soil biota have been so widespread until fairly recently (Lavelle 2000). For example, in a survey of 163 farmers from the state of Veracruz, Mexico,

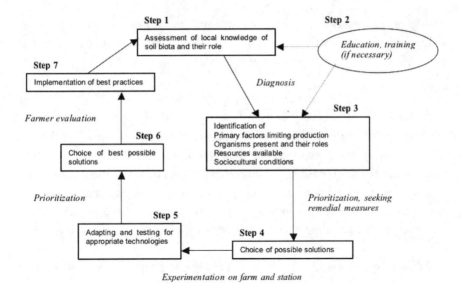

FIGURE 9.3. The 7-step process to optimum soil biological management and conservation (modified after Swift 1997).

55% ignored the role of earthworms in soil fertility, and 11% considered them harmful, mainly because they confused them with intestinal parasites (Ortiz et al. 1999). Lack of knowledge can lead to abuses to the soil ecosystem (e.g., contamination of surface water and groundwater, erosion, loss of biodiversity) as well as the underuse of the benefits derived from biological soil management.

When limited knowledge and need for alternative management are detected, awareness and capacity building should be targeted to farmers, extension agents, local communities, service providers, politicians, and industries responsible for promoting particular land uses and management practices that disregard the importance of soil biota and their functions (Step 2). In commercial agriculture, knowledge of the role of some soil biota may be even poorer than in smallholder subsistence systems because product-oriented, intensive management practices often bypass biological mechanisms and interactions through their use of external inputs (notably pesticide or herbicide use rather than biocontrol of pests and weeds, and chemical fertilizers rather than OM restoration). Increasingly, traditional knowledge systems on how to maintain and restore a healthy soil and sustainable crop, crop–livestock, or agroforestry system are being lost. For intensive systems, alternatives must be demonstrated that make better use of ecological processes and reduce the medium- to long-term risks and potential damage of conventional practices such as monocultures, deep and frequent plowing, and high chemical inputs.

The identification of local conditions and the resources available, both biotic (e.g., human, plants, organic materials, soil biota) and abiotic (e.g., draft or mechanical traction, cash or credit, external inputs, soil nutrient contents), is essential for determining which soil biological management practices realistically can be used. This diagnostic process (Step 3, figure 9.3) should create an understanding of potential constraints, opportunities, and needs at various levels.

Reflecting increased attention to ecological principles and to human management considerations, several minimum datasets for assessing soil and environmental resources and their quality have been proposed (Doran and Jones 1996). These generally include characterization of the current farming system and practices of different farmer groups, such as the available human resources, organic resources, and biological indicators of soil quality and function (box 9.4). The particular advantage of bioindicators rests in the fact that they are generally able to detect changes (for better or worse) in the agroecosystem more quickly than traditional chemical or physical indicators of soil quality.

Relative to physical and chemical soil quality indicators, there are few biological soil quality indicators in datasets hitherto proposed, and often they are not independent measures (e.g., microbial biomass, potential nitrogen mineralization, soil respiration, and the ratio of respiration to microbial biomass, proposed by Doran and Parkin 1994). Reduction of redundant indicators to one key or a few integrated indicators such as potential nitrogen mineralization (Keeney and Nelson 1982) does simplify matters, but it does not solve the essential shortcoming that these indicators are related mostly to element transformations, not to soil structure or hydrological and biological properties of the topsoil. Therefore, the challenge is to identify a minimum set of soil quality bioindicators that can be related to nutrients, contaminants, soil structure, and topsoil hydrological properties (Brussaard et al. 2004) and that have the additional purpose of

- Signaling changes in soil quality earlier or more precisely than chemical and physical indicators. An example is shown in figure 9.4, where microbial

Box 9.4 Soil Quality Indicators: What Are They, and Why Use Them?

Soil quality indicators are biological, physical, and chemical properties and processes that can be measured to monitor changes in soil function (Muckel and Mausbach 1996). They are quantitative tools to assess the health of soil and provide an early warning of system collapse, allowing land managers to react before irreversible damage occurs (Pankhurst et al. 1997). Indicators must be rapid reactors, robust yet sensitive (can be detected above background noise), meaningful and predictive (good relationship between the indicator and the function), and easy to measure and interpret. Examples of indicators associated with biological activity in soils include the following (from Brown 1991; Stork and Eggleton 1992; Doran et al. 1994; Oades and Walters 1994; Doran and Jones 1996; Pankhurst et al. 1997; van Straalen 1998; Paoletti 1999):

- Biodiversity at the molecular, genetic, taxonomic, and functional levels
- Organisms and their properties (presence–absence, biomass and density at species, genera, community, or functional group levels), such as some bacteria and fungi, nematodes, protozoa, earthworms, termites, ants, some beetles, isopods, millipedes, spiders, flies, collembolans, mites, roots, weed seed numbers, plant pathogens and root feeders, microbial biomass carbon and nitrogen
- Soil processes affected by biological activity such as compaction, aggregation and aggregate stability, erosion, water infiltration, potentially mineralizable carbon and nitrogen, nitrogen fixation, nitrification and denitrification, soil respiration, decomposition rates, enzyme activities, and ergosterol
- The ability of the soil to support and sustain plant growth, the ultimate indicator of soil quality and health in agroecosystems (Pankhurst 1994)

biomass carbon is shown to indicate changes in soil OM at an earlier stage and more precisely than changes in total soil carbon content.
- Giving an integrated assessment of changes in physical, chemical, and biological properties. Examples are plentiful in ecotoxicology, where water or soil organisms are used for integrated assessment of multiple contaminant effects on growth, reproduction, and longevity of organisms and associated biological processes. Similarly, earthworms can indicate OM availability and hence soil nutrients and water-holding capacity, as well as porosity, aggregation, and aerobic microbiological activity.

It is important that individual farmers can actually work with soil quality indicators. Therefore, visual assessment of soil quality is an essential

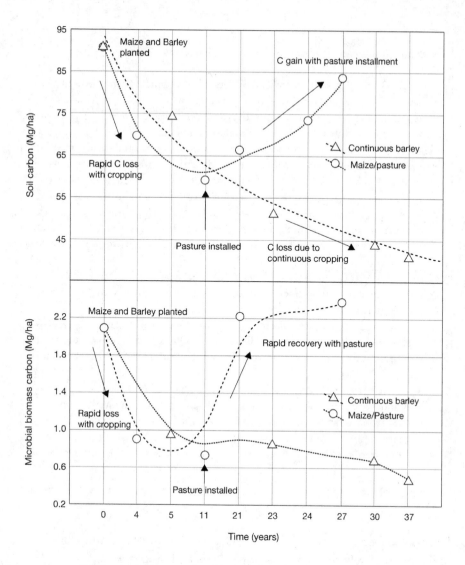

FIGURE 9.4. Patterns in the loss or gain with time of total soil carbon and microbial biomass carbon in the topsoil (0–20 cm) under maize and continuous barley cropping and in a pasture installed after 11 years of maize. Note the more rapid recovery of biomass carbon (compared with total soil carbon) after pasture installment (T. G. Shepherd, pers. comm., 2002).

starting point, such as the method developed by Shepherd (2000) in New Zealand. This method is very simple and requires only that the farmer visually inspect a spade of soil in terms of structure, porosity, color, mottles, and earthworm number. Assessment values are recorded on a soil score card and integrated to a total value on a scale from poor to good. The farmers' visual measures can be backed up by and correlated with specific laboratory measurements of chemical, physical, and biological soil quality. More than 90% of farmers and scientists who were asked to work with this system considered it both practical and scientifically sound.

A visual soil assessment toolbox is being developed by FAO and includes soil management guidelines for the prevention and amelioration of soil degradation and the sustainable management of farms (Benites, pers. comm., 2005). The FAO Programme for Community Integrated Pest Management in Asia also released a very useful booklet with a series of training exercises on integrated soil management (Settle 2000), and a soil fauna and biological quality assessment manual will also be available shortly from FAO.

However, existing soil quality assessment tools (e.g., farmer interviews, surveys, and soil health kits) still need to be adapted to the specific conditions present in smallholder farming contexts in tropical humid and semiarid regions, to allow their use by farmers and extensionists rather than soil scientists. Simple methods and measurements such as those described earlier are the most useful and most likely to be widely adopted.

OVERCOMING LIMITATIONS

Once the main biotic and abiotic constraints have been identified, they should be hierarchically organized, and the potential alternatives—adapted to the local human, climate, soil, and agroecosystem conditions—should be chosen. At this point, an understanding of how limitations to agricultural production at various levels (social, cultural, economic, political, agronomic, biological, environmental, edaphic, genetic) can be overcome using local or imported resources, knowledge, and capacity and how agricultural practices affect soil biota and their activity is essential to predict possible management options and other solutions.

Unfortunately, information on the effects of various different agricultural practices on the soil biota is not available for all soil organisms, and there can be important differences in the effects of different practices on the same organism or of the same practices on different organisms. Some organisms are susceptible to certain land management practices and become locally extinct, whereas others respond positively and take advantage of the modified conditions to increase their abundance, biomass, and activity.

To adequately assess the effect of an individual species on a given soil function and the effect of management practices on its populations and activity, sample measurements must be undertaken within the spatiotemporal scales relevant to that particular species; that is, they must be taken within the functional domain of that species, often a particularly difficult methodological challenge.

Despite the complexity of this task, there are a few general rules that apply and can be used with farmers at local level in predicting management effects and /choosing potential solutions. Some of these are presented in figure 9.5 and table 9.3, together with the description of some of the major constraints on different management practices and their effects on soil function.

Cropping drastically affects the soil environment and hence the number and kinds of organisms present, especially the process of turning the soil through plowing. In general, when a forest or grazing land is converted for cultivation, the quantity and quality of plant residues and the

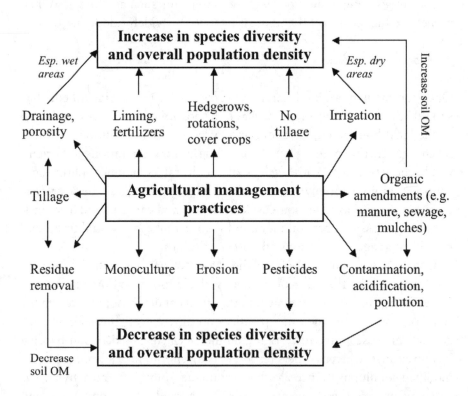

FIGURE 9.5. The effect of different agricultural management practices on soil biota (modified from Hendrix et al. 1990).

Table 9.3. Constraints on different agricultural management practices and their effects on soil biota and soil function.

Management Practice	Constraints to Use	Effect on Biota and Function
Tillage	Labor, tools and machinery, cost, soil-borne diseases, sloping lands	More rapid decomposition of OM, higher ratio of bacteria to fungi, lower populations of macrofauna and mesofauna, short-term increase in nutrient availability but increase in long-term losses, better root growth in tilled layer, higher erosion risks
No tillage	Machinery, cost, soil compaction and heavy textures, pest management	Higher populations of macrofauna, mesofauna, and microfauna; greater ratio of fungi to bacteria; organic matter accumulation on soil surface; nutrient conservation; lower runoff and erosion; increase in presence and incidence of pests and diseases associated with litter layer
Organic matter input	Availability, labor, livestock presence, cost	Changes in decomposition rates and organism populations (some increase, others decrease, depending on the type of material); increased nutrient availability, storage, and exchanges; improved soil physical structure and water relations; reduction in acidity and Al toxicity; greater microbial and faunal activity, especially detritivores
Fertilization	Availability, cost	Usually reduction in mycorrhization and N_2 fixation (with P and N, respectively), mineralization–immobilization balance changes, increased plant production and organic matter inputs, increases in populations of some organisms through greater food supply
Pesticides	Cost, environmental and health impacts	Reduced incidence of diseases, pests, parasites, and pathogenic organisms but negative effects on nontarget biota such as beneficial insects and earthworms; improved plant production but often creation of dependence; destabilization of nutrient cycles; loss of soil structure; long-term increased resistance of target biota

Table 9.3. continues to next page

Table 9.3. *continued*

Management Practice	Constraints to Use	Effect on Biota and Function
Irrigation or flooding	Cost, slope, labor, tools, available water	Increased water availability, pH neutralization, changes in nutrient cycling (often higher anaerobic processes) and availability, higher asymbiotic N_2 fixation, increased populations of drought-stressed biota, lower numbers of sensitive biota, lower organic matter decomposition rates, depression of soilborne diseases and weeds
Crop rotations	Social acceptability, opportunity costs, agroecosystem compatibility, climate, soil conditions	Rotation effect, with improved productivity and pest and disease management; more efficient soil nutrient use; greater diversity aboveground and belowground; higher populations, biomass, and activity of most organisms (especially with legumes); improved soil aggregation and infiltration; reduced bulk density; higher organic matter
Inoculation of selected soil biota (e.g., rhizobia, mycorrhizae, earthworms, rhizobacteria, antagonists, biocontrol agents)	Cost, availability, environmental adapt-ability, competition with or replacement of native biota, adequate soil conditions	Increased N fixation, nutrient availability in soil, water uptake, and efficiency of nutrient acquisition by plants; higher yields; increased heavy metal tolerance; better resis-tance to plant diseases, pests, and parasites; increased soil porosity, aeration, aggregate stability, water infiltration, and holding capacity; faster decomposition rates and nutrient cycling

Source: Expanded from Swift (1997).

number of species of higher plants are greatly reduced, thereby reducing the range of habitats and food sources for soil organisms. The ratio of different organisms and their interactions is also significantly altered. In general, moldboard plowing, monoculture, pesticide use, erosion, and soil contamination or pollution have negative effects on most organisms, and their effects should be observed and the practices adapted, avoided, or minimized to the extent possible. On the other hand, practices such as the application of organic wastes, moderate fertilizer use, crop rotations, and irrigation in dry and drainage in wet areas generally have positive impacts on soil organism densities, diversity, and activity. In most cases these practices can also be improved for better resource use efficiency.

However, it is not only the biophysical factors that affect farmers' decisions (Step 4, figure 9.3) but also socioeconomic considerations. Some of the most common constraints to the use of different soil biological management practices include the monetary cost (purchased inputs), labor and time costs, the availability of the resources, and the tools to implement them (table 9.3).

ADAPTIVE MANAGEMENT: CHOOSING THE BEST SOLUTIONS

After a number of different possible solutions have been chosen from best farmers' practices, innovations, and new technologies, these should be tested using an iterative and participatory screening process of adaptive experimentation (Step 5 in figure 9.3). The Tropical Soil Biology and Fertility Institute of the International Center for Tropical Agriculture has developed an approach for the adaptive management of soil biological processes, emphasizing an iterative, cooperative interaction between farmers, extension agents, local community facilitators, and scientists (TSBF 2000). In this adaptive process, different treatments and techniques are tested simultaneously and repeated over several cropping cycles to identify the most adaptable, economic, practical, and socially acceptable practices.

The Farmer Field School approach is being used by FAO together with partners in eastern Africa and Asia to promote such experiential learning by farmer groups in soil and water management, including the development of training modules for a dynamic farmer-driven process and a toolbox of practical exercises. In addition, FAO has produced practical training materials on conservation agriculture through its Land and Water Digital Media Series (see CD Nos. 27 and 22 at www.fao.org/landandwater/lwdms.stm).

It is the farmers and other stakeholders in the decision-making process who finally select the desired or appropriate technologies for implementation at various levels (Step 6, figure 9.3). The final decision by the stakeholder groups as to which practices are to be implemented may be substantially different for small-scale and large-scale farmers and for resource-poor and resource-rich farmers.

The adoption of integrated soil biological management (Step 7, figure 9.3) is a long-term participatory learning process resulting from the diagnosis, analysis of options, prioritization, choice, testing, adaptation, discussion, agreement, and choice of the best soil biological management options. The final step in the cycle is the farmers' evaluation of the best options in the field and decision whether to implement these practices on larger and long-term scales or to revert to their traditional management strategies. This is a critical step in which all the hard work of the previous steps is often at stake. Certain support services may be needed, such as the provision of seed of selected crop species or varieties, the supply of fertilizers at appropriate prices and quantities, training of artisans for manufacture of adapted tools, and further farmer training in, for example, livestock management for OM and fertilizer applications.

Biological Management of Soil Fertility: Some Examples

Importance of OM Management

Of the various successful practices available, the most interesting for conservation of soil biota and long-term maintenance of productivity generally have been associated with stimulation or maintenance of active pools of OM in soils. Through the manipulation of the whole cropping system, the appropriate combination of crops, the appropriate pattern in space and time, and appropriate soil management practices, OM quantity and quality can be increased, promoting a cascading effect on all soil life and physicochemical functions. This phenomenon often is observed when degraded ecosystems are undergoing recovery. Once plants are established, roots begin to penetrate the soil, and a protective litter layer is formed on the soil surface, a synergistic effect of increased carbon availability, microclimatic changes to the soil environment, and biological activities help to speed ecosystem recovery. In drier environments soil moisture is critical to this restoration process and buildup of soil OM. Soil moisture retention can be enhanced through a protective cover of crops or mulch and through no-till or minimum-tillage which retains plant root biomass and soil OM.

Indirect Soil Biological Management

A good example of the application of the management of biological soil recovery processes and the ecosystem approach is the case of the farmers of the Grupo Vicente Guerrero (GVG) in Tlaxcala, Mexico (Ramos 1998). Soils in the state of Tlaxcala have been cultivated for thousands of years using traditional practices (Gliessman 1990). However, intensification of the fragile, easily erodible soils of the state deeply scarred the landscape, leading to statewide soil erosion, siltation, and water catchment problems. In response to these concerns, more than 20 years ago peasant farmers in the small village of Vicente Guerrero initiated a program (along with the Quaker House of Friends) to generate, share, and promote experiences that might improve their quality of life and that of their neighbors.

The motivating force behind the success of GVG is a profound respect for the environment, evidenced in an evolving, integrated use of local natural resources and the firm conviction that sharing their discoveries with other farmers is an undeniable moral obligation. This has allowed the group to patiently put into practice, and successfully refine, a farmer-to-farmer model for transmitting to other neighboring farmers the knowledge given to them by rural development facilitators and technical experts. Members of GVG have trained more than 2,000 peasant farmers in Mexico and elsewhere in Latin America in the past two decades. Some of the successful management practices adopted by the group are shown in box 9.5. The success of this case study highlights the importance of integrated, multilateral (not just top-down) approaches to farming system development in order to guarantee long-lasting results.

In the state of Paraná, Brazil, in the last 20 years, a similar process of cooperative technology development, adaptation, and extension has taken place, with the resulting widespread adoption of conservation agriculture practices, especially no-till farming. In the 1970s and most of the 1980s, after the abandonment of coffee and the adoption of conventional tillage annual cropping (especially soybean and wheat), much of the state experienced

Box 9.5 Adaptive Management and Conservation Methods Adopted by the GVG, Tlaxcala, Mexico

The GVG counts the following as some of its main successes:

- Significant reduction of agrochemical use by many farmers who initially rejected organic fertilizers and total elimination of agrochemicals in some farmer fields.
- Increased adoption of soil and water conservation measures and soil fertility restoration efforts by local farmers.
- Greater incorporation of stubble and crop residues into the soil.
- Increased agricultural productivity. One farmer in the group won first prize in a statewide competition for improved yield of dryland maize with a 5.5-Mg/ha grain yield (much higher than average yields in the state).
- Formal recognition of their efforts by the government of the state of Tlaxcala.
- Increased group capacity to organize and attract outside funding, thanks to collective experience and well-earned prestige.

Management and conservation methods adopted include the following:

- Grain production using techniques that enhance soil biodiversity and biological functions
- Crop rotations, leguminous cover crops, improved local seed varieties, and diversified crop associations to broaden agroecosystem resilience and improve yields
- Low-impact tillage methods to reduce disturbances to soil structure and soil biota
- Organic fertilizer production using stubble, harvest residues, livestock manure, and green manure
- Soil cover conservation measures to maintain soil structure and moisture content
- Land management that favors plant and animal diversity and its association with soil biological activity
- Mosaics of different crops and land uses
- Capture and conservation of rainwater for plants, animals, and people
- Incorporation of backyard animals (native races of chickens, turkeys, and rabbits), whose excrements are used in home gardens
- Restoration of agricultural biodiversity by planting native crops, medicinal plants, and tree species.

Participatory methods and various tools include the following:

- Visits to farmer fields
- Field demonstrations of crop and soil management techniques
- On-farm experimentation
- Rapid participatory diagnostics
- Workshops, talks, courses, didactic games, and community theater shows

Further information on the GVG and this case study can be found in Ramos (1998) and at www.fao.org/ag/AGL/agll/soilbiod/cases/caseD1.pdf.

problems similar to those of the GVG in Mexico. Widespread gullies, silted rivers, floods, water quality problems, and severely damaged crops prompted farmers to seek alternatives to the traditional soil preparation practices (disk and moldboard plowing). Driven by farmers' experimentation, adaptation, and demand, a partnership was formed with the industrial sector and farmer cooperatives, aided by government subsidies, to develop no-till planters that could be used manually, with animal traction and tractors. Simultaneously, management practices were developed based on crop rotations, cover crops for weed management and soil protection, and reduced traffic to minimize soil compaction. Consequently, no-tillage is now used on almost 20 million ha in Brazil, 5.5 million of which are in the state of Paraná (25% of the state's surface area). These techniques are particularly interesting from the biological standpoint because they avoid soil disturbance, build up OM in soil (mostly on the surface; Sá 1993), and permit the recovery of soil biological activity, enhancing their role in soil fertility (House and Parmelee 1985; Hendrix et al. 1990; Brown et al. 2002b).

Complementary Direct Biological Management Technologies

Although higher-level interventions in the agroecosystem are more likely to succeed and influence the system through cascading effects at lower levels and down the soil food chain, specific technologies that directly manipulate soil biota are also useful and can complement indirect interventions through OM and agroecosystem management. Nonetheless, they usually have more limited application and should be adopted under particular conditions, depending on the agroecosystem characteristics. The following sections contain some examples of soil biological management techniques using microorganisms and macrofauna, the perspectives and benefits of their use, and some of the problems that must be overcome to allow their widespread adoption.

BENEFICIAL SOIL MICROORGANISMS

Beneficial microorganisms include those that create symbiotic associations with plant roots, promote nutrient mineralization and availability, produce plant growth hormones, and are antagonists of plant pests, parasites, or diseases. Many of these organisms are naturally present in the soil, although in some situations it may be beneficial to increase their populations either by inoculation or by applying various agricultural management techniques that enhance their abundance and activity.

The role of the N_2-fixing Rhizobiaceae bacterial family in agricultural production probably is the most successful and familiar form of direct biological management (box 9.6). The rhizobia infect plant roots, creating nodules where N_2 is fixed, providing the plant with most of the nitrogen needed for its development. Well-nodulated plants with an efficient symbiosis may fix up to several hundred kilograms of nitrogen per hectare annually. Some of this nitrogen is added to the soil during plant growth by leaky roots, although much of the nitrogen is exported in the grains (grain legumes) or remains in plant tissues and is released during residue decomposition, to the benefit of subsequent crops or the intercrop. Previous colonization of the legume roots by mycorrhizae may greatly enhance nodulation by rhizobia, ultimately increasing the potential growth benefits. Yield increases with inoculation have been well documented, and some of the main limitations are discussed by Giller (2001) and Montáñez (2002). Nevertheless, despite the obvious benefits of rhizobial inoculation or management, the widespread use of this technique to enhance legume yields continues to be limited by extensive promotion of nitrogen fertilizers, lack of market incentives to grow legumes, lack of understanding of the importance of N_2 fixation or adoption of inoculants by farmers, environmental constraints (e.g., low soil phosphorus, drought), low quality and limited availability of inoculants, low genetic compatibility of the host legume with the bacteria, and lack of appropriate political and economic incentives and infrastructure (Giller et al. 1994; Hungria et al. 1999).

Biological nitrogen fixation (BNF) is crucial for agricultural sustainability but is often constrained by the absence in soil of efficient and competitive nitrogen-fixing microorganisms. There is an obvious need to improve the availability, quality, and delivery of such microorganisms because of their importance for crop production. Research on BNF has expanded greatly over the last decades, improving knowledge of the process. However, application of BNF technologies and their impact in agricultural systems has been less than desired. Inoculants do not play a major role in the production of some of the most important food legumes, and many of the inoculants produced in the world are still of poor quality (FAO 1991). Spontaneous adoption can be assured only when farmers have seen and are convinced of the benefits of BNF and are able to overcome constraints, in partnership with researchers, the private sector, and policymakers.

Opportunities to enhance BNF inputs are available across different agroecosystems and socioeconomic conditions through the following means:

- Altering the number of effective symbiotic or associated organisms in the system (inoculation)
- Enhancing inoculation methods and technology
- Screening and selecting the most appropriate crops and microbial strains
- Management practices that enhance N_2 fixation and recycling of net N inputs into the cropping system (e.g., rotation, green manure application, no tillage, strategic use of legumes; Montáñez 2002).

Further information on this topic can be found in Giller (2001) and at www.fao.org/ag/AGL/agll/soilbiod/cases/caseB1.pdf.

Mycorrhizae are highly evolved mutualistic associations between soil fungi and plant roots. The plant donates carbon to the mycorrhizae in exchange for a greater ability to use native soil resources. More than 90% of the world's plants are mycorrhizal, with varying degrees of dependence and benefits derived from this association. The most well-known and perhaps the most common mycorrhizal symbioses involve arbuscular mycorrhizae (many crop species) and ectomycorrhizae (only woody species, mostly trees and shrubs), although several other types also exist (Allen et al. 1995). The positive role of mycorrhizae in plant production is well documented, with many cases of growth and yield enhancement, particularly in highly dependent, susceptible plants. The plant response can result from various factors, although in most cases it results from an increase in effective root area for water and nutrient extraction because the mycorrhizal hyphal network works as a natural extension of the plant root system. Other benefits of the mycorrhizal association include enhanced protection against pathogens, improved tolerance to pollutants, and greater resistance to water stress, high soil temperature, adverse soil pH, and transplant shock.

However, the widespread use of mycorrhizal inoculants in agroecosystems has been hampered by the difficulty in cultivating arbuscular mycorrhizae and producing sufficient inocula at affordable prices. Furthermore, the efficiency of the symbiosis decreases with increasing soil fertility status (particularly phosphorus content) or with high phosphorus fertilizer application. It appears that the most practical current uses of mycorrhizae involve land restoration and reclamation efforts and arbuscular and ecto-mycorrhizal inoculation of tree and crop seedlings in nurseries. Nonetheless, enhancement of naturally occurring mycorrhizal populations in agricultural fields (and their potential benefits to the growing crops) is feasible, and important benefits can arise through the adoption of various management practices that enhance mycorrhizal populations and activity, such as reduced tillage, crop rotations, lower fertilizer (especially nitrogen and phosphorus) applications, and choice of suitable hosts to increase the infectivity of the soil before sowing of the main crop (Abbott and Robson 1994). Thus the potential for improved benefits of mycorrhizal symbioses appears to be particularly worthwhile in reduced external input agroecosystems and organic agriculture.

BENEFICIAL SOIL MACROFAUNA

Direct biological management practices can also involve inoculation or enhancement of the activity of soil ecosystem engineers. A very successful example of this technique was developed in India, using earthworms and organic fertilizers in tea gardens of Tamil Nadu (Giri 1995; Lavelle et al. 1998; Senapati et al. 1999, 2002).

Tea is a high-value plantation crop in India with a long history (many estates are more than 100 years old). In recent years, green tea production has stabilized, despite increased application of external inputs such as fertilizers and pesticides. The long-term exploitation of soil under the tea gardens has led to important changes in various soil physical, chemical, and biological conditions, decreasing OM content, cation exchange, water-holding capacity, soil macrofauna populations (reduced up to 70%), and pH, simultaneously increasing concentrations of toxic aluminum.

In response to these limitations on tea production, a patented technology called Fertilisation Bio-Organique dans les Plantations Arborées (FBO) was developed by Parry Agro Industries Ltd., in association with the French Institut de Recherche pour le Développement and Sambalpur University (Orissa, India). This technology aims at improving physical, chemical, and biological soil conditions by inoculating a mixture of low- and high-quality organic materials (tea prunings and manure) and earthworms into trenches dug between the rows of tea plants. Measurements performed at two sites since 1994 have shown that this technique is much more effective than 100% organic or 100% inorganic fertilization alone, increasing yields on average by up to 276% and profits by an equal percentage (from around US$2,000/ha using conventional techniques to about US$7,600/ha using FBO) in the first year of application. The technique has been extended to other countries, and the principles of its application may be useful for other perennial plantation crops as well. Details of the method are given in the patent document (ref. PCT/FR 97/01363; see also www.fao.org/ag/AGL/agll/soilbiod/cases/caseA1.pdf).

A similar but indirect manipulation of ecosystem engineer populations through the application of OM on crusted Sahelian soils increased termite activity and resulted in the restoration of soil structure and consequent improvement in plant production (Mando et al. 1997, 2002; see also www .fao.org/ag/AGL/agll/soilbiod/cases/caseA2.pdf). The extension of bare and crusted soils in the Sahel has increased in recent decades, seriously degrading the landscape and reducing crop production. However, when mulch was placed on crusted and bare soil in northern Burkina Faso, termites migrating from nearby areas invaded the organic substrate and the topsoil, significantly changing its physical structure. Many galleries were opened to the soil surface, reducing surface sealing. Throughout the topsoil profile, macropores with irregular shapes and sizes were created, reducing soil compaction and increasing water infiltration and drainage so that crops could again be planted. Furthermore, termites increased the decomposition and mineralization of the mulch, releasing nutrients for plant uptake. In mulched plots where termites were artificially excluded, cowpea yields were less than 1% as much as in plots where termites were present and active. This work demonstrated again that termites, far from being pests in agroecosystems, can be extremely important in plant production and ecosystem function and that it is possible to manage their activities for human benefit in some cases.

In much of Africa, farmers clean their fields of any OM because of the fear of associated pests and diseases, particularly of termites, which, if there is no available food source, will indeed resort to feeding on crops, although they prefer dry materials. Bringing about a change in behavior depends on convincing farmers, for example through study plots, of the value of mulch in enhancing biological activity and water infiltration, reducing evaporation, and providing essential plant nutrients and moisture.

Soil Biodiversity Accidents

The aforementioned examples are all planned interventions targeting the improvement of agricultural management practices through biological means. Occasionally, however, accidents occur (Lavelle 2000) that offer opportunities to test the principles of biological management. In these accidents, the loss of key functional groups of soil biota in a particular site, generally caused by human interference, can have dramatic effects (generally negative) on ecosystem function. One such example involves the destruction of soil structure and pasture degradation in kaolinitic soils of the Amazon Basin (Chauvel et al. 1997, 1999; Barros et al. 2004).

In the Brazilian Amazon, 95% of the deforested area is converted into pastures, and of these, about 50% can be considered degraded because of mismanagement, phytosanitary problems, poor soil fertility, and soil structural modification (linked to faunal activity). The kaolinitic soils that predominate in the Amazonian region have a favorable yet fragile microaggregate structure because of its low oxy-hydroxide metal content. When the forest is converted to pasture, first machines and later cattle trampling can cause severe soil compaction, particularly in the 5- to 10-cm layer (Chauvel et al. 1997).

But, even more important, the native forest soil macrofaunal communities are radically altered, with disappearance of most of the native taxa. The opportunistic invading earthworm species *Pontoscolex corethrurus* occupies the empty niches, reaching a biomass of more than 450 kg/ha (nearly 90% of total soil faunal biomass). This species produces annually more than 100 Mg/ha of castings, dramatically decreasing soil macroporosity down to a level equivalent to that produced by the action of heavy machinery on the soil ($2.7\,cm^3/100\,g$). During the rainy season these casts plug up the soil surface, saturating the soil and producing a thick, muddy layer in which anaerobic conditions prevail (simultaneously increasing both methane emission and denitrification). In the dry season, desiccation cracks the surface, blocking root growth and hindering their ability to extract water from the soil. The plants then wilt and die, leaving bare patches in the pasture (Chauvel et al. 1997).

An experiment performed near Manaus, Brazil, demonstrated the role of *P. corethrurus* and a diverse assemblage of soil organisms (e.g., other earthworm species, termites, millipedes, isopods, ants) in destroying and recreating soil structure (Barros et al. 2004). Soil monoliths 25 by 25 cm square were removed from the pasture and placed in the forest; similar blocks were also taken from the forest and placed in the pasture. After one year, the structure of the compacted pasture soil was completely restored to levels typical of native forest soils by the action of the diverse community of forest soil invertebrates. Meanwhile the macroaggregate structure of the forest soil was completely destroyed by *P. corethrurus*, reaching compaction and porosity levels similar to those of the degraded pasture.

This research highlights not only the extremely important role of a diverse assemblage of macroinvertebrates in the maintenance of soil structure (especially in these kaolinitic soils) but also the problems associated with management practices that are not well adapted to the environment (extensive pastures on problem soils after deforestation) and the role of invading species in ecosystem properties and processes. Such research findings must be made available for training farmers and extensionists because the sharing of experiences between farmers and researchers will help catalyze innovation and adaptive management and provide feedback on constraints that must be addressed by researchers and policymakers.

Conclusion

We have seen that the soil biota represent a substantial proportion of earth's biological diversity. They also contribute significantly to human welfare through their role in the production of goods and services ranging from agricultural products to climate regulation and groundwater quality. Yet this group of organisms remains largely unknown by the public, largely ignored in scientific evaluations of biodiversity, and neglected in farming system development. The intimate and complex links between soil organisms and those above ground, particularly plants, make this an extremely worrying omission. The capacity for knowledge-based management of these organisms remains limited, but significant strides have been made in developing principles and methods. The development and exploitation of these approaches deserves to be considered as one of the most important challenges of the century ahead.

If sustainable and more productive agricultural systems are to be realized, the impact of land management change on the short- and long-term functioning of soil ecosystems should be clarified. This entails the development of appropriate indicators to improve understanding of land use and soil biodiversity interactions and to assist in monitoring and assessment of the trends and impacts and the progress in promoting conservation and sustainable use of agroecosystems and their components. Such indicators should facilitate monitoring at various spatial scales and provide a tool for adequate management of land resources and biodiversity both locally and on a national level and for regional and global overviews of biodiversity and natural resource status and trends.

Soil, crop, and pest management practices often are developed as separate technologies, and their impacts on the function of other parts of the ecosystem are ignored. The development of ecosystem-oriented management strategies entails an integrated system approach rather than component and reductive studies. If soil processes are addressed simultaneously through a system approach, taking into account soil–water–crop–livestock–human management interactions, strategies and recommendations can be developed that more effectively address multiple goals of farmers and livestock managers. Various cases are available showing both positive and negative effects of soil biological management practices for improved agricultural productivity and agroecosystem sustainability. When management strategies are not placed in an ecosystem context, or lack of knowledge does not permit adequate assessment of potential risks or constraints, the consequences of inappropriate practices or technologies can be disastrous. On the other hand, when the specific ecosystem characteristics and opportunities and limitations of the farming system are considered, interventions are more likely to succeed but are not guaranteed. Integrated soil biological and agroecosystem management demands knowledge of the soil organisms, their interactions and needs, the effect of various practices on their populations and functions, and the soil, plant, livestock, agroecosystem, climate, socioeconomic, and human contexts.

Acknowledgments

The authors would like to thank the FAO, through the Netherlands Partnership Programme, and the Conselho Nacional de Desenvolvimento Cientifico e Tecnológico (Profix) for their support during the development of this chapter. We also thank J. Benites, D. Cooper, J. J. Jiménez, and an anonymous reviewer for their comments and especially D. Jarvis and L. Seers for their hard editing work, without which this chapter would not have been published. For further information on this topic, see FAO's Soil Biodiversity Portal (www.fao.org/ag/AGL/agll/soilbiod/).

References

Abbott, L. K. and A. D. Robson. 1994. The impact of agricultural practices on mycorrhizal fungi. In C. E. Pankhurst, B. M. Doube, V. V. S. R. Gupta, and P. R. Grace, eds., *Soil Biota: Management in Sustainable Farming Systems*, 88–95. East Melbourne, Australia: CSIRO.

Allen, E. B., M. F. Allen, D. J. Helm, J. M. Trappe, R. Molina, and E. Rincon. 1995. Patterns and regulation of mycorrhizal plant and fungal diversity. *Plant Soil* 170:47–62.

Amman, R. and W. Ludwig. 2000. RNA ribosomal-targeted nucleic acid probes for studies in microbial ecology, *Federation of European Microbiological Societies Microbiology Reviews* 24:555–565.

Anderson, J. M. 1994. Functional attributes of biodiversity in land use systems. In D. J. Greenland and I. Szabolcs, eds., *Soil Resilience and Sustainable Land Use*, 267–290. Wallingford, UK: CAB International.

Anderson, J. M. 1995. Soil organisms as engineers: Microsite modulation of macroscale processes. In C. G. Jones and J. H. Lawton, eds., *Linking Species and Ecosystems*, 94–106. New York: Chapman and Hall.

Anderson, J. M. 2000. Foodweb functioning and ecosystem processes: Problems and perceptions of scaling. In D. C. Coleman and P. F. Hendrix, eds., *Invertebrates as Webmasters in Ecosystems*, 3–24. Wallingford, UK: CAB International.

André, H. M., M.-I. Noti, and P. Lebrun. 1994. The soil fauna: The other last biotic frontier. *Biodiversity and Conservation* 3:45–56.

Barros, M. E., M. Grimaldi, M. Sarrazin, A. Chauvel, D. Mitja, T. Desjardins, and P. Lavelle. 2004. Soil physical degradation and changes in macrofaunal communities in Central Amazon. *Applied Soil Ecology* 26:157–168.

Bater, J. E. 1996. Micro- and macro-arthropods. In G. S. Hall, ed., *Methods for the*

Examination of Organismal Diversity in Soils and Sediments, 163–174. Wallingford, UK: CAB International.

Brown, G. G., I. Barois, and P. Lavelle. 2000. Regulation of soil organic matter dynamics and microbial activity in the drilosphere and the role of interactions with other edaphic functional domains. *European Journal of Soil Biology* 36:177–198.

Brown, G. G., C. A. Edwards, and L. Brussaard. 2004. How earthworms affect plant growth: Burrowing into the mechanisms. In C. A. Edwards, ed., *Earthworm Ecology,* 13–49. Boca Raton, FL: CRC Press.

Brown, G. G., M. Hungria, L. J. Oliveira, S. Bunning, and A. Montañez. 2002a. *Programme, Abstracts and Related Documents of the International Technical Workshop on Biological Management of Soil Ecosystems for Sustainable Agriculture.* Série Documentos Vol. 182. Londrina, Brazil: Embrapa Soja.

Brown, G. G., A. Pasini, N. P. Benito, A. M. de Aquino, and M. E. F. Correia. 2002b. Diversity and functional role of soil macrofauna communities in Brazilian no-tillage agroecosystems. In *Proceedings of the International Symposium on Managing Biodiversity in Agricultural Ecosystems,* November 8–10, 2001, CD-ROM, 1–20. Montreal: UNU/CBD.

Brown, K. S. Jr. 1991. Conservation of neotropical environments: Insects as indicators. In N. M. Collins and J. A. Thomas, eds., *The Conservation of Insects and Their Habitats,* 349–403. London: Academic Press.

Brussaard, L., V. M. Behan-Pelletier, D. E. Bignell, V. K. Brown, W. Didden, P. Folgarait, C. Fragoso, D. Wall-Freckman, V. V. S. R. Gupta, T. Hattori, D. L. Hawksworth, C. Klopatek, P. Lavelle, D. W. Malloch, J. Rusek, B. Söderström, J. M. Tiedje, and R. A. Virginia. 1997. Biodiversity and ecosystem functioning in soil. *Ambio* 26:563–570.

Brussaard, L., T. W. Kuyper, W. A. M. Didden, R. G. M. de Goede, and J. Bloem. 2004. Biological soil quality from biomass to biodiversity: Importance and resilience to management stress and disturbance. In P. Schjønning, S. Emholt, and B. T. Christensen, eds., *Managing Soil Quality: Challenges in Modern Agriculture,* 139–161. Wallingford, UK: CAB International.

CBD (Convention on Biological Diversity). 2002. *Action for a Sustainable Future: Decisions from the Sixth Meeting of the Conference of the Parties to the Convention on Biological Diversity.* Montreal, Canada: Secretariat of the Convention on Biological Diversity.

Chambers, R. 1991. Farmer first: A practical paradigm for the third agriculture. In M. Altieri and S. B. Hecht, eds., *Agroecology and Small Farm Development,* 237–244. Boca Raton, FL: CRC Press.

Chauvel, A., E. M. Barbosa, E. Blanchart, M. Grimaldi, J. Ferraz, P. D. Martins, O. Topall, E. Barros, T. Desjardins, N. F. Filho, I. P. A. Miranda, M. Sarrazin, and D. Mitja. 1997. Mise en valeur de la fôret et modifications écologiques. In H. Théry, ed., *Environnement et développement en Amazonie Brésilienne*, 42–75. Paris: Editions Berlin.

Chauvel, A., M. Grimaldi, E. Barros, E. Blanchart, M. Sarrazin, and P. Lavelle. 1999. Pasture degradation by an Amazonian earthworm. *Nature* 389:32–33.

Decaëns, T., L. Mariani, N. Betancourt, and J. J. Jiménez. 2001. Earthworm effects on permanent soil seed banks in Colombian grasslands. In J. J. Jiménez and R. J. Thomas, eds., *Nature's Plow: Soil Macroinvertebrate Communities in the Neotropical Savannas of Colombia*, 274–293. Cali, Colombia: CIAT.

Doran, J. W., D. C. Coleman, D. F. Bezdicek, and B. A. Stewart. 1994. *Defining Soil Quality for a Sustainable Environment*. SSSA Special Publication 35. Madison, WI: ASA.

Doran, J. W. and A. J. Jones. 1996. *Methods for Assessing Soil Quality*. SSSA Special Publication 49. Madison, WI: ASA.

Doran, J. W. and T. B. Parkin. 1994. Defining and assessing soil quality. In J. W. Doran, D. C. Coleman, D. F. Bezdicek, and B. A. Stewart, eds., *Defining Soil Quality for a Sustainable Environment*, 3–21. Madison, WI: ASA.

FAO (Food and Agriculture Organization of the United Nations). 1991. *Expert Consultation on Legume Inoculant Production and Quality Control*. Rome: FAO.

FAO (Food and Agriculture Organization of the United Nations). 2002. *World Agriculture: Towards 2015/2030, Summary Report*. Rome: FAO.

FAO (Food and Agriculture Organization of the United Nations). 2003. *Biological Management of Soil Ecosystems for Sustainable Agriculture*. World Resources Soil Reports 101. Rome: FAO.

Giller, K. E. 2001. *Nitrogen Fixation in Tropical Cropping Systems*, 2nd ed. Wallingford, UK: CAB International.

Giller, K. E., M. H. Beare, P. Lavelle, A.-M. N. Izac, and M. J. Swift. 1997. Agricultural intensification, soil biodiversity and agroecosystem function. *Applied Soil Ecology* 6:3–16.

Giller, K. E., J. F. McDonagh, and G. Cadish. 1994. Can biological nitrogen fixation sustain agriculture in the tropics? In J. K. Syers and D. L. Rimmer, eds., *Soil Science and Sustainable Land Management in the Tropics*, 173–191. Wallingford, UK: CAB International.

Giri, S. 1995. *Short Term Input Operational Experiment in Tea Garden with Application of Organic Matter and Earthworm*. M.Phil. thesis, Sambalpur University, Jyvoti Vihar, India.

Gliessman, S. R. 1990. Understanding the basis of sustainability for agriculture in the tropics: Experiences in Latin America. In C. A. Edwards, R. Lal, P. Madden, R. H. Miller, and G. House, eds., *Sustainable Agricultural Systems*, 378–390. Ankeny, IA: SWCS.

Hågvar, S. 1998. The relevance of the Rio Convention on Biodiversity to conserving the biodiversity of soils. *Applied Soil Ecology* 9:1–7.

Hawksworth, D. L. 1991. The fungal dimension of biodiversity: Magnitude, significance, and conservation. *Mycological Research* 95:641–655.

Hawksworth, D. L. and M. T. Kalin-Arroyo. 1995. Magnitude and distribution of biodiversity. In V. H. Heywood, ed., *Global Biodiversity Assessment*, 107–191. Cambridge, UK: Cambridge University Press.

Hawksworth, D. L. and L. A. Mound. 1991. Biodiversity databases: The crucial significance of collections. In D. L. Hawksworth, ed., *The Biodiversity of Microorganisms and Invertebrates: Its Role in Sustainable Agriculture*, 17–29. Wallingford, UK: CAB International.

Hendrix, P. F., D. A. Crossley Jr., J. M. Blair, and D. C. Coleman. 1990. Soil biota as components of sustainable agroecosystems. In C. A. Edwards, R. Lal, P. Madden, R. H. Miller, and G. House, eds., *Sustainable Agricultural Systems*, 637–654. Ankeny, IA: SWCS.

Hendrix, P. F., R. W. Parmelee, D. A. Crossley Jr., D. C. Coleman, E. P. Odum, and P. M. Groffman. 1986. Detritus food webs in conventional and non-tillage agroecosystems. *BioScience* 36:374–380.

Hillel, D. 1991. *Out of the Earth: Civilization and the Life of the Soil*. Berkeley: University of California Press.

House, G. J. and R. W. Parmelee. 1985. Comparison of soil arthropods and earthworms from conventional and no-tillage agroecosystems. *Soil Tillage Research* 5:351–360.

Hungria, M., M. A. T. Vargas, D. de S. Andrade, R. J. Campo, L. M. de O. Chueire, M. C. Ferreira, and I. C. Mendes. 1999. Fixação biológica do nitrogênio em leguminosas de grãos. In J. O. Siqueira, F. M. S. Moreira, A. S. Lopes, L. R. G. Guilherme, V. Faquin, A. E. Furtani Neto, and J. G. Carvalho, eds., *Inter-relação fertilidade, biologia do solo e nutrição de plantas*, 597–620. Lavras, Brazil: UFLA.

Ingham, E. R. 1999. The food web and soil health. In A. J. Tugel and A. M. Lewandowski, eds., *Soil Biology Primer*, B1–B10. Ames, IA: NRCS Soil Quality Institute.

Ingham, R. E., J. A. Trofymow, E. R. Ingham, and D. C. Coleman. 1985. Interactions of bacteria, fungi, and their nematode grazers: Effects on nutrient cycling and plant growth. *Ecological Monographs* 55:119–140.

Jones, C. G., J. H. Lawton, and M. Shachak. 1994. Organisms as ecosystem engineers. *Oikos* 69:373–386.

Keeney, D. R. and D. W. Nelson. 1982. Nitrogen: Inorganic forms. In C. A. Black, D. D. Evans, L. E. Ensminger, J. L. White, and F. E. Clark, eds., *Methods of Soil Analysis,* Part 2, 682–687. Madison, WI: ASA.

Kevan, D. K. M. 1985. Soil zoology, then and now—mostly then. *Quaestiones Entomologicae* 21:371.7–472.

Lavelle, P. 1996. Diversity of soil fauna and ecosystem function. *Biology International* 33:3–16.

Lavelle, P. 2000. Ecological challenges for soil science. *Soil Science* 165:73–86.

Lavelle, P. 2002. Functional domains in soils. *Ecological Research* 17:441–450.

Lavelle, P., I. Barois, E. Blanchart, G. G. Brown, L. Brussaard, T. Decaëns, C. Fragoso, J. J. Jimenez, K. Ka Kajondo, M. A. Martínez, A. G. Moreno, B. Pashanasi, B. K. Senapati, and C. Villenave. 1998. Earthworms as a resource in tropical agroecosystems. *Nature and Resources* 34:28–44.

Lavelle, P., D. Bignell, M. Lepage, V. Wolters, P. Roger, P. Ineson, O. W. Heal, and S. Ghillion. 1997. Soil function in a changing world: The role of invertebrate ecosystem engineers. *European Journal of Soil Biology* 33:159–193.

Lewinsohn, T. M. and P. I. Prado. 2005. How many species are there in Brazil? *Conservation Biology* 19:619–624.

Lewinsohn, T. M. and P. I. Prado. 2006. *Sintese do conhecimento da biodiversidade brasileira*, Vol. I. Brasilia: Ministério do Meio Ambiente, Secretaria do Biodiversidade e Florestas, 21–109.

Lowdermilk, W. C. 1978. *Conquest of the Land Through 7,000 Years.* Agriculture Information Bulletin 99. Washington, DC: USDA.

Mando, A., L. Brussaard, and L. Stroosnijder. 1997. Termite- and mulch-mediated rehabilitation of vegetation on crusted soil in West Africa. *Restoration Ecology* 7:33–41.

Mando, A., L. Brussaard, L. Stroosnijder, and G. G. Brown. 2002. Managing termites and organic resources to improve soil productivity in the Sahel. In G. G. Brown, M. Hungria, L. J. Oliveira, S. Bunning, and A. Montañez, eds., *Program, Abstracts and Related Documents of the International Technical Workshop on Biological Management of Soil Ecosystems for Sustainable Agriculture,* Série Documentos Vol. 182, 191–203 (also available at www.fao.org/ag/AGL/agll/soilbiod/cases.stm). Londrina, Brazil: Embrapa Soja.

McNeely, J. A., M. Gadgil, C. Levèque, C. Padoch, and K. Redford. 1995. Human influences on biodiversity. In V. H. Heywood, ed., *Global Biodiversity Assessment,* 711–821. Cambridge, UK: Cambridge University Press.

Montáñez, A. 2002. Overview and case studies on biological nitrogen fixation: Perspectives and limitations. In G. G. Brown, M. Hungria, L. J. Oliveira, S. Bunning, and A. Montañez, eds., *Program, Abstracts and Related Documents of the International Technical Workshop on Biological Management of Soil Ecosystems for*

Sustainable Agriculture, Série Documentos Vol. 182, 204–224. Londrina, Brazil: Embrapa Soja. Available at www.fao.org/ag/AGL/agll/soilbiod/cases.stm.

Moreira, F. M. S., J. O. Siqueira, and L. Brussaard. 2006. Soil organisms in tropical ecosystems: A key role for Brazil in the global quest for the conservation and sustainable use of biodiversity. In F. M. S. Moreira, J. O. Siqueira, and L. Brussaard, eds., *Soil Biodiversity in Amazonian and Other Brazilian Ecosystems,* 1–12. Wallingford, UK: CABI.

Muckel, G. B. and M. J. Mausbach. 1996. Soil quality information sheets. In J. W. Doran and A. J. Jones, eds., *Methods for Assessing Soil Quality.* SSSA Special Publication 49, 393–400. Madison, WI: ASA.

Myers, R. J. K., C. A. Palm, E. Cuevas, I. U. N. Gunatilleke, and M. Brossard. 1994. The synchronisation of nutrient mineralisation and plant nutrient demand. In P. L. Woomer and M. J. Swift, eds., *The Biological Management of Tropical Soil Fertility,* 81–116. Chichester, UK: Wiley.

Oades, J. M. and L. J. Walters. 1994. Indicators for sustainable agriculture: Policies to paddock. In C. E. Pankhurst, B. M. Doube, V. V. S. R. Gupta, and P. R. Grace, eds., *Soil Biota: Management in Sustainable Farming Systems,* 219–223. East Melbourne, Australia: CSIRO.

Ortiz, B., C. Fragoso, I. Mboukou, B. Pashanasi, B. K. Senapati, and A. Contreras. 1999. Perception and use of earthworms in tropical farming systems. In P. Lavelle, L. Brussaard, and P. F. Hendrix, eds., *Earthworm Management in Tropical Agroecosystems,* 239–252. Wallingford, UK: CAB International.

Palm, C. A., K. E. Giller, P. L. Mafongoya, and M. J. Swift. 2001. Management of organic matter in the tropics: Translating theory into practice. *Nutrient Cycling in Agroecosystems* 61:63–75.

Palm, C. A., K. E. Giller, and M. J. Swift. 2000. Synchrony: An overview. In *The Biology and Fertility of Tropical Soils. Tropical Soil Biology and Fertility Programme Report 1997–1998,* 18–20. Nairobi, Kenya: TSBF.

Pankhurst, C. E. 1994. Biological indicators of soil health and sustainable productivity. In D. J. Greenland and I. Szabolcs, eds., *Soil Resilience and Sustainable Land Use,* 331–351. Wallingford, UK: CAB International.

Pankhurst, C. E., B. M. Doube, and V. V. S. R. Gupta. 1997. *Biological Indicators of Soil Health.* Wallingford, UK: CAB International.

Paoletti, M. G. 1999. *Invertebrate Biodiversity as Bioindicators of Sustainable Landscapes: Practical Use of Invertebrates to Assess Sustainable Land Use.* Amsterdam: Elsevier.

Piearce, T. G., N. Roggero, and R. Tipping. 1994. Earthworms and seeds. *Journal of Biological Education* 28:195–202.

Pimentel, D., C. Wilson, C. McCullum, R. Huang, P. Dwen, J. Flack, Q. Tran,

T. Saltman, and B. Cliff. 1997. Economic and environmental benefits of biodiversity. *BioScience* 47:747–757.

Puentes, R. and M. J. Swift. 2000. Tropical soil ecology: Matching research opportunities with farmers' needs. In M. J. Swift, ed., *Managing the Soil Biota for Sustainable Agriculture: Opportunities and Challenges*. Nairobi, Kenya: TSBF.

Ramos, S. F. J. 1998. *Grupo Vicente Guerrero de Españita, Tlaxcala. Dos decadas de promoción de campesino a campesino*. México City, Mexico: Red de Gestión de Recursos Naturales and Rockefeller Foundation.

Sá, J. C. M. 1993. *Manejo da fertilidade do solo no plantio direto*. Ponta Grossa, Brazil: Fundação ABC.

Sanchez, P. A. 1994. Tropical soil fertility research: Towards the second paradigm. In *Transactions of the 15th World Congress of Soil Science*, Vol. 1, 65–88. Acapulco, Mexico: ISSS.

Sanchez, P. A. 1997. Changing tropical soil fertility paradigms: From Brazil to Africa and back. In A. C. Moniz, ed., *Plant–Soil Interactions at Low pH*, 19–28. Lavras, Brazil: Brazilian Soil Science Society.

Senapati, B. K., P. Lavelle, S. Giri, B. Pashanasi, J. Alegre, T. Decaëns, J. J. Jiménez, A. Albrecht, E. Blanchart, M. Mahieux, L. Rousseaux, R. Thomas, P. K. Panigrahi, and M. Venkatachalan. 1999. In-soil technologies for tropical ecosystems. In P. Lavelle, L. Brussaard, and P. F. Hendrix, eds., *Earthworm Management in Tropical Agroecosystems*, 199–237. Wallingford, UK: CAB International.

Senapati, B. K., P. Lavelle, P. K. Panigrahi, S. Giri, and G. G. Brown. 2002. Restoring soil fertility and enhancing productivity in Indian tea plantations with earthworms and organic fertilizers. In G. G. Brown, M. Hungria, L. J. Oliveira, S. Bunning, and A. Montañez, eds., *Program, Abstracts and Related Documents of the International Technical Workshop on Biological Management of Soil Ecosystems for Sustainable Agriculture*, Série Documentos Vol. 182, 172–190. Londrina, Brazil: Embrapa Soja. Available at www.fao.org/ag/AGL/agll/soilbiod/cases.stm.

Settle, W. 2000. *Living Soils: Training Exercises for Integrated Soils Management*. Jakarta, Indonesia: FAO Programme for Community IPM in Asia.

Shepherd, T. G. 2000. *Visual Soil Assessment*, Vol. 1, *Field Guide for Cropping and Pastoral Grazing on Flat to Rolling Country*, 84. Palmerston North, New Zealand: Horizon.mw & Landcare Research.

Stork, N. E. and P. Eggleton. 1992. Invertebrates as determinants and indicators of soil quality. *American Journal of Alternative Agriculture* 7:38–47.

Swift, M. J. 1997. Biological management of soil fertility as a component of sustainable agriculture: Perspectives and prospects with particular reference to tropical regions. In L. Brussaard and R. Ferrera-Cerrato, eds., *Soil Ecology in Sustainable Agricultural Systems*, 137–159. Boca Raton, FL: Lewis Publishers.

Swift, M. J. 1999. Towards the second paradigm: Integrated biological management of soil. In J. O. Siqueira, F. M. S. Moreira, A. S. Lopes, L. R. G. Guilherme, V. Faquin, A. E. Furtani Neto, and J. G. Carvalho, eds., *Inter-relação fertilidade, biologia do solo e nutrição de plantas*, 11–24. Lavras, Brasil: UFLA.

Swift, M. J., L. Bohren, S. E. Carter, A. M. Izac, and P. L. Woomer. 1994. Biological management of tropical soils: Integrating process research and farm practice. In P. L. Woomer and M. J. Swift, eds., *The Biological Management of Tropical Soil Fertility*, 209–227. New York: Wiley.

Swift, M. J., J. Vandermeer, P. S. Ramakrishnan, J. M. Anderson, C. K. Ong, and B. A. Hawkins. 1996. Biodiversity and agroecosystem function. In H. A. Mooney, J. H. Cushman, E. Medina, O. E. Sala, and E.-D. Schulze, eds., *Functional Roles of Biodiversity: A Global Perspective*, 261–298. New York: Wiley.

Torsvik, V., J. Goksøyr, F. L. Daae, R. Sørheim, J. Michalsen, and K. Salte. 1994. Use of DNA analysis to determine the diversity of microbial communities. In K. Ritz, J. Dighton, and K. E. Giller, eds., *Beyond the Biomass: Composition and Functional Analysis of Soil Microbial Communities*, 39–48. Chichester, UK: Wiley.

Torsvik, T. and L. Ovreas. 2002. Microbial diversity and function in soil: From genes to ecosystems. *Current Opinion in Microbiology* 5:240–245.

TSBF (Tropical Soil Biology and Fertility Institute). 1999. *Managing the Soil Biota for Sustainable Agricultural Development in Africa: A Collaborative Initiative. A Proposal to the Rockefeller Foundation*. Nairobi, Kenya: TSBF.

TSBF (Tropical Soil Biology and Fertility Institute). 2000. *The Biology and Fertility of Tropical Soils. Tropical Soil Biology and Fertility Programme Report 1997–1998*. Nairobi, Kenya: TSBF.

Usher, M. B., P. Davis, J. Harris, and B. Longstaff. 1979. A profusion of species? Approaches towards understanding the dynamics of the populations of microarthropods in decomposer communities. In R. M. Anderson, B. D. Turner, and L. R. Taylor, eds., *Population Dynamics*, 359–384. Oxford: Oxford University Press.

Vandermeer, J., M. van Noordwijk, J. M. Anderson, C. Ong, and I. Perfecto. 1998. Global change and multi-species agroecosystems: Concepts and issues. *Agriculture, Ecosystems Environment* 67:1–22.

van Straalen, N. M. 1998. Evaluation of bioindicator systems derived from soil arthropod communities. *Applied Soil Ecology* 9:429–437.

Wall, D. H. and J. C. Moore. 1999. Interactions underground: Soil biodiversity, mutualism, and ecosystem processes. *BioScience* 49:109–117.

Willems, J. H. and K. G. A. Huijsmans. 1994. Vertical seed dispersal by earthworms: A quantitative approach. *Ecography* 17:124–130.

Wilson, E. O. 1985. The biological diversity crisis: A challenge to science. *BioScience* 35:700–706.

10 ✤ Diversity and Pest Management in Agroecosystems

Some Perspectives from Ecology

A. WILBY AND M. B. THOMAS

At a time when biodiversity is being lost at an unprecedented rate because of human activity, much research effort has been spent on assessing the importance of biodiversity for the functioning and stability of ecosystems and for the delivery of ecosystem services. Pest control has been identified on numerous occasions as a valuable ecosystem service delivered by biodiversity (Pimentel 1961; Horn 1988; Altieri 1991; Mooney et al. 1995a, 1995b; Naylor and Ehrlich 1997; Naeem et al. 1999; Schläpfer et al. 1999), one that is at risk from human activity (Naylor and Ehrlich 1997). There is much evidence that as agricultural production systems are intensified by increased use of external inputs to increase yield and change the landscape structure, they tend to lose biodiversity and become destabilized, with greater frequency and extent of pest outbreaks (Pimentel 1961; Andow 1991; Kruess and Tscharntke 1994; Swift et al. 1996; Knops et al. 1999). However, we know little about the ecological mechanisms that result in this destabilization or how important natural enemy diversity is in maintaining pest control. The aim of this chapter is to explore how insights from ecology can facilitate investigation of these mechanisms and contribute to development of a framework for examining and understanding the role of biodiversity in maintaining pest control and how this role is shaped by different management practices. Through this, and drawing on insights from some of our earlier work (see Wilby and Thomas 2002a, 2002b), we identify a number of hypotheses and recommendations for future research into the role and management of agrobiodiversity for sustainable pest management.

In order for us to predict consequences of human-induced species loss for pest control, we must improve our understanding of two linked issues of agroecosystem ecology. First, we need to identify and characterize the mechanisms whereby agroecosystem management affects the diversity and species composition of pest and natural enemy assemblages. Second, we need to understand the consequences of these effects for pest control (Wilby and Thomas 2002a). In approaching these questions we use existing ecological theories concerning community assembly and function of biodiversity. The latter, in particular, has made major advances recently despite some controversy. Initially, we examine this controversy and ask how the lessons learned should influence our study of the diversity–pest control relationship.

Biodiversity and Ecosystem Functioning

Characterization of the relationship between biodiversity and ecosystem functioning has been a major research goal of ecology for the past decade (see chapter 9). Numerous theoretical and empirical studies have been undertaken on many ecosystem properties, including biomass production (producers, consumers, and decomposers), nutrient uptake and retention, decomposition, soil pH, soil water and organic matter content, and community respiration (Schläpfer et al. 1999). Although the majority of studies have revealed a saturating positive relationship between diversity and ecosystem functioning (Schwartz et al. 2000), several issues have made the interpretation of diversity–ecosystem functioning relationships contentious. For instance, controversy surrounds the relative merit of observational and experimental evidence in characterizing the diversity–ecosystem functioning relationship. Experimental studies have been criticized because the species composition and species abundance distributions in experimental communities often do not adequately resemble natural communities and because the effects of species identity and species diversity sometimes are confounded (Huston 1997; Wardle 1999; Wardle et al. 2000). Furthermore, where diversity effects have been shown to occur, there has been controversy over whether they are caused by complementary functioning of different taxa or functional groups, positive interactions between species, or the sampling effect, that is, the increased probability of including a highly influential species as diversity increases (Huston 1997; Tilman et al. 1997). The first two mechanisms are viewed as true diversity effects because they are emergent properties arising from diversity, whereas the latter is a stochastic effect resulting from species composition, which can be viewed as a real diversity effect only if the experimental probabilities of inclusion of species match those seen in nature. Methods have been developed to separate true diversity effects from sampling or selection effects. For instance, replication of diversity levels with differing composition removes the confounding of diversity and identity, and the Loreau–Hector equation (Loreau and Hector 2001) enables separation of the sampling effect of diversity from the effect of complementarity and positive species interactions if the study has the right design features.

The problems with experimental approaches have led some researchers to promote observational studies. These too have been criticized because they do not control for variables correlated with diversity, and therefore they cannot be used reliably to determine the importance of biodiversity in ecosystem functioning (Naeem et al. 1999; Naeem 2000). However, it has been recognized that observational studies are necessary in order to identify diversity patterns that exist in nature, a vital step in the design of more realistic experiments (Wardle et al. 2000).

In addition to these problems with the interpretation of the results of biodiversity–ecosystem functioning studies, there are also difficulties in applying emerging theory to real ecosystems. Notably, we suggest that the context and extent of studies can strongly affect the shape of the observed relationship between diversity and functioning, and therefore it is unclear how results of experiments can be extrapolated across scales or to different environmental contexts (Fridley 2001).

Central to hypotheses concerning the relationship between biodiversity and ecosystem functioning is the concept of complementarity of taxonomic or functional biodiversity elements. The shape of the relationship is determined by the extent to which biodiversity elements (e.g., genotypes, species, feeding guilds) are functionally similar in terms of what they do and where, or how, they do it. If there is a significant amount of complementarity between elements with respect to a particular function, then the rate of the associated process will decrease with the loss of each element. Conversely, if there is a large amount of redundancy between elements, then initial biodiversity loss will tend not to affect the ecosystem process rate.

To date, theory concerning functional complementarity between biodiversity elements has focused on the ecological properties of the elements themselves. However, it is likely that the ecological context of the study will also have an important influence on functional complementarity. Complementarity arises if elements use exclusive portions of space or time or if they exclusively carry out particular functions. Consequently, the potential for complementarity is determined in part by the range of available spatial and temporal niches and the extent of the process under study. In fixing the bounds and subject of a study, an experimenter determines each of these attributes. As the extent along temporal, spatial, and process scales increases, so will the number of niches and the number of elements needed for maximal function. This is illustrated with a hypothetical example in figure 10.1a, which represents the role of elements of biodiversity (shown as lowercase letters with subscripts) in the transition between states (boxes) within a process. For a given position in time and space (i.e., a particular ecological context), the number of elements of biodiversity necessary to fulfill ecosystem function depends on the extent of the process in question. If the extent of the process of interest is simply the transition from state A to state B, then just one biodiversity element is

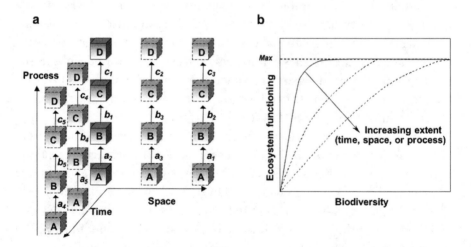

FIGURE 10.1. The influence of process and spatial and temporal extent on biodiversity–ecosystem functioning relationships. (a) Hypothetical example of a process in which the transition between individual states (boxes with capital letters) is mediated by activity of individual biodiversity elements (lowercase letters with subscripts). Increasing extent of process, time, or space from the baseline set of ecological conditions (represented by the solid, heavily shaded boxes), encompasses more niches and increases the number of elements of biodiversity needed for maximum function. (b) This changes the shape of the biodiversity–ecosystem functioning relationship.

needed (e.g., a_2). However, if we are interested in a process involving transition from state A to state D, then three elements are needed for full function (e.g., a_2, b_1, and c_1). As the temporal or spatial extent of a study increases, the number of elements necessary for maximal functioning increases further because different elements are likely to be the most efficient at different points in space and time. Therefore to move from state A to state B across space in this example entails the activity of three biodiversity elements (a_1, a_2, and a_3). For the full extent of the process A to D across all combinations of time and space, a maximum of 27 elements is needed for full function. Thus although a saturating function may be a common biodiversity–ecosystem functioning relationship (Schwartz et al. 2000), the minimum diversity at which maximal functioning is obtained will depend, in part, on the spatial–temporal–process extent of the study (figure 10.1b).

Of course, this conceptual model is an extreme simplification, and we do not know, for instance, how niches accumulate across space, time, and process scales. It is also clear that niche expression depends to some extent on interspecific interactions. However, there is evidence of spatial extent effects due to niche specialization along environmental gradients in space (Tilman et al. 1997; Fridley 2001; Wellnitz and Poff 2001) and temporal extent effects due to phenological differences between species (Hooper 1998). Also, the insurance hypothesis of species diversity (Naeem and Li 1997; Petchey et al. 1999; Yachi and Loreau 1999) suggests that niches in time and space will interact as environment changes, further increasing the likelihood of complementary function in time among species. In discussing the definition of ecosystem functioning, Ghilarov (2000) points out the importance of the process scale. For example, we expect very different relationships between ecosystem functioning and biodiversity if by *ecosystem functioning* we mean the total consumption of CO_2 by all plants rather than the production and consumption of all compounds used by all organisms. These scale and context effects make it difficult to extrapolate experimental results to other points on temporal, spatial, or process scales, and we must therefore be very careful in using experimental results to guide, for example, agricultural (or conservation) policy until scaling relationships are adequately understood.

What are the implications of these issues for the development of ecological frameworks for the study of the relationships between agricultural management, biodiversity, and pest control? Given the problems concerning the use of unrealistic communities in experimental studies and the problems of determining causation in observational studies, we suggest that future research should attempt to link the effects of changing biodiversity on pest control with the expected effects of agricultural management on biodiversity. In ecological terms, this means linked study of assembly and function of biodiversity. In taking this approach, we negotiate the problems associated with unnatural species assemblages and unrealistic patterns of species loss. As a consequence of scale and context dependence in biodiversity–ecosystem functioning relationships and in the absence of models to predict niche accumulation along scales, experimental tests should also focus on the full ecosystem process of interest and should be undertaken at scales appropriate to normal agricultural management. In the case of pest control by natural enemies, this is likely to be the field scale across a single or several cropping seasons.

It is clear that both diversity and species composition effects can have a strong influence on several ecosystem processes, and it is unlikely that pest control will be an exception. In our view, a profitable path for research may be to elucidate the biological properties of pests and natural enemies that affect the relative importance of diversity and composition as determinants of pest control functioning. In cases where species composition effects are strong, it is imperative that we understand the mechanisms whereby species diversity is lost and compare the ecological or biological traits that determine the probability of species loss with those that determine function. By following such a method we also avoid the problem of sampling effects because estimating the probability of including a particular species in an assemblage is an inherent goal of the study and is not assumed to be random. In the following sections we propose ecological frameworks for the study of agroecosystem management and its impact on arthropod community assembly and for natural enemy diversity and pest control functioning. We then discuss how these can profitably be linked in order to predict pest emergence during agricultural intensification and extensification.

Agroecosystem Management and Arthropod Community Assembly

Numerous studies report a range of effects of agroecosystem management on abundance, distribution, and diversity of arthropods. For example (and with particular reference to pest control), Letourneau and Goldstein (2001), comparing the effects of organic and conventional production on pest damage and arthropod community structure in tomatoes, found that whereas herbivore abundance did not differ between production systems (i.e., pest problems were not greater where pesticides were restricted), organic farms had greater species richness of all functional groups of arthropods and higher abundance of natural enemies than did conventional farms. These differences were associated with specific on-farm practices and landscape features, particularly fallow management, surrounding habitat, and transplant date of the crop. In another U.S. system, Menalled et al. (1999) examined whether agricultural landscape structure affected parasitism and parasitoid diversity. They found that at some sites, complex landscapes comprising cropland intermixed with mid- and late-successional noncrop habitats had higher rates of parasitism and parasitoid diversity than simple (primarily cropland) landscapes (see chapter 11). However, this pattern was not consistent across all sites, and so no clear effect of landscape complexity on parasitism could be determined. This ambiguous result contrasts with a study by Thies and Tscharntke (1999), who found that complex landscapes did promote parasitism and control of rape pollen beetle. Yet other studies, such as that by Weibull et al. (2003), indicate that although species richness generally increases with landscape heterogeneity at the farm scale, changes in diversity do not clearly lead to effects on natural pest control.

From the aforementioned studies it is clear that if we are to better understand the effect of agricultural management on arthropod diversity (and then the effects of any changes in diversity for pest control), we need a framework that describes the ecological mechanisms whereby species inhabit particular areas (Wilby and Thomas 2002a). Assembly rules have a long history in community ecology, where the objective has been to predict which species will occur in a particular habitat (Keddy 1992; Kelt et al. 1995; Belyea and Lancaster 1999).

In order for a species to be present at a particular site, it must first be capable of arriving at a site. Dispersal constraints govern which species from the regional species pool are included in the local species pool, that is, the species that are able to disperse to the site in question (figure 10.2). Of species in the local species pool, only the species that are able to overcome environmental constraints at the site occupy the ecological species pool. Finally, internal community dynamics, including intraspecific and interspecific processes govern composition of the actual species pool. In addition to governing the presence of species in communities, these processes also influence species abundance. This allows the framework to be used to describe the important processes driving species abundance distributions in addition to species richness.

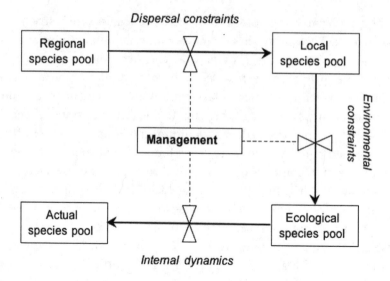

FIGURE 10.2. Community assembly processes highlighting sequential filters of dispersal, environment, and internal community dynamics in the transition from the regional species pool to the actual species pool. Agroecosystem management can modify each of these processes.

Although this assembly framework is applicable to any community, certain characteristics of agroecosystems may change the relative importance of the filters in governing species diversity compared with other ecosystems. Disturbance, for example, tends to increase the importance of nonequilibrium processes, such as dispersal and colonization, relative to equilibrium processes, such as interspecific interactions (Chapin et al. 1997). Because agroecosystems are regularly disturbed, for example by harvest, cultivation, or pesticides, the importance of dispersal constraints may be inflated because many species are forced to recolonize after these disturbances. The value of using such a framework to describe the community assembly process is that our understanding of the impact of agricultural management practices on diversity can be understood in terms of the way in which management modifies one or more of the filters comprising the assembly process.

Dispersal constraints, for example, are affected largely by structural changes in the landscape that occur as a result of land use change. Such effects are certainly qualitative (i.e., changes in the relative abundance of different species) and potentially also quantitative (i.e., changes in diversity or total abundance). Typically, a process of agricultural development results in an increase in cropped area relative to uncropped area. A region will consequently have greater source area for crop-associated insects, such that any given agricultural site is likely to be more easily colonized by specialist arthropods. However, the opposite effect may be true for more generalist arthropods if they need resources not available in the cropped area. Agricultural development also tends to increase average field size, so the centers of fields become distanced from colonization sources. The influence of adjoining vegetation on the type and number of natural enemies occupying a site has been stated on many occasions (Wainhouse and Coaker 1981; Altieri and Schmidt 1986a, 1986b; Thomas et al. 1991, 1992; Landis et al. 2000; Gurr et al. 2003), and there is evidence that the abundance and species richness of generalist insects are greater in diverse landscapes (Carmona and Landis 1999). Thus species diversity and composition are altered by land use changes through changes in the number and isolation of colonization sources. Importantly, we already know of biological traits of arthropods, such as diet breadth or body size, that will determine response of particular species to land use change. If we can also understand the functional significance of these traits for pest control, we are well on the way to predicting pest control response to land use change.

Agricultural management also has a large impact on the environmental conditions at sites, the second filter in our scheme. For example, intensification of rice production systems often increases the number of cropping cycles per year. (From here on we will tend to draw on examples from rice production because this is one of our main study systems. However, the implications of our ideas and interpretations clearly extend to other systems.) This move to multicropping allows insects specializing in the rice ecosystem to persist in a local site between crops (Loevinsohn 1994). This contrasts with more traditional farming practices that include prolonged fallow periods, in which a large proportion of the arthropod community is forced to recolonize a site after the fallow period. The marked effect of a fallow period is exhibited by rice-associated insect communities in Indonesia, where the insect community dynamics vary widely between the first and second crops (Settle et al. 1996) because the first crop follows a prolonged fallow period, whereas the second crop follows soon after the first.

Like fallow periods, insecticides can also change environmental constraints that force recolonization of a large proportion of the insect community, albeit on a smaller time scale. Differential recolonization rates of generalist predators and pest species after insecticide use are the principal cause of brown planthopper (*Nilaparvata lugens*) problems across vast areas of Southeast Asia (Heong 1991; Cohen et al. 1994; Settle et al. 1996).

Agricultural management also affects internal community dynamics, the third filter of our framework. Changes in food quality, caused by fertilization, for example, have been shown to increase abundance and damage by a number of pest groups, and these effects have been attributed to greater survival rates, faster growth, and greater fecundity of pest species (Ooi and Shepard 1994). Evidence suggests that these effects cascade through the food chain; de Kraker et al. (2000) showed that herbivores and natural enemies increased in abundance with nitrogen fertilization of rice fields.

In addition to variability in vegetation through time associated with the cropping cycle, the diversity of vegetation within a crop affects the diversity of arthropods within the crop. Generally, the change from diverse vegetation to virtual monoculture, which occurs with increased weed management, is associated with a decrease in diversity but not necessarily of abundance of arthropod species (Andow 1991; Tonhasca and Byrne 1994). Qualitatively, the presence of weeds in a rice crop increases the abundance of generalist herbivores relative to specialists. Afun et al. (1999) showed in West African rice that the abundance of generalist herbivores and predators is positively correlated with weed biomass, whereas specialist herbivore abundance was positively correlated with rice biomass.

This brief analysis of community assembly with respect to agroecosystems shows how agricultural management can influence each of the assembly filters governing species diversity of arthropods. The challenge ahead, if we are to predict impacts of management on pest control, is to further elucidate generalized trait characteristics that explain response to management practices and to investigate the functional significance of these or linked traits for pest control functioning.

Natural Enemy Diversity and Pest Control Function

Having discussed an ecological framework for the study of arthropod community assembly in agroecosystems and shown how certain traits may be instrumental in governing response to agricultural management, we turn to the second question addressed by this chapter: how species diversity and composition affect natural pest control functioning.

A few studies have identified some broad patterns predicting natural enemy responses to herbivores in natural and managed systems. For example, Dyer and Gentry (1999) provide evidence that specialist and gregarious lepidopteran larvae may be better controlled by parasitoids, whereas predators might better control smooth and cryptic larvae. Similarly, Hawkins et al. (1997) present data to suggest that predators and pathogens may cause greater mortality of external feeding herbivores, whereas certain endophytic herbivores suffer greater parasitoid-induced mortality. In general, however, our understanding of the relationship between biodiversity and pest control functioning remains poor, and the mechanisms through which natural enemies interact to determine the extent and stability of pest control are unclear. For example, in a recent study of the effect of landscape, habitat diversity, and management on species diversity in cereal systems, Weibull et al. (2003) revealed that there was no straightforward relationship between species richness of carabids, rove beetles, and spiders, at the farm level or in individual cereal fields, and biological control. They concluded that species richness in itself is not as important as a high diversity of different guilds of predators, such as ground and foliage predators, spring and summer breeders, and day and night active species, for the overall efficiency of biological control. That is, the key to effective natural control is in maximizing functional complementarity between the natural enemies of pest species. Unfortunately, our understanding of complementarity and the factors determining the emergent properties of multispecies predator assemblages is limited (Schmidt et al. 2003). Although there is evidence of significant niche partitioning across microhabitats and functional complementarity between spider species (Sunderland 1999), for example, few other studies have shown significant complementarity between natural enemies (Snyder and Wise 1999). Similarly, although examples of synergistic interactions between predators exist (e.g., foliar predators eliciting dropping responses in aphid prey, which increases their vulnerability to ground-foraging predators; Losey and Denno 1998), processes such as intraguild predation can severely disrupt biological control (Rosenheim et al. 1995; Snyder and Ives 2001; Finke and Denno 2004).

Given the types of complexity outlined here, we believe there is value in exploring in more detail the ecological factors determining the extent of complementarity between natural enemies and the nature of the relationship between biodiversity and ecosystem functioning. Consider a hypothetical saturating positive relationship between biodiversity and ecosystem functioning, similar to the relationship commonly found in empirical studies, where the slope of the relationship depends on the amount of complementarity between species (figure 10.3). If all species have an equal and nonoverlapping effect on a particular process (perfect complementarity), then a linear reduction in the rate of the process would occur as species richness decreased. At the other extreme, if there is no complementarity (species are redundant with respect to the function concerned [Walker 1992; Lawton and Brown 1993]), then remaining species would be able to compensate for lost species, and a sudden and complete loss of function would occur as the final functional species was lost. Aside from the shape of the relationship between mean function rate and diversity, it was shown earlier that species composition often plays a strong role in determining function rate. Species composition effects increase the variance about the relationship between species richness and process rate. In the extreme case of a single species having a much greater impact on process rate than other species (e.g., a keystone species), the observed relationship would take any

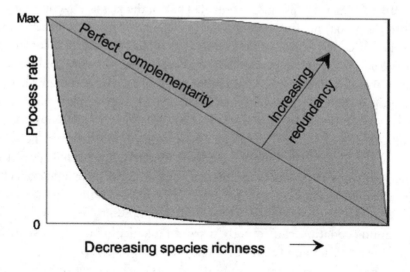

FIGURE 10.3. Hypothesized response of ecosystem function to a decrease in species richness (diversity). If species act in a redundant manner, then the response tends to a threshold behavior in which significant reductions in function occur only when the threshold is reached. If species act in a complementary manner, with each having nonoverlapping roles, then the response tends toward linear, with an incremental decrease in function with each species lost. If species identity is important, then the response can follow a broad range of trajectories (represented by the shaded region) depending on the order of species loss.

trajectory within a broad envelope of response, depending on the order of species loss (Sala et al. 1996). In a case of significant negative interactions between species, such as intraguild predation, functioning may even increase as species richness declines. In such cases of strong compositional effects, species richness would be a poor predictor of process rate, and further exploration of trait differences between species would be needed. With reference to pest control functioning, recent evidence suggests that life history characteristics of the species concerned sometimes may be used to predict the shape of the relationship between biodiversity and ecosystem functioning and whether strong species composition effects (high variance in response) are likely to occur (Wilby and Thomas 2002b).

For example, major differences in pest life history, such as whether they are enopterygote or exopterygote insects, can have a major influence on the number and diversity of natural enemies that attack them (Wilby and Thomas 2002b). Endopterygote insects undergo dramatic morphological and behavioral changes between egg, larva, pupa, and adult, and these life stages often play very different ecological roles, using different food and occupying different sites. In contrast, exopterygote insects undergo gradual changes between life stages, which often occupy similar sites and use similar food. As a consequence of these differences, we expect differences in the structure of links between these insect types and their natural enemies. The natural enemy complex interacting with endopterygote insects should be largely segregated into groups attacking particular life stages of the herbivore; most of the enemies feeding on adults would not be expected to also feed on larvae or pupae. There is evidence that this may be the case for certain endopterygote insects (e.g., Barrion et al. 1991; Mills 1994; Luna and Sánches 1999). However, because of the similarity in site occupancy and behavior, natural enemies feeding on exopterygote insects are expected to feed on both nymphs and adults. Because of the life cycle structure, the natural enemy complex feeding on endopterygote insects should exhibit more complementarity between species than that feeding on exopterygote insects. Simulation studies have shown that the higher complementarity between natural enemies of endopterygote insects may result in gradual loss of pest control function with decreasing natural enemy species diversity, whereas control of exopterygotes should be more resistant, with a sudden loss of pest control function after extreme loss of natural enemy diversity. These predictions appear to be consistent with pest emergence patterns during intensification of Asian rice production systems (Wilby and Thomas 2002b).

Other aspects of herbivore life history are likely to influence the richness or characteristics of the natural enemy complex, with implications for pest control function. For example, morphological or behavioral concealment tends to promote tight linkage of population dynamics between the herbivore and natural enemy. Thus, with concealed herbivores, specialist parasitoids are more important than generalist predators (Memmott et al. 2000), although concealed herbivores may be attacked by more species of generalist parasitoid than specialist parasitoid (Hawkins 1990; Hawkins and Gross 1992). In essence, tight links between more specialist natural enemies and pests inflate the importance of species composition effects and increase the uncertainty about response to random species loss.

The life span of herbivorous insects is also likely to have important consequences for the richness and characteristics of the natural enemy complex. The shorter the life span of a herbivorous species, the smaller the number of natural enemy species that overlap phenologically with the herbivore. This effect has been shown to affect the richness of parasitoid species attacking herbivores (Cornell and Hawkins 1993). Consequently, control of a pest with a shorter life span should depend on a group of natural enemies with less redundancy, either across the life cycle in the case of exopterygote insects or within life cycle stages in the case of endopterygote insects. In either case, control of short-lived insects should decrease earlier, on average, in response to natural enemy species loss.

What these examples show is that basic life history characteristics of the organisms involved can lead to different predictions about the relative importance of diversity and species composition in determining pest control functioning. This enables us to move beyond arguments of whether species composition or diversity is the most important attribute of the natural enemy community to a position where we can predict which types of pest are likely to warrant particular natural enemy species for control and which are likely to be better controlled by a diverse assemblage of natural enemies. To understand how agricultural management is likely to affect pest occurrence, this information must be linked with our previous discussion of arthropod community assembly in agroecosystems.

Pest Occurrence and Agroecosystem Management

How might linking of the two ecological frameworks help us predict natural pest control functioning in response to different types of agroecosystem management? For unconcealed exopterygote pests such as planthoppers (Delphacidae) and leafhoppers (Cicadellidae), we might predict control by natural enemies to be resistant to natural enemy species loss. However, certain management techniques such as insecticide use have a large impact on natural enemy diversity and may result in a nonfunctional natural enemy assemblage. Such an effect is exemplified by the well-studied case of the brown planthopper (*Nilaparvata lugens*) in Asian irrigated rice ecosystems. *N. lugens* usually is well controlled by natural enemies but often becomes a pest because of resurgence after insecticide use (Kenmore et al. 1984; Heong 1991). The mechanism for this phenomenon has been well elucidated, and it seems that the nearly complete removal of a guild of generalist predators, which are maintained by detritivorous insects at times when *N. lugens* is scarce, is the cause of the emergence of *N. lugens* as a pest (Settle et al. 1996). *N. lugens* is able to escape control partly because it survives insecticide treatment to some extent, because it is highly dispersive and can recolonize rice fields quickly, and because it has a high population growth rate (Heinrichs and Mochida 1984). A combination of being more successful in dealing with dispersal, environmental, and biological constraints than its natural enemies allows *N. lugens* to escape control. Thus effective control can be attained largely by limiting the use of insecticides and by ensuring alternative food supplies for predators at times when pest species are scarce.

Control of endopteran herbivores is predicted to be sensitive to loss of natural enemy species diversity. The emergence of endopteran pests in the early stages of intensification of rice may be avoided if management ensures the maintenance of natural enemy diversity. As indicated previously, there is a large literature on the effects of uncultivated land on natural enemy abundance and diversity. In terms of our framework, uncultivated areas close to crops dampen dispersal constraints in the natural enemy species that spend part of their life cycle outside the crop habitat. At the same time, close proximity of crop habitat increases the chance of colonization of the more specialist insects. Of course, this means herbivorous species, too, but we work on the assumption that ordinarily pest emergence entails disruption of natural enemy control, and therefore colonization of potential pest species is unlikely to be a problem unless colonization of their natural enemies is impeded.

For concealed herbivores, more detailed use of our assembly framework is needed in order to predict the impact of management on particular species or guilds of natural enemies. Concealed herbivores tend to be attacked by parasitoids more than by predators, and this has a number of implications for the persistence of a functional natural enemy assemblage. Parasitoids tend to be more specialist than predators, and therefore they are more likely to depend on a small number of host species. With specialists there is inevitably a lag between the increases in host abundance, as would occur during emergence of a pest, for example, and the response of the natural enemy. This increases the likelihood that pests will reach damaging densities before the natural enemy limits pest abundance. Therefore it is an important management goal to maintain stable parasitoid–host dynamics. This may entail provision of supplementary food for the parasitoids, such as nectar sources provided by weeds in or near the crop (relaxation of biological constraints), changes in fallow management so as to facilitate survival of parasitoids over fallow periods (relaxation of environmental constraints), and landscape management (small field sizes, asynchrony of cropping cycles, and pesticide use) to ensure local colonization sources (relaxation of dispersal constraints). The latter is likely to be particularly important for specialist insects that are known to be more sensitive to habitat fragmentation by virtue of their trophic position and body size (Tscharntke and Brandl 2004).

An alternative or complementary practice might be to ensure the presence of an effective complex of generalist natural enemies or to use plant varieties conferring some resistance to the target pest (Thomas 1999), which may slow or delay population buildup (basically by providing some additional density-independent mortality), allowing the key parasitoid species to establish at an earlier point in the pest population cycle. Such an approach has been discussed with respect to control of African rice gall midge (*Orseolia oryzivora*), for example, where although two key parasitoid species are known to cause substantial declines in gall midge numbers, their regulatory effect often occurs too late in the season to prevent economic damage (F. Nwilene, pers. comm., 2002). Interestingly, this supplementary role of generalist natural enemies increases the importance of natural enemy diversity in the system overall, even though pest population regulation may rely on just a few specialists. This confirms the importance of framing ecological insights in an appropriate context for the system in which they are being applied (in this case an economic pest control context and not necessarily a population dynamic context).

Conclusion

This application of ecological theories of assembly and function of biodiversity to pest–natural enemy interactions has led to a range of hypotheses about pest control response to agroecosystem management. An examination of simplified food webs leads to the predictions about how pests with certain traits will respond to loss of natural enemy diversity. Our analysis suggests that pests controlled largely by generalists would be expected to show threshold or gradual loss of control as diversity of natural enemies decreases. For these species, management techniques designed to maintain natural enemy diversity through modification of biological, environmental, and dispersal constraints may be generally adequate to prevent pest emergence. In contrast, the herbivores controlled by specialist natural enemies may exhibit unpredictable responses to decreases in natural enemy diversity. In these cases, an understanding of dispersal and colonization processes of pests and natural enemies is needed to predict response of pest control to loss of natural enemy diversity. Already, biological generalizations about specialist natural enemies allow us to predict the effect of agricultural management on dispersal, environmental, and biotic constraints that determine their abundance.

In our view, only by linking the ecological processes of assembly and function of biodiversity can we effectively answer questions about the likely impact of human activity on ecosystem functioning via biodiversity effects. We have shown how certain traits of species may link their response to human-induced ecosystem changes and their functional characteristics. Although we have limited our discussion to pest control, we believe that similar approaches may allow more accurate prediction of management impact on other ecosystem processes. Examination of the biological characteristics and detail of an ecosystem process and the organisms involved may allow us to move away from arguments about whether diversity effects or composition effects predominate to a prediction of which circumstances lead to a predominance of composition or diversity effects.

Acknowledgments

This chapter is an output from a research project funded by the United Kingdom Department for International Development (DFID) for the benefit of developing countries (R7570 Crop Protection Research Programme). The views expressed are not necessarily those of the DFID.

References

Afun, J. V. K., D. E. Johnson, and A. Russell-Smith. 1999. Weeds and natural enemy regulation of insect pests in upland rice: A case study from West Africa. *Bulletin of Entomological Research* 89:391–402.

Altieri, M. A. 1991. Increasing biodiversity to improve insect pest management in agroecosystems. In D. Hawksworth, ed., *The Biodiversity of Microorganisms and Invertebrates: Its Role in Sustainable Agriculture.* Wallingford, UK: CAB International.

Altieri, M. A. and L. L. Schmidt. 1986a. Cover crops affect insect and spider populations in apple orchards. *California Agriculture* 40:15–17.

Altieri, M. A. and L. L. Schmidt. 1986b. The dynamics of colonizing arthropod communities at the interface of abandoned, organic and commercial apple orchards and adjacent woodland habitats. *Agriculture, Ecosystems and Environment* 16:29–43.

Andow, D. A. 1991. Vegetational diversity and arthropod population response. *Annual Review of Entomology* 36:561–586.

Barrion, A. T., J. A. Litsinger, E. B. Medina, R. M. Aguda, J. P. Bandong, P. C. Pantua Jr., V. D. Viajante, C. G. de la Cruz, C. R. Vega, J. S. Soriano Jr., E. E. Camañg, R. C. Saxena, E. H. Tyron, and B. M. Shepard. 1991. The rice *Cnaphalocricis* and *Marasmia* (Lepidoptera: Pyralidae) leaffolder complexes in the Philippines: Taxonomy, bionomics and control. *Philippines Entomologist* 8:987–1074.

Belyea, L. R. and J. Lancaster. 1999. Assembly rules within a contingent ecology. *Oikos* 86:402–416.

Carmona, D. M. and D. A. Landis. 1999. Influence of refuge habitats and cover crops on seasonal activity-density of ground beetles (Coleoptera: Carabidae) in field crops. *Environmental Entomology* 28:1145–1153.

Chapin, F. S., B. H. Walker, R. J. Hobbs, D. U. Hooper, J. H. Lawton, O. E. Sala, and D. Tilman. 1997. Biotic control over the functioning of ecosystems. *Science* 277:500–504.

Cohen, J. E., K. Schoenly, K. L. Heong, H. Justo, G. Arida, A. T. Barrion, and J. A. Litsinger. 1994. A food-web approach to evaluating the effect of insecticide spraying on insect pest population-dynamics in a Philippine irrigated rice ecosystem. *Journal of Applied Ecology* 31:747–763.

Cornell, H. V. and B. A. Hawkins. 1993. Accumulation of native parasitoid species on introduced herbivores: A comparison of hosts as natives and hosts as invaders. *American Naturalist* 141:847–865.

de Kraker, J., R. Rabbinge, A. van Huis, J. C. van Lenteren, and K. L. Heong. 2000. Impact of nitrogenous-fertilization on the population dynamics and natural control of rice leaffolders (Lep.: Pyralidae). *International Journal of Pest Management* 46:219–224.

Dyer, L. A. and G. Gentry. 1999. Predicting natural-enemy responses to herbivores in natural and managed systems. *Ecological Applications* 9:402–408.

Finke, D. L. and R. F. Denno. 2004. Predator diversity dampens trophic cascades. *Nature* 429:407–410.

Fridley, J. D. 2001. The influence of species diversity on ecosystem productivity: How, where, and why? *Oikos* 93:514–526.

Ghilarov, A. M. 2000. Ecosystem functioning and intrinsic value of biodiversity. *Oikos* 90:408–412.

Gurr, G. M., S. D. Wratten, and J. M. Luna. 2003. Multi-function agricultural biodiversity: Pest management and other benefits. *Basic and Applied Ecology* 4:107–116.

Hawkins, B. A. 1990. Global patterns of parasitoid assemblage size. *Journal of Animal Ecology* 59:57–72.

Hawkins, B. A., H. V. Cornell, and M. E. Hochberg. 1997. Predators, parasitoids, and pathogens as mortality agents in phytophagous insect populations. *Ecology* 78:2145–2152.

Hawkins, B. A. and P. Gross. 1992. Species richness and population limitation in insect parasitoid–host systems. *American Naturalist* 139:417–423.

Heinrichs, E. A. and O. Mochida. 1984. From secondary to major pest status: The case of insecticide-induced rice brown planthopper, *Nilaparvata lugens*, resurgence. *Protection Ecology* 1:201–218.

Heong, K. L. 1991. Management of the brown planthopper in the Tropics. In *Migration and Dispersal of Agricultural Insects*, 269–279, Tsukuba, Japan, September 25–28, 1991.

Hooper, D. U. 1998. The role of complementarity and competition in ecosystem responses to variation in plant density. *Ecology* 79:704–719.

Horn, D. J. 1988. *Ecological Approach to Pest Management*. New York: Guilford.

Huston, M. A. 1997. Hidden treatments in ecological experiments: Re-evaluating the ecosystem function of biodiversity. *Oecologia* 110:449–460.

Keddy, P. A. 1992. Assembly and response rules: Two goals for predictive community ecology. *Journal of Vegetation Science* 3:157–164.

Kelt, D. A., M. L. Taper, and P. L. Meserve. 1995. Assessing the impact of competition on community assembly: A case-study using small mammals. *Ecology* 76: 1283–1296.

Kenmore, P. E., F. O. Cariño, C. A. Perez, V. A. Dyck, and A. P. Gutierrez. 1984. Population regulation of the rice brown planthopper (*Nilaparvata lugens* Stål) within rice fields in the Philippines. *Journal of Plant Protection in the Tropics* 1:19–37.

Knops, J. M. H., D. Tilman, N. M. Haddad, S. Naeem, C. E. Mitchell, J. Haarstad, M. E. Ritchie, K. M. Howe, P. B. Reich, E. Siemann, and J. Groth. 1999. Effects of plant species richness on invasion dynamics, disease outbreaks, insect abundances and diversity. *Ecology Letters* 2:286–293.

Kruess, A. and T. Tscharntke. 1994. Habitat fragmentation, species loss, and biological control. *Science* 264:1581–1584.

Landis, D. A., S. D. Wratten, and G. M. Gurr. 2000. Habitat management to conserve natural enemies of arthropod pests in agriculture. *Annual Review of Entomology* 45:175–201.

Lawton, J. H. and V. K. Brown. 1993. Redundancy in ecosystems. In E.-D. Schulze and H. A. Mooney, eds., *Biodiversity and Ecosystem Function,* 255–270. Berlin: Springer-Verlag.

Letourneau, D. K. and B. Goldstein. 2001. Pest damage and arthropod community structure in organic vs. conventional tomato production in California. *Journal of Applied Ecology* 38:557–570.

Loevinsohn, M. E. 1994. Rice pests and agricultural environments. In E. A. Heinrichs, ed., *Biology and Management of Rice Insects,* 487–513. New Delhi: Wiley Eastern.

Loreau, M. and A. Hector. 2001. Partitioning selection and complementarity in biodiversity experiments. *Nature* 412:72–76.

Losey, J. E. and R. F. Denno. 1998. Positive predator–predator interactions: Enhanced predation rates and synergistic suppression of aphid populations. *Ecology* 79:2143–2152.

Luna, M. and N. Sánches. 1999. Parasitoid assemblages of soybean defoliator Lepidoptera in north-western Buenos Aires province, Argentina. *Agricultural and Forest Entomology* 1:255–260.

Memmott, J., N. D. Martinez, and J. E. Cohen. 2000. Predators, parasites and pathogens: Species richness, trophic generality and body sizes in a natural food web. *Journal of Applied Ecology* 69:1–15.

Menalled, F. D., P. C. Marino, S. H. Gage, and D. A. Landis. 1999. Does agricultural landscape structure affect parasitism and parasitoid diversity? *Ecological Applications* 9:634–641.

Mills, N. J. 1994. Parasitoid guilds: Defining the structure of the parasitoid communities of endopterygote insect hosts. *Environmental Entomology* 23:1066–1083.

Mooney, H., J. Lubchenco, R. Dirzo, and O. Sala. 1995a. Biodiversity and ecosystem functioning: Basic principles. In V. Heywood, ed., *Global Biodiversity Assessment,* 279–323. Cambridge, UK: Cambridge University Press.

Mooney, H., J. Lubchenco, R. Dirzo, and O. Sala. 1995b. Biodiversity and ecosystem functioning: Ecosystem analyses. In V. Heywood, ed., *Global Biodiversity Assessment,* 347–452. Cambridge, UK: Cambridge University Press.

Naeem, S. 2000. Reply to Wardle et al. *Bulletin of the Ecological Society of America* 81:241–246.

Naeem, S., F. S. Chapin III, R. Costanza, P. R. Ehrlich, F. B. Golley, D. U. Hooper, J. H. Lawton, R. V. O'Neill, H. A. Mooney, O. E. Sala, A. J. Symstad, and D. Tilman. 1999. Biodiversity and ecosystem functioning: Maintaining natural life support processes. Ecological Society of America. *Issues in Ecology* 4:1–12.

Naeem, S. and S. B. Li. 1997. Biodiversity enhances ecosystem reliability. *Nature* 390:507–509.

Naylor, R. L. and P. R. Ehrlich. 1997. Natural pest control services and agriculture. In G. C. Daily, ed., *Nature's Services*, 151–174. Washington, DC: Island Press.

Ooi, P. A. C. and B. M. Shepard. 1994. Predators and parasitoids of rice insect pests. In E. A. Heinrichs, ed., *Biology and Management of Rice Insects*, 586–612. New Delhi: Wiley Eastern.

Petchey, O. L., P. T. McPhearson, T. M. Casey, and P. J. Morin. 1999. Environmental warming alters food-web structure and ecosystem function. *Nature* 402:69–72.

Pimentel, D. 1961. Species diversity and insect populations outbreaks. *Annals of the Entomological Society of America* 54:76–86.

Rosenheim, J. A., H. K. Kaya, L. E. Ehler, J. J. Marois, and B. A. Jaffee. 1995. Intra-guild predation among biological-control agents: Theory and evidence. *Biological Control* 5:303–335.

Sala, O. E., W. K. Lauenroth, S. J. McNaughton, G. Rusch, and X. Zhang. 1996. Biodiversity and ecosystem functioning in grasslands. In H. A. Mooney, J. H. Cushman, E. Medina, O. E. Sala, and E. D. Schulze, eds., *Functional Roles of Biodiversity: A Global Perspective*, 129–149. New York: Wiley.

Schläpfer, F., B. Schmid, and I. Seidl. 1999. Expert estimates about effects of biodiversity on ecosystem processes and services. *Oikos* 84:346–352.

Schmidt, M. H., A. Lauer, T. Purtauf, C. Thies, M. Schaefer, and T. Tscharntke. 2003. Relative importance of predators and parasitoids for cereal aphid control. *Proceedings of the Royal Society of London Series B-Biological Sciences* 270:1905–1909.

Schwartz, M. W., C. A. Bringham, J. D. Hoeksema, K. G. Lyons, M. H. Mills, and P. J. van Mantgem. 2000. Linking biodiversity to ecosystem function: Implications for conservation ecology. *Oecologia* 122:297–305.

Settle, W. H., H. Ariawan, E. T. Astruti, W. Cahyana, A. L. Hakim, D. Hindayana, A. S. Lestari, and P. Sartanto. 1996. Managing tropical pests through conservation of generalist natural enemies and alternative prey. *Ecology* 77:1975–1988.

Snyder, W. E. and A. R. Ives. 2001. Generalist predators disrupt biological control by a specialist parasitoid. *Ecology* 82:705–716.

Snyder, W. E. and D. H. Wise. 1999. Predator interference and the establishment of generalist predator populations for biocontrol. *Biological Control* 15:283–292.

Sunderland, K. 1999. Mechanisms underlying the effects of spiders on pest populations. *Journal of Arachnology* 27:308–316.

Swift, M. J., J. Vandermeer, P. S. Ramakrishnan, J. M. Anderson, C. K. Ong, and B. A. Hawkins. 1996. Biodiversity and agroecosystem function. In H. A. Mooney, J. H. Cushman, E. Medina, O. E. Sala, and E. D. Schulze, eds., *Functional Roles of Biodiversity: A Global Perspective*, 261–297. New York: Wiley.

Thies, C. and T. Tscharntke. 1999. Landscape structure and biological control in agroecosystems. *Science* 285:893–895.

Thomas, M. B. 1999. Ecological approaches and development of "truly integrated" pest management. *Proceedings of the National Academy of Sciences of the USA* 96:5944–5951.

Thomas, M. B., S. D. Wratten, and N. W. Sotherton. 1991. Creation of island habitats in farmland to manipulate populations of beneficial arthropods: Predator densities and emigration. *Journal of Applied Ecology* 28:906–917.

Thomas, M. B., S. D. Wratten, and N. W. Sotherton. 1992. Creation of island habitats in farmland to manipulate populations of beneficial arthropods: Predator densities and species composition. *Journal of Applied Ecology* 29:524–531.

Tilman, D., C. L. Lehman, and K. T. Thomson. 1997. Plant diversity and ecosystem productivity: Theoretical considerations. *Proceedings of the National Academy of Sciences of the USA* 94:1857–1861.

Tonhasca, A. and D. N. Byrne. 1994. The effects of crop diversification on herbivorous insects: A meta-analysis approach. *Ecological Entomology* 19:239–244.

Tscharntke, T. and R. Brandl. 2004. Plant–insect interactions in fragmented landscapes. *Annual Review of Entomology* 49:405–430.

Wainhouse, D. and T. H. Coaker. 1981. The distribution of carrot fly (*Psila rosae*) in relation to the fauna of field boundaries. In J. M. Thresh, ed., *Pests, Pathogens and Vegetation: The Role of Weeds and Wild Plants in the Ecology of Crop Pests and Diseases*, 263–272. London: Pitman.

Walker, B. H. 1992. Biodiversity and ecological redundancy. *Conservation Biology* 6:18–23.

Wardle, D. A. 1999. Is "sampling effect" a problem for experiments investigating biodiversity–ecosystem function relationships? *Oikos* 87:403–407.

Wardle, D. A., M. A. Huston, J. P. Grime, F. Berendse, E. Garnier, W. K. Lauenroth, H. Setälä, and S. D. Wilson. 2000. Biodiversity and ecosystem function: An issue in ecology. *Bulletin of the Ecological Society of America* 81:235–239.

Weibull, A. C., O. Östman, and A. Granqvist. 2003. Species richness in agroecosystems: The effect of landscape, habitat and farm management. *Biodiversity and Conservation* 12:1335–1355.

Wellnitz, T. and N. Poff. 2001. Functional redundancy in heterogeneous environments: Implications for conservation. *Ecology Letters* 4:177–179.

Wilby, A. and M. B. Thomas. 2002a. Are the ecological concepts of assembly and function of biodiversity useful frameworks for understanding natural pest control? *Agricultural and Forest Entomology* 4:237–243.

Wilby, A. and M. B. Thomas. 2002b. Natural enemy diversity and natural pest control: Patterns of pest emergence with agricultural intensification. *Ecology Letters* 5:353–360.

Yachi, S. and M. Loreau. 1999. Biodiversity and ecosystem productivity in a fluctuating environment: The insurance hypothesis. *Proceedings of the National Academy of Sciences of the* USA 96:1463–1468.

11 ✤ Managing Crop Disease in Traditional Agroecosystems

Benefits and Hazards of Genetic Diversity

D. I. JARVIS, A. H. D. BROWN, V. IMBRUCE, J. OCHOA, M. SADIKI, E. KARAMURA, P. TRUTMANN, AND M. R. FINCKH

For millennia, farmers have contended with pest and disease outbreaks that threaten their crops and livelihoods. Their legacy of domesticated varieties or landraces is notably diverse genetically, both between and within populations. The question that naturally arises is whether the maintenance of diversity on farm, particularly for genes that affect host–pathogen interactions, has given farmers an effective strategy against disease, or, conversely, whether it has provided the opportunity for the evolution of adverse diversity in pathogen populations. In other words, is crop genetic diversity a benefit in reducing disease in time, or is it a hazard in giving scope for the emergence of pathogen super-races?

Two conflicting hypotheses summarize the essential issue to be resolved for the best management of genetic diversity on farm. They can be starkly spelled out in terms of whether a farmer relies on a diverse planting, achieved as a mixture of genotypes differing in resistance structure,[1] or plants a monoculture of a crop variety[2] that is protected by one form of resistance.

Under the *diversity benefit hypothesis,* a diverse genetic basis of resistance is beneficial for the farmer because it allows a more stable management of disease pressure than a monoculture allows. This is because theory and experience indicate how readily the resistance of a monoculture can break down and the whole population succumb. The genetically diverse field will need the much less likely event of different types of resistance to break down in the same place for comparable disease damage.

The alternative or *diversity hazard hypothesis* argues that a monoculture of a variety that carries multigenic, or indeed a combined form of several genetically different major gene resistances, is the better, more stable option because pathogen populations are kept very low. The joint double or multiple mutation needed to overcome all resistances should be extremely rare. In stark contrast, this hypothesis predicts that mixed host populations that have genotypes differing in resistance to different sets of pathotypes will allow diverse pathogen populations to build up and the potential of new super-race pathotypes to arise by single-step mutation, or recombination. The theory behind these arguments is subject to much discussion (Mundt 1990, 1991; Kolmer et al. 1991), and it is difficult to determine experimentally the threat from super-races.

In this chapter we discuss the evidence that bears on whether local crop cultivar diversity reduces genetic susceptibility to pathogens. The ultimate aim of such research is to discover when and how the use of local crop varieties and genotypes has a beneficial effect for farmers on pest and disease incidence. We discuss what type of research is necessary to decide between the two hypotheses and determine the optimal use of diversity to manage pathogen pressures. Finally, we note that the farmer is at the center of the host–pathogen–environment triangle and that local crop cultivars (landraces) managed in long extant, low-input agricultural systems are reservoirs of genetic variation resulting from a dynamic interaction between host, pest, environment, and farmer.

Genetic Vulnerability and Genetic Uniformity

As early as the 1930s, agricultural scientists recognized the potentially damaging consequences of planting large areas to single, uniform crop cultivars (Marshall 1977). This situation is known as increased genetic vulnerability because it increases the risk of disease epidemics.[3] The expected reduction in vulnerability caused by genetically heterogeneous plantings is in line with the diversity benefit hypothesis. On the other hand, diseases severely affect production, especially in developing countries. Much of the 30% of the world's annual harvest lost to disease and pests occurs in developing countries (Oerke et al. 1994). Superficially, the diversity hazard hypothesis predicts that traditional varieties are prone to such losses and explain the severity of disease in developing countries. However, inappropriate or limited strategies of resistance gene deployment that ignore environmental and agronomic complexities in traditional systems may lie at the root of these generalizations.

The Irish potato famine in the wake of the introduction of the late blight pathogen (*Phytophthora infestans*) in the 1840s is a dramatic example of genetic vulnerability accompanying genetic uniformity and leading to the devastating loss of the crop (Schumann 1991). Another is the 1979–1980 rust attack on Cuba's sugarcane (caused by *Puccinia melanocephala*), where one cultivar covered 40% of the sugarcane area, which resulted in US$500 million in losses (FAO 1998:32). The southern corn leaf blight (caused by *Cochliobolus carbonum*) destroyed $1 billion worth of corn in the United States in the 1970s (Ullstrup 1972). Susceptibility of the five major commercial cultivars of banana to the fungal disease black sigatoka (caused by *Mycosphaerella fijiensis*) resulted in Central American countries losing about 47% of their banana yield (FAO 1998). Although measure to control the disease are available, over the course of eight years they cost Central America, Colombia, and Mexico US$350 million and caused serious human health problems through exposure to pesticides. Cassava mosaic virus causes yield losses of up to 40% in some parts of Africa, where many depend on cassava as an important nutritional resource (Otim-Nape and Thresh 1998). Most rubber clones grown throughout the world derive from crosses based on very limited genetic variation (Oldfield 1989). South American leaf blight, caused by *Microcyclus ulei,* has a history of devastating rubber plantations in South America and remains the main obstacle to the development of rubber there because of the high variability of leaf blight (Rivano 1997). The real threat of rubber tree leaf blight is in Asia, where 90% of rubber is produced. At present this region is free of the disease, but clones are considered very susceptible (Compagnon 1998; Kennedy and Lucks 1999).

Much damage results from the evolution of new races of pests and pathogens that overcome resistance genes currently deployed over large areas. When new cultivars are produced that carry new resistance genes, these resistances may protect for only a few cropping seasons as new pathotypes emerge. However, gene deployment can also increase pathogen complexity. For instance, in a scenario more suggestive of the diversity hazard hypothesis, some landraces of quinoa in Ecuador were resistant to low-virulence isolates of downy mildew that were common before gene deployment. However, with the increased planting of resistant landraces, pathogen strains developed that were virulent to all the hypersensitive resistance deployed (Ochoa et al. 1999; box 11.1). The real epidemiological

Box 11.1 Deployment of New Resistant Varieties and Shifts in Pathogenicity in Ecuador

Evolution in the wheat–yellow rust, the quinoa–downy mildew, the bean rust, and anthracnose pathogens has been followed in some detail in Ecuador.

The population structure of the yellow rust pathogen (*Puccinia striiformis* f. sp. *tritici*) was very simple in the early 1970s, including a fraction avirulent on all the differential cultivars used for race characterization (INIAP 1974). In a 1991 survey, Ochoa et al. (1998) identified virulence to the yellow rust resistance genes (Yr1, Yr2, Yr3, Yr6, Yr7, YrA). Since then, virulence to Yr9 and other genes has been identified in the Ecuadorian population. Currently all major resistance genes available to breeders have been overcome by pathogens.

Quinoa breeding based on line selection began in the early 1980s and continued until the early 1990s. Local and introduced germplasm from Peru and Bolivia was tested at several locations, and four cultivars were released: Cochasqui, Imbaya, Tunkahuan, and Ingapirca. Resistance to downy mildew caused by *Peronospora farinosa* f. sp. *chenopodii* was a major selection criterion in this program. In a study of the population structure of *P. farinosa* in 1994–1995, four groups of pathotypes were found, apparently differing in virulence by successive single steps. The avirulent isolate (V1-group) was found only once in a local landrace in Otavalo. Such avirulent isolates probably were more common in the former subsistence quinoa system before breeding began. Cultivar Imbaya apparently carries resistance factor R1, cultivar Ingapirca carries resistance factor R2 (Peru and Bolivia origin), and the most recently released cultivar Tunkahuan lacks any resistance factor. Resistance factor R1 is common in landraces, and R3 is more common in advanced lines. Germplasm screening for resistance to pathogen isolates in the V4 group has so far been unsuccessful (Ochoa et al. 1999).

A rapid evolutionary process has taken place in the quinoa–downy mildew pathosystem in this short period of breeding improvement. Isolates of low virulence that appear to be less aggressive and less complex are postulated to have been common in traditional agroecosystems. In contrast, virulent isolates are more common in modern quinoa, possibly because of higher levels of aggressiveness. In the quinoa–downy mildew pathogen, adaptation appears as quickly and efficiently as in other biotrophic specialists.

Bean rust (*Uromyces appendiculatus*) and anthracnose (*Colletotrichum lindemuthianum*) are serious constraints on bush bean cultivation in Ecuador. Pathogen structure and host resistance have been studied for both these diseases. From 21 isolates selected for their variability, 17 different rust pathotypes were identified. Fourteen out of 20 differentials were susceptible. However, local cultivars and landraces were more useful in discriminating beween pathotypes,

Box 11.1 continues to next page

Box 11.1 continued

which indicates coevolution of host plant and pathogen. Most of the modern commercial cultivars were found to be susceptible to rust (Ochoa et al. 2002).

Akin to the results for rust, the formal differentials were less efficient in discriminating between anthracnose pathotypes. Six races were found using the differential set. However, 12 different patterns were discriminated when local cultivars and landraces were included. In common with bean rust, most of the commercial cultivars were found susceptible (Falconi et al. 2003).

Although resistance to rust and anthracnose is an important bean breeding objective, grain quality is the predominant objective at time of release. The most widespread variety (Paragachi) is very susceptible to both rust and anthracnose. The rust-resistant cultivar Gema is not adapted to the low valleys where rust is a constraint, but it is grown in areas prone to anthracnose, to which it is susceptible. This apparent contradiction occurs because bean breeding and selection for resistance have been done outside the country, and only adaptation and yield potential were tested before cultivar release. Instead, breeding programs are needed that develop multiple resistances in varieties suitable to local conditions.

consequences of this interference are difficult to establish because the extent of cultivation of resistant landraces is unknown.

Adaptation of Landraces to the Pathogen Environment

Different types of resistance appear to be widespread in local crop landraces (Teshome et al. 2001). This is attributed to the long-term coevolution between pest and host species in primary and secondary centers of diversity. For many crop species, it is likely that centers of crop genetic diversity and those of pest or pathogen diversity coincide (Leppik 1970; Allen et al. 1999).

As humans have moved around the globe with their crops, so have resistant germplasm and virulent races of pathogens. Resistance genes evolve in response to new pathogens, but there may also be remnants of resistance already present in a region if the crops had historically been in contact with the disease. This phenomenon has resulted in the occurrence of resistance outside the primary center of diversity, an example being resistance to chocolate spot (caused by *Botrytis fabae*) in faba bean (*Vicia faba*) in the Andes (Hanounik and Robertson 1987). This crop first reached the Americas several hundred years ago; its center of diversity is the Fertile Crescent.

Marked geographic patterns of host resistance in relation to pest and disease presence can suggest the operation of coevolution. In a screening of world barley collections, Qualset (1975) found resistance to barley yellow dwarf virus (BYDV) to be highly localized in Ethiopia, a center of diversity. Qualset concluded that the mutation for BYDV resistance happened in Ethiopia, and the presence of the disease is reason to believe that natural selection favored resistant barleys. Subrahmanyam et al. (1989) screened a global peanut collection for resistance to rust caused by *Puccinia arachnidas* and to leaf spot caused by *Phaeoisariopsis personata*. They found that 75% of the resistant accessions came from the Tarapoto region of Peru. Peru is a secondary center of diversity for peanut that developed from the primary center of domestication in southern Bolivia.

There is evidence that landraces are adapted to their biotic environment, which includes pests and pathogens. Leppik (1970), Harlan (1977), and Buddenhagen (1983) noted that the greatest numbers of disease resistance genes usually come from landraces in which host and pathogen have coexisted for long periods of time. Although some of these populations may be low yielding, the genetic variability for resistance within and between them has provided some degree of insurance against the hazard of epidemics.

Other selective forces combine with pathogen pressure and the relative importance of a disease in the host's environment to determine the intensity of selection for resistance. For example, occasional epidemics of rice blast (caused by *Pyricularia grisea*) can be devastating at the high altitudes of Bhutan, locally eradicating whole crops. This suggests that blast is a strong selective pressure. Yet cold resistance is a vital trait and may in fact be the dominant selective force in the system (Thinlay 1998).

Box 11.2 discusses recent research on local varieties of faba bean in Morocco as sources of resistance to the crop's major foliar diseases: chocolate spot and ascochyta blight. Of key interest in this work is that much of the screening was done with local isolates of the pathogens under both laboratory and field conditions. The host populations were found to be polymorphic for resistance, which genetic analysis has indicated is multigenic and partial in the case of chocolate spot of faba bean.

In addition to resistance genes themselves, the resistance responses in landraces can result from morphological differences, correlated traits, or

Box 11.2 Local Moroccan Varieties as Sources of Multigene Resistance

Chocolate spot, caused by the fungus *Botrytis fabae,* is the most destructive leaf disease of faba bean (*Vicia faba* L.) crops in Morocco. This pathogen can reduce yields by up to 80% annually under optimum conditions for the disease development. Bouhassan et al. (2003a) identified and reported sources of resistance in the local germplasm. Therefore screening of 136 faba bean local accessions for resistance to *B. fabae* was conducted. Response under field conditions and on detached leaves was tested with artificial inoculation using a local strain of *Botrytis.* Significant differences were detected between genotypes for reaction to the disease in both tests. Nine accessions were clearly resistant in both the field and in vitro, and two were highly resistant. However, no complete resistance was observed, and the authors concluded that these genotypes have partial resistance, presumably under multigenic control.

Bouhassan et al. (2003b) analyzed the epidemiological components of this partial resistance to chocolate spot using five different lines developed from five different local faba bean varieties, which showed different levels of field susceptibility to the disease. They found that the components lesion diameter, latent period, and number of spores per leaflet were significantly involved in the characterization of the partial resistance. Incubation period did not appear to play a role. The work was based on local isolates of the fungus.

Ascochyta blight, caused by *Ascochyta fabae* Speg., is one of the major fungal diseases of faba beans worldwide. The fungus can damage all aerial parts of the plant and cause severe loss in both quality and quantity of the product. Genetic resistance is one of the major components of integrated control of the disease. Through a collaborative network (FRYMED), the local germplasm of North Africa was screened for sources of resistance to this pathogen in order to develop a resistant genepool (Kharrat et al. 2002). In total, 309 accessions (of which 106 originated from Morocco) have been screened in the field under inoculation with the local pathogen isolate FRY AFT04. The most resistant lines have been retested for confirmation in the field and in the growth chamber under artificial conditions against two virulent isolates (FRY AFT04 and FRY AFT37). These tests resulted in the identification of 18 resistant faba bean accessions. Some accessions showed better resistance on stems than on leaves and were retained to keep the genetic basis of the resistance as broad as possible. Almost all accessions identified as resistant or partially resistant belong to small- and medium-seeded types, but they have a large variability for cycle length and some other morphological traits. These resistant genotypes were introduced in the Ascochyta Disease Specific Gene Pool collection, held by IAV Hassan II Institute, Rabat, Morocco.

indirect effects. For example, the solid-stemmed types in Turkish wheat landraces were resistant to sawfly, whereas the hollow-stemmed types were not (Damania et al. 1997). In East Africa, response to selection for tolerance to heavy rain was correlated with resistance to anthracnose (Trutmann et al. 1993).

Composite crosses or bulk populations that are highly variable genetically are interesting experimental systems that can portray how host populations evolve to meet pressures from varying pathogen populations (Brown 1999). Allard (1990) analyzed temporal trends in resistance to scald (caused by *Rhynchosporium secalis*) in barley composite crosses and inferred that not all resistance alleles are useful, some being detrimental to yield, reproductive capacity, and adaptability. He also concluded that pathotypes differ in their ability to overcome various resistance alleles, to infect and damage the host. Several aspects of the pathosystem are interrelated in ways that affect population dynamics of host and pathogen, including frequencies of resistance alleles in the host population and virulence alleles in the pathogen population.

Several mechanisms may contribute to changes in disease incidence or severity (usually a reduction) in host populations that are diverse for resistance (Wolfe and Finckh 1997). Seven such mechanisms are listed here, the first four of which apply to all mixtures and variable populations irrespective of whether pathogen specialization to the host in question is present. The last three apply to host–pathogen systems with specific resistance.

- The *increased distance* between plants of the more susceptible genotypes in the population reduces spore density and the probability that a virulent spore will land on a susceptible host.
- Resistant plants act as *barriers* to pathogen spread.
- *Selection in the host population* for the more competitive or more resistant genotypes can reduce overall disease severity.
- *Increased diversity of the pathogen population* per se can in some cases decrease disease (Dileone and Mundt 1994).
- Where pathogen specialization for host genotypes occurs, the *resistance reactions* that avirulent spores induce may prevent or delay infection by adjacent virulent spores (e.g., for powdery mildew of barley mixtures [Chin and Wolfe 1984] and for yellow rust of wheat [Lannou et al. 1994; Calonnec et al. 1996]).

- *Interactions between pathogen races* (e.g., competition for available host tissue) may reduce disease severity.
- *Barrier effects are reciprocal,* that is, plants of one host genotype will act as a barrier for the pathogen specialized to a different genotype, and plants of the latter will act as a barrier for the pathogen specialized to the first genotype.

These mechanisms apply to airborne, splash-borne, and some soil-borne diseases. Thus mixtures of host genotypes that vary in response to a range of plant diseases tend to show an overall response to those diseases that is correlated with the disease levels on the more resistant components in the population. In addition, when particular genotypes are affected by disease, the yields of the other, more resistant individuals generally compensate for them.

Pathogen Evolution in Response to Host Resistance Management

The biotic environment of landraces differs in degree from the abiotic environment in at least two ways. First, it is potentially a responsive moving target, able to change to meet new evolutionary opportunities and match changes in the host. Second, the pathogen component is partially hidden in that potential diseases that currently are under control in the population may not be evident as threats. Thus the presence of a serious pathogen requires disease developed on specific host plants to be evident, whereas edaphic or climatic stresses are apparent in an area from either physical or biological data.

A serious concern is the potential for genetically heterogeneous host populations to select for resistance to super-races, which could lead to the simultaneous loss of all resistances. However, the approach to dominance of a pathogen race able to attack all genotypes will slow with increasingly complex host populations because the selective advantage of being able to attack one more host decreases as the number of different genotypes increases (Wolfe and Finckh 1997). On the other hand, increasing the diversity of resistance responses may lower the adaptation or the use or value of the crop population to the farmer. Therefore there is likely to be an optimum in host complexity.

There are other strategies to delay the evolution of super-races. For example, some researchers suggest that within local populations the optimum evolutionary strategy may be the development of complementary patterns of genetic variation for resistance in the host and virulence in the pathogen (McDonald et al. 1989). A substantial theoretical and empirical literature exists investigating such strategies for deliberate mixtures; much less is known about this question in traditional landraces.

The long-term effects of resistance gene deployment on the genetic structure of pathogen populations are widely debated. Many studies directed toward coevolutionary models in agricultural systems have stressed the importance of fitness costs associated with resistance and virulence. However, such costs are hard to document. If virulence does have a fitness cost to the pathogen, then mixtures carrying different resistance genes will slow the rate of evolution of the pathogen, and simple races will dominate the pathogen population. However, recent models indicate that mechanisms other than the cost of virulence might act to the same effect (Lannou and Mundt 1996; Finckh et al. 1998).

As farmers have manipulated genetic diversity in their crops, how have the pathogens responded? This important question probably has as many answers as there are cropping systems, but one overriding generalization is that evolutionary shifts in the pathogen are the rule. Box 11.1 gives some examples from recent research in Ecuador that emphasize the complex situations that arise in resistance gene deployment. Suboptimal use of resistant varieties can cause unintended and untoward shifts in pathogen virulence that must be met with the use of further resistance sources.

Using Genetic Diversity to Manage Diseases

Both farmers and plant breeders have selected for and used genotypes that are resistant to the pests and pathogens of their crops and have developed farming systems that reduce the damage they cause (Frankel et al. 1995; Finckh and Wolfe 1997; Thinlay et al. 2000a). Here we discuss three kinds of genotype use: direct use by farmers, use of resistance in mixtures, and use in breeding programs.

Direct Use by Farmers

Traditional farmers often are aware of and exploit intervarietal differences in susceptibility to major pathogens. Box 11.3 provides an example of farmer use of genotypic diversity to cope with a suite of diseases and

Box 11.3 Managing Leaf Spot Diseases in East African Highland Banana Production Systems

Banana cultivar diversity in the Great Lakes region of East Africa is estimated at 100–150 cultivars (Karamura and Karamura 1995). Banana cultivation is so closely intertwined in the sociocultural fabric of the communities that every part of the plant is used in the households; different cultivars are used as medicine and in the execution of cultural functions such as birth, death, and marriage. In an ethnobotanical study, Karamura et al. (2003) reported seven criteria farmers use for their selection breeding, and five of these were related to pest and diseases. In addition, cultural practices such as desuckering, deep planting, and the uprooting of postharvest stumps are measures practiced to manage pest and diseases in subsistence banana systems.

The East African highland bananas, AAA-EAHB (Karamura 1999), are a group unique to the Great Lakes region of East Africa, which is now regarded as a secondary center of banana diversity (Karamura et al. 1999). Although this group dominates the crop in the region (78%), other banana groups including bluggoes (ABB), dessert bananas (AAA-Gros Michel), AB (Sukali Ndizizi), and plantain (AAB-Gonja) are grown in mixtures with the AAA-EAHB, ranging from 30 to 40 different cultivars per farm.

In this region a host of viral, fungal, and bacterial diseases and pests attack the crop, all of which elicit a variety of responses from the crop. Chief among these stresses is a complex of leafspots: black sigatoka, caused by *Mycosphaerella fijiensis* Morelet; *Cladosporium* speckle, caused by *Cladosporium musae* Mason; and yellow sigatoka, caused by *Mycosphaerella musicola* Leach. Occasionally, in areas of warm and humid conditions, the crop may be attacked by the eye spot disease (*Drechslera* sp.).

Tushemereirwe (1996) studied the incidence and distribution of the leaf spot diseases in the Great Lakes region, with specific emphasis on the highland bananas. His results showed a range of responses across plant populations with respect to different leaf spot diseases. Box table 11.3 summarizes those for *M. musicola,* for which the AAA-EAHB varieties in the trial (Entundu, Mbwazirume, and Nakitembe) had the lowest incidence, whereas the "beer" banana, Kayinja cultivar, had the highest. In an average farm in areas where the disease is prevalent, this cultivar normally constitutes less than 5% of the stand (Karamura and Karamura 1995). This may help keep the disease inoculum low in the garden and minimize the farmers' losses. The response for black sigatoka disease (*M. fijiensis*) contrasts with yellow sigatoka. The ABB cultivars display a high level of resistance, whereas AAA-EAHB appears to be very susceptible.

The results described here imply that intraspecific diversity can contribute to the management of the leaf spots in bananas. By growing several cultivars, farmers guard against total yield losses that may result from variability or change in

Box 11.3 continued

BOX TABLE 11.3. Incidence of yellow sigatoka and reaction to black sigatoka of banana genomes

Cultivar	Genome	M. musicola	Incidence	Black Sigatoka	Response[*]
Kayinja	ABB	72%	Susceptible	7.1±0.1	Resistant
Gros Michel	AAA	19%	Resistant	5.2±0.3	
3 cultivars	AAA-EAHB	7%		—	
Many cultivars	AAA-EAHB	—		4.7±0.0	Susceptible
Sukali Ndiizi	AB	—		5.4±0.1	
Plantain	ABB	—		4.8±0.2	

Source: Tushemereirwe (1996).

[*]Response measured as the youngest spotted leaf (± standard error of mean), counting the last funnel or unexpanded leaf as zero. In susceptible cultivars, symptoms appear quickly on young leaves, whereas in resistant cultivars only the older leaves show symptoms.

Box 11.3 continues to next page

Box 11.3 continued

the pathogen population, thereby ensuring food security and household income. In the Great Lakes region, farmers address the disease problem at two levels. First, they take advantage of the variation between genomes. The ABB cultivars are susceptible to yellow sigatoka but resistant to black sigatoka. The opposite is true for the East African highland bananas. The ranges of the two diseases are also modified by temperature, with the cooler highlands heavily infested by yellow sigatoka and the warmer lowlands by black sigatoka.

Second, farmers may use the variation within the subgroup such as Lujugira-Mutika, where the most susceptible ones are also the early maturing ones (9–12 months), whereas the most resistant tend to be big-bunched and late maturing (12–15 months). Early maturing cultivars will escape at least one humid season during which leaf spots tend to proliferate, and yields are higher than expected. At a cropping system level, farmers at high altitudes tend to grow susceptible but early maturing cultivars, whereas lowland farmers largely grow resistant or tolerant cultivars.

pests in bananas in Uganda. Disease susceptibility often joins a complex list of criteria that determine farmers' choice of seed. The choice reflects a compromise between conflicting criteria, or farmers may select several varieties to meet distinct needs.

Multilines and Mixtures for Disease Control

In many regions of the world, farmers have local preferences for growing mixtures of cultivars that provide resistance to local pests and diseases and enhance yield stability (Trutmann et al. 1993). Thus within-crop diversity (through variety mixtures, multilines, or the planned deployment of different varieties in the same production environment) can reduce damage by pests and diseases (box 11.4).

Another approach available to farmers is to use mixtures of traditional and resistant modern varieties to achieve reduced pest and disease damage and thus retain and use traditional varieties on farm (Zhu et al. 2000; chapter 12). Pyndji and Trutmann (1992) and Trutmann and Pyndji (1994) showed over three seasons that adding a resistant variety to 25–50% of a local bean mixture that was susceptible to angular leaf spot (ALS) (caused by *Colletotrichum lindemuthianum*) both protected the susceptible components in the local mixture and increased yields significantly above expected.

Box 11.4 East Africa: Farmers' Use of Common Bean Genetic Diversity to Reduce Disease

The Great Lakes Region in Africa is a secondary center of diversity of a major local food crop, the common bean (*Phaseolus vulgaris*). Beans are grown as genetic mixtures, which are preferred for reasons of higher yield and greater stability of production (Voss 1992). Farmers play a central part in developing and manipulating the available genetic diversity to optimize production in highly variable environments. Traditionally, a mixture for each field is selected and kept separately, each unique for the slope, sun exposure, soil exposure to rain, and other factors. At first, when farmers settle an area or cultivate a new field, they develop a mixture by sowing as many variable sources of seed as possible in each field, harvesting the survivors, and repeating the process over seasons and years. Eventually, other selection criteria are added to satisfy other targets such as family tastes, color, and cooking preferences. New varieties are selectively added to mixtures only at a later stage, and only after they are tested separately. Without farmer selection, the composition of mixtures changes rapidly. Therefore, the makeup of farmer mixtures is partly the result of natural selection and partly of farmer management. Substantial levels of resistance to local pathogens are inherent in these mixtures, and the level of resistance increases in zones more favorable to pathogens (Trutmann et al. 1993). In particular, under controlled conditions varieties have resistance to local races of *Colletotrichum lindemuthianum,* the causal agent of an often lethal disease called anthracnose. Yet the farmer mixtures vary in both the number of different seed types (the richness diversity of the mixture) and the percentage of component types (evenness diversity), depending on the zone. Resistance to local pathotypes of *C. lindemuthianum* of varieties from zones with more favorable conditions for anthracnose increases with altitude, as do the number of varieties with high levels of resistance. Additional ways in which farmers manage resistance to diseases include the use of plant architecture, the removal of blemished seed during selection, and varying the use of genetic diversity in temporal and spatial settings.

Varieties have to resist rain. Resistance to rain and yield are the most important farmer criteria for varietal selection. Although diseases on the whole usually are not recognized individually, they are related to rain. Rain is associated with the rotting of leaves or roots (as seen from the farmers' perspective) and with causing floral abortion (Trutmann et al. 1996). Plant architecture that enables plants to escape the effects of rain is preferred, and certain types of plant vigor are selected depending on the conditions. Farmers also deploy their genetic diversity, using different mixtures in the first and second rainy season. Traditionally, seed for each season is kept for each field. This strategy is interwoven with rotations. In addition, fields are kept small, and beans often are intercropped

Box 11.4 continues to next page

Box 11.4 *continued*

with other crops such as banana, sweet potato, and maize. The overall effect is that genetic variation to manage diseases is enhanced by variation of its placement, frequency or density, and timing. In these ways local farmers enhance the use of the available genetic diversity beyond the within-crop deployment of genes that directly confer resistance to local pathogens.

However, no yield benefit accrued without disease pressure. Angular leaf spot (ALS) is an important factor limiting yield, and new sources of resistance can have a major impact on yields of traditional mixtures. Such new resistances and their use in mixtures can help conserve traditional varieties and reduce their displacement by monocultures.

The story is more complex. In multilocation trials, the yield benefits from new resistant mixtures were not as clear cut as severity ratings had indicated. In these sites the probable interacting factor was another disease, floury leaf spot (caused by *Ramularia phaseoli*), to which the ALS-resistant variety was susceptible. These results underline the typical difficulties that breeders meet as they have to select for multiple disease resistance among other traits. Wolfe (1985) proposed that cultivar mixtures might help to achieve this goal more efficiently because it will suffice if different components in the mixture are resistant to different diseases.

Multilines are mixtures of genetically similar lines or varieties that differ only in their resistances to different pathotypes. They are in use in cereals in the United States (Finckh and Wolfe 1997) and in coffee (*Coffea arabica*) in Colombia. There the variety Colombia is a multiline of coffee lines differentially resistant to rust (caused by *Hemilera vastatrix*) and grown on more than 360,000 ha (Moreno-Ruiz and Castillo-Zapata 1990; Browning 1997).

Epidemiological studies of pathogen populations in experimental varietal mixtures and multilines provide an empirical test of whether the resistance heterogeneity in a landrace population might also reduce the spread of disease. Wolfe (1985) reviewed more than 100 observations from such experimental evidence and found that the infection rate in the more susceptible component in binary mixtures was only 25% of the infection rate in pure plots. The overall infection rate in varietal mixtures approached that of the resistant component grown alone. Also, he found that mixtures generally are more effective than multilines because of their higher level of genetic heterogeneity.

Another line of argument that supports the adaptive properties of multiple resistance is its prevalence in wild plant populations. Burdon (1987) reviewed the evidence from eight herbaceous and forest tree species as well as *Avena, Glycine,* and *Trifolium* showing that natural plant populations often are polymorphic in their response to pathogens. In the wild *Linum marginale–Melampsora linii* system, the more resistant natural plant populations harbor more virulent rust populations (Thrall and Burdon 2003). Yet in this system, disease is generally less prevalent in host populations with greater genetic diversity of resistance. Very similar observations were made for rice landraces and rice blast (Thinlay et al. 2000b).

Competition and compensation are the most important intergenotypic interactions occurring in plant populations, both of which influence yield and yield stability. In the absence of disease, mixtures tend to yield around the mean of the components and overall average slightly more than the mean (Finckh and Wolfe 1997). Yield increases in genotype mixtures may arise partially from niche differentiation among the components (Finckh and Mundt 1992). Allelopathy and synergisms of unknown origin might also play a role.

Disease levels in mixtures are almost always lower than the average levels of their components (Burdon 1987; Burdon and Jarosz 1989). In the presence of disease, mixtures of cultivars often yield more than the mean of the components grown as pure stands (Finckh and Wolfe 1997). Although the correlation between disease severity and yield often is clear in pure stands, it is not always so in mixtures (Finckh et al. 1999). This is because the correlation between disease severity and yield of the individual component plants of a mixture often is poor. One important reason for this is the effects of disease on the competitive interactions between cultivars (Finckh and Mundt 1992; Finckh et al. 1999).

Breeding

Because of the value of resistance genes to breeding programs, many researchers have screened genebank samples of landraces and wild crop relatives as well as newly collected samples from the field. When interpreting the results of such studies it is important to keep in mind when the genebank samples were originally collected and what pathotypes were used to test resistance (Teshome et al. 2001). The temporal factor is important because pathogen and host populations change over time in the field. Comparison collections made at different times may show a diversity of response that is misleading as to what level of diversity may be present at any one time. Although the use of nonlocal pathotypes in testing for resistance response in landraces is relevant to specific breeding goals, data of this type may not be useful for the study of coevolutionary processes in situ.

Because landraces often are diverse for resistance, it is also important to use sufficiently large samples for screening against multiple pathogen races. Often only a certain fraction of a landrace carries resistance (Thinlay et al. 2000b). In addition even predominantly inbreeding crops will outcross to a certain degree when maintained as diverse landraces and therefore may be segregating and show changes in resistance over time (Finckh 2003).

Breeders' use of resistances in landraces typically begins with germplasm screening. For example, Negassa (1987) screened Ethiopian wheat landraces for response to leaf rust (caused by *Puccinia recondita*) and found moderate resistance to an isolate that was virulent on six genes. Subsequently Dyck and Sykes (1995) tested whether such resistance was transferable in a wheat breeding program. In tests using crosses and backcrosses, they demonstrated that resistance in Ethiopian tetraploid and hexaploid wheat to leaf rust and to stem rust (caused by *P. graminis* f. sp. *tritici*) was usable.

In Ethiopian landraces of barley, Alemayehu and Parlevliet (1996) found a near absence of race-specific, major resistance and a high frequency of moderate levels of partial resistance to *Puccinia hordei*. Breeding with quantitative, partial, or multigenic resistances poses difficulties in modern plant breeding, which sometimes can be helped with linked genetic markers. Alternatively, dispersed breeding efforts in participatory schemes that involve farmers selecting in their fields are encouraging, as box 11.5 reports.

Resistance whose genetic basis is complex can be handled in ways other than by pedigree breeding. Ever since pathogens were recognized as "shifting enemies" (Stakman 1947), many breeders have advocated the use of resistance gene diversity to cope with, if not to forestall, evolving pathogen populations (e.g., Suneson's 1956 "evolutionary plant breeding" approach; Le Boulc'h et al. 1994). Among other breeding concepts, population selection, composite crosses, top crosses, and multilines all make use of within-crop diversity (Finckh and Wolfe 1997).

Box 11.5 Response of Local Varieties to Participatory Recurrent Selection in Morocco

In Morocco, germplasm enhancement based on recurrent selection has proved to be an efficient approach for improving faba bean populations, particularly for quantitative traits (Sadiki et al. 2000). This strategy is attractive as a method of participatory plant selection for improving local germplasm of faba bean. Three cycles of half-sib family multitrait selection were completed for yield components and resistance to *Botrytis fabae* under natural infestation in a broadly based population developed from local varieties (Sadiki et al. 2000). Evaluation of response to selection showed that significant gain was achieved for yield and that resistance to *Botrytis* improved by 54%. The first cycle induced the largest response to selection for all traits. This approach demonstrates that local farmers may improve their varieties locally by increasing the frequency of the disease-resistant genotypes that combine resistance genes. Nevertheless, the improved populations are still appreciably diverse for visible traits and for the reaction to disease itself. The improved populations are selected against local populations of the pathogen.

Farmers' Role in Shaping Coevolved Genetic Diversity

Farmers manipulate the genetic composition of their crops and the biotic and abiotic environment in and around their fields, creating distinct selective pressures in agricultural systems. Four kinds of genetic management are notable.

Selection of Crop Genetic Diversity

The choices of planting materials that farmers make clearly have a major effect on pathogen populations. Crops differ in the extent to which farmers' selection criteria for seed explicitly or effectively include the avoidance of pathogen damage. For many (e.g., faba bean, box 11.2; banana, box 11.3; phaseolus beans, box 11.4), disease response criteria rank highly in farmers' decisions. In other crops without obvious disease symptoms, resistance selection is indirect via pragmatic selection for yield.

The effects of farmers' seed selections depend on their access to genetic resources and the history of cropping in the region. Landrace crops growing in regions where the species was domesticated still can interact with their wild progenitors and relatives along with weeds and shared pests, pathogens, and beneficial organisms. On the other hand, crops that have traversed continents and are separated from their origin may retain less genetic diversity and display a variety of relationships with their pests. Outcomes for any particular situation are difficult to predict. Most crops, away from the constraints of their coevolved pests, can flourish. In some cases, crops have developed resistance outside their center of domestication (e.g., *Vicia faba*), presumably involving farmer selection.

Field Size and Position

Field location affects the interaction of crop species with populations in other farmers' fields and wild alternative hosts in the surrounding natural vegetation. Small, isolated fields are more likely to diverge from one another than larger fields, so that in many traditional systems small fields become mosaics of diversity that may reduce the chance of large-scale epidemics. Adjacent fields have increased opportunity for gene-flow between populations of both host and pathogen. Natural populations of wild relatives can support pathogen evolution and the potential of pathogens to overcome crop resistance (Allen et al. 1999). An extreme example is the movement of virulent rust strains from wild relatives of wheat in the Himalayas to cultivated wheat in India and Pakistan, resulting in epidemics (Joshi 1986).

Within-Field Spatial Arrangement of Crop Genetic Diversity

Farmers may grow their crops as varietal monocultures or as species mixtures and in various intercropping patterns. Each of these strategies affects the rate and level of host–pathogen interaction, as discussed earlier.

Temporal Variables

Seasonality in temperature and rainfall in relation to harvest and planting affects plant–pathogen interactions. Farmer practices such as fallowing, rotation, adjustment of sowing date, use of cultivars of different duration, use of trap crops, and temporal deployment of specific resistances can build on seasonality for managing pests (Thurston 1992).

Crop rotations are fundamental in improving crop health in various ways (Finckh 2003). These can be divided into time effects to outlast residual pathogen propagules in soil or on crop residue, indirect effects via soil microbial activity, and direct suppressive effects of certain crops on certain pathogens. Although the presence of a pathogen is needed to cause disease, the absence of a pathogen is not necessarily needed for a healthy crop. In fact, the balance between beneficial and detrimental organisms usually determines the outcome.

Yet as this chapter has shown, the need is to complement and extend such integrated pest management strategies as rotations by using and managing the intraspecific diversity of local crop cultivars as a key resource. For resource-poor farmers in developing countries, local crop diversity and its management may be one of the few resources and options available to combat pest and disease pressures. Thus biodiversity benefits that will accrue through application of this approach, in addition to the conservation of agrobiodiversity, will include less environmental damage, conservation of insects, fungi, soil microorganisms, and aquatic biodiversity of adjacent ecosystems.

Discussion and Research Challenges

Although it is known that crop genetic diversity can be used to reduce pest and disease pressures, it is also known that this approach is not appropriate in all circumstances. The challenge is to develop criteria that determine when and where diversity can play or is playing a key role in managing pest and disease pressures. These criteria will form the basis for tools and decision-making procedures for farmers and development workers to enable the appropriate adoption of diversity-rich strategies to manage pests and diseases.

The key questions for research to yield such guidelines in the use of crop genetic diversity are as follows:

• Host resistance diversity: Between and within traditional crop cultivars, what genetic variation for resistance exists against the pathogen populations they harbor?
• Diversity and field resistance: Does the resistance diversity present in a crop actually reduce pest and disease pressure and vulnerability, at least in the short term?
• Biotype diversity: How does the population structure of pathogens vary across systems and in space?

The answers to these questions will be based on data collected to characterize hosts, pests, pathogens, and surrounding environments from direct field measurements in conjunction with information from farmers.

In general, the development of disease in plant populations and the co-evolution of resistance and virulence is the outcome of interaction between three factors: the host, the pest or pathogen, and the environment, depicted as the disease triangle (Burdon 1987). Host–pathogen coevolution in traditional cropping systems can also be portrayed as a triangle in common with natural communities or with composite crosses. However, for landraces in traditional systems, it is important to add the farmers to this model because of the crucial role they play in selection (Finckh and Wolfe 1997).

Conclusion

Understanding the interconnecting forces at work between farmers, their crops, the environment, and host and pest species in agroecosystems is critical to developing effective mechanisms for combating diseases based on optimal maintenance and management of crop genetic diversity in highly variable environments. Resource-poor farmers depend on the diversity of local crop cultivars to cope with all the factors that lower yield. Development of alternative strategies to meet their needs, such as highly bred homogeneous varieties that combine several resistances ("pyramid breeding"), is costly. They are unlikely to be adapted to marginal or highly variable environments. Inevitably, such varieties must be replaced as new diseases or pathotypes arise to attack them. Most developing countries are not able to finance such continued maintenance breeding. The public sector is shrinking, the environments often are highly variable, and the climate is optimal for most pathogens. Therefore, it is essential that resistance diversity be both maintained and used optimally on farm to ensure present production and future options for farmers. Cases of inappropriate deployment do not rule out this fundamental principle. Diversity is not a hazard in itself, nor is it necessarily a benefit. Rather, the task is to determine the key genetic, environmental, and agronomic parameters that will affect when farmers will benefit from its use and reduce the vulnerability of their crops to disease and pests.

Acknowledgments

The authors would like to thank the United National Environmental Programme Global Environmental Facility, the Food and Agriculture Organization of the United Nations, and the governments of Switzerland (Swiss Agency for Development and Cooperation) and Germany (Bundesministerium für Wirtschaftliche Zusammenarbeit/Deutsche Gesellschaft für Technische Zusammenarbeit) for their financial support of some of the studies in this chapter.

Notes

1. For simplicity, we do not include multiple species as a strict diversity strategy because a component species of a multicrop system may be genetically homogeneous and be host to entirely different diseases yet immune to others that afflict the other component. Resistance benefits from such a strategy arise from physical effects (e.g., spore trapping, host density) rather than genetic effects (e.g., differential resistance).

2. The term *monoculture* usually refers to the continuous use of a single crop species over a large area. For the pathologist, however, *monoculture* alone is inadequate because it can apply at the level of species, variety, or gene. If all varieties available within the species possess the same resistance gene, then the system is effectively a resistance gene monoculture (Finckh and Wolfe 1997).

3. Genetic vulnerability is defined as "the condition that results when a widely planted crop is uniformly susceptible to a pest, pathogen or environmental hazard as a result of its genetic constitution, thereby creating a potential for widespread crop losses" (FAO 1998:30). Thus vulnerability reflects a potential for damage rather than actual damage.

References

Alemayehu, F. and J. E. Parlevliet. 1996. Variation for resistance to *Puccinia hordei* in Ethiopian barley landraces. *Euphytica* 90:365–370.

Allard, R. W. 1990. The genetics of host–pathogen coevolution: Implications for genetic resource conservation. *Journal of Heredity* 81:1–6.

Allen, D. J., J. M. Lenne, and J. M. Walker. 1999. Pathogen biodiversity: Its nature, characterization and consequences. In D. Wood and J. Lenne, eds.,

Agrobiodiversity. Characterization, Utilization and Management, 123–153. Wallingford, UK: CAB International.

Bouhassan, A., M. Sadiki, and B. Tivoli. 2003a. Evaluation of a collection of faba bean (*Vicia faba* L.) genotypes originating from the Maghreb for resistance to chocolate spot (*Botrytis fabae*) by assessment in the field and laboratory. *Euphytica* 135:55–62.

Bouhassan, A., M. Sadiki, B. Tivoli, and N. El Khiati. 2003b. Analysis by detached leaf assay of components of partial resistance of faba bean (*Vicia faba* L.) to chocolate spot caused by *Botrytis fabae Sard. Phytopathologia Mediterranea* 42:183–190.

Brown, A. H. D. 1999. The genetic structure of crop landraces and the challenge to conserve them in situ on farm. In S. Brush, ed., *Genes in the Field: On Farm Conservation of Crop Diversity*, 29–48. Boca Raton, FL: Lewis Publishers.

Browning, J. A. 1997. A unifying theory of the genetic protection of crop plant populations from diseases. In I. Wahl, G. Fischbeck, and J. A. Browning, eds., *Disease Resistance from Crop Progenitors and Other Wild Relatives*. Berlin: Springer Verlag.

Buddenhagen, I. W. 1983. Breeding strategies for stress and disease resistance in developing countries. *Annual Review of Phytopathology* 21:385–409.

Burdon, J. J. 1987. *Diseases and Plant Population Biology*. Cambridge, UK: Cambridge University Press.

Burdon, J. J. and A. M. Jarosz. 1989. Disease in mixed cultivars, composites, and natural plant populations: Some epidemiological and evolutionary consequences. In A. H. D. Brown, M. T. Clegg, A. L. Kahler, and B. S. Weir, eds., *Plant Population Genetics, Breeding and Genetic Resources*, 215–228. Sunderland, MA: Sinauer Associates.

Calonnec, A., H. Goyeau, and C. de Vallavieille-Pope. 1996. Effects of induced resistance on infection efficiency and sporulation of *Puccinia striiformis* on seedlings in varietal mixtures and on field epidemics in pure stands. *European Journal of Plant Pathology* 102:733–741.

Chin, K. M. and M. S. Wolfe. 1984. The spread of *Erysiphe graminis* f. sp. *hordei* in mixtures of barley varieties. *Plant Pathology* 33:89–100.

Compagnon, P. 1998. *El caucho natural, Biologia-Cultivo-Producción*, 142–1559. Paris: Consejo Mexicano del Hulei y CIRAD.

Damania, A., B. L. Pecetti, C. O. Qualset, and B. O. Humeid. 1997. Diversity and geographic distribution of stem solidness and environmental stress tolerance in a collection of durum wheat landraces from Turkey. *Genetic Resources and Crop Evolution* 44:101–108.

Dileone, J. A. and C. C. Mundt. 1994. Effect of wheat cultivar mixtures on populations of *Puccinia striiformis* races. *Plant Pathology* 43:917–930.

Dyck, P. L. and E. E. Sykes. 1995. Inheritance of stem rust and leaf rust resistance in some Ethiopian wheat collections. *Euphytica* 81:291–297.

Falconi, E., J. B. Ochoa, E. Peralta, and D. Daniel. 2003. *Virulence Pattern of* Colletrotrichum lindemuthianum *in Common Bean in Ecuador*. Bean Improvement Cooperative (BIC). East Lansing: Michigan State University.

FAO (Food and Agriculture Organization of the United Nations). 1998. *The State of the World's Plant Genetic Resources for Food and Agriculture*. Rome: FAO.

Finckh, M. R. 2003. Ecological benefits of diversification. In T. W. Mew, D. S. Brar, S. Peng, D. Dawe, and B. Hardy, eds., *Rice Science: Innovations and Impact for Livelihood*, Proceedings of the International Rice Research Conference, September 16–19, 2002, 549–564. Beijing: International Rice Research Institute, Chinese Academy of Engineering and Chinese Academy of Agricultural Sciences.

Finckh, M. R., E. S. Gacek, H. J. Czembor, and M. S. Wolfe. 1999. Host frequency and density effects on disease and yield in mixtures of barley. *Plant Pathology* 48:807–816.

Finckh, M. R., E. S. Gacek, H. J. Nadziak, and M. S. Wolfe. 1998. Suitability of cereal cultivar mixtures for disease reduction and improved yield stability in sustainable agriculture. *Sustainable Agriculture for Food, Energy and Industry* 1:571–576.

Finckh, M. and C. Mundt. 1992. Plant competition and disease in genetically diverse wheat populations. *Oecologia* 91:82–92.

Finckh, M. and M. S. Wolfe. 1997. The use of biodiversity to restrict plant diseases and some consequences for farmers and society. In L. E. Jackson, ed., *Ecology in Agriculture*, 203–237. San Diego, CA: Academic Press.

Frankel, O. H., A. H. D Brown, and J. J. Burdon. 1995. *The Conservation of Plant Biodiversity*. Cambridge, UK: Cambridge University Press.

Hanounik, S. B. and L. D. Robertson. 1987. New sources of resistance in *Vicia faba* L. to chocolate spot caused by *Botrytis faba*. *Plant Disease* 72:696–698.

Harlan, J. R. 1977. Sources of genetic defense. *Annals of New York Academy of Sciences* 287:345–356.

INIAP. 1974. *Annual Report*. Instituto Nacional Autónomo de Investigaciones Agropecuarias (INIAP). Quito, Ecuador: Department of Plant Pathology.

Joshi, L. M. 1986. Perpetuation and dissemination of wheat rusts in India. In L. Joshi, D. Singh, and K. D. Srivastava, eds., *Problems and Progress of Wheat Pathology*. New Delhi: South Asia Malhorta Press.

Karamura, D. A. 1999. *Numerical Taxonomic Studies of the East African Highland Bananas (Musa AAA–East Africa) in Uganda*. PhD thesis, University of Reading, UK.

Karamura, D. A., S. Mgenzi, E. Karamura, and S. Sharrock. 2003. Exploiting indigenous knowledge for the management and maintenance of *Musa* diversity. *African Crop Science Journal* 12:67–74.

Karamura, E., E. Frison, D. Karamura, and S. Sharrock. 1999. Banana production systems in eastern and southern Africa. In C. Picq, E. Foure, and E. Frison, eds., *Bananas and Food Security*, 401–412. International Symposium, November 10–14, 1998, Cameroon. Montpellier, France: INIBAP.

Karamura, E. B. and D. A. Karamura. 1995. Banana morphology. Part II. The aerial shoot: In S. Gowen, ed., *Bananas and Plantains*, 190–205. London: Chapman and Hall.

Kennedy, D. and M. Lucks. 1999. Rubber, blight, and mosquitoes: Biogeography meets the global economy. *Environmental History* 4:369–383.

Kharrat, M., M. Sadiki, R. Esnault, B. Tivoli, A. Porta Puglia, and M. R. Hajlaoui. 2002. *Identification of Sources of Resistance to* Ascochyta *Blight in Faba Bean*. Grain Legumes in the Mediterranean Agriculture (LEGUMED). Paris: AEP.

Kolmer, J. A., P. L. Dyck, and A. P. Roelfs. 1991. An appraisal of stem rust resistance in North American hard red spring wheats and the probability of multiple mutations to virulence in populations of cereal rust fungi. *Phytopathology* 81:237–239.

Lannou, C. and C. C. Mundt. 1996. Evolution of a pathogen population in host mixtures: Simple race–complex race competition. *Plant Pathology* 45:440–453.

Lannou, C., C. de Vallavieille-Pope, and H. Goyeau. 1994. Induced resistance in host mixtures and its effect on disease control in computer-simulated epidemics. *Plant Pathology* 44:478–489.

Le Boulc'h, V., J. L. David, P. Brabant, and C. de Vallavieille-Pope. 1994. Dynamic conservation of variability: Responses of wheat populations to different selective forces including powdery mildew. *Genetics Selection Evolution* 26:221–240.

Leppik, E. E. 1970. Gene centers of plants as a source of disease resistance. *Annual Review of Phytopathology* 8:323–344.

Marshall, D. R. 1977. The advantages and hazards of genetic homogeneity In P. Day, ed., The genetic basis of epidemics in agriculture. *Annals of the New York Academy of Sciences* 287:1–20.

McDonald, B. A., J. M. McDermott, S. B. Goodwin, and R. W. Allard. 1989. The population biology of host–parasite interactions. *Annual Review of Plant Pathology* 27:77–94.

Moreno-Ruiz, G. and J. Castillo-Zapata. 1990. The variety Colombia: A variety of coffee with resistance to rust (*Hemileia vastatrix* Berk. & Br.), *Cenicafe Chinchiná–Caldas. Colombia Technical Bulletin* 9:1–27.

Mundt, C. C. 1990. Probability of mutation to multiple virulence and durability of resistance gene pyramids. *Phytopathology* 80:221–223.

Mundt, C. C. 1991. Probability of mutation to multiple virulence and durability of resistance gene pyramids: Further comments. *Phytopathology* 81:240–242.

Negassa, M. 1987. Possible new genes for resistance to powdery mildew, *Septoria,* glume blotch and leaf rust of wheat. *Plant Breeding* 98:37–46.

Ochoa, J., H. D. Frinking, and T. H. Jacobs. 1999. Postulation of virulence groups and resistance factors in the quinoa/downy mildew pathosystem using material from Ecuador. *Plant Pathology* 48:425–430.

Ochoa, J., J. Lowers, and L. Broers. 1998. Analysis of virulence and evolution of the Ecuadorian population of stripe rust in wheat. *Fitopatología* 33:160–164.

Ochoa, L. B., E. Cruz, and D. Daniel. 2002. *Physiological Variation of Bean Rust in Ecuador.* Bean Improvement Cooperative (BIC). East Lansing: Michigan State University.

Oerke, E. C., H. W. Dehne, F. Schönbeck, and A. Weber. 1994. *Crop Production and Crop Protection, Estimated Losses in Major Food and Cash Crops.* Amsterdam: Elsevier.

Oldfield, M. L. 1989. *The Value of Conserving Genetic Resources.* Sunderland, MA: Sinauer Associates.

Otim-Nape, G. W. and J. M. Thresh. 1998. The current pandemic of cassava mosaic virus disease in Uganda. In D. G. Jones, ed., *The Epidemiology of Plant Diseases,* 423–443. Dordrecht, The Netherlands: Kluwer.

Pyndji, M. M. and P. Trutmann. 1992. Managing angular leaf spot development on common bean in Africa by supplementing farmer mixtures with resistant varieties. *Plant Disease* 76:1144–1147.

Qualset, C. O. 1975. Sampling germplasm in a center of diversity: An example of disease resistance in Ethiopian barley. In O. H. Frankel and J. G. Hawkes, eds., *Crop Genetics Resources for Today and Tomorrow,* 81–96. Cambridge, UK: Cambridge University Press.

Rivano, F. 1997. South American leaf blight of Hevea. 1. Viability of *Microcyclus ulei* pathogenicity. *Plantations, Recherche, Developpement* 4:104–114.

Sadiki, M., L. Belqadi, S. Mehdi, and A. El Alami. 2000. Sélection de la fève pour la résistance polygénique aux maladies par voies d'amélioration des populations. *Petria* 10:203–262.

Schumann, G. L. 1991. *Plant Diseases: Their Biology and Social Impact.* St. Paul, MN: APS Press.

Stakman, E. C. 1947. Plant diseases are shifting enemies. *American Scientist* 35:321–350.

Subrahmanyam, P., V. Ramanatha Rao, D. McDonald, J. P. Moss, and R. Gibbons. 1989. Origins of resistances to rust and late leaf spot in peanut (*Arachis hypogea*, Fabaceae). *Economic Botany* 43:444–455.

Suneson, C. A. 1956. An evolutionary plant breeding method. *Agronomy Journal* 48:188–191.

Teshome, A., A. H. D. Brown, and T. Hodgkin. 2001. Diversity in landraces of cereals and legume crops. *Plant Breeding Reviews* 21:221–260.

Thinlay, X. 1998. *Rice Blast, Caused by* Magnaporthe grisea, *in Bhutan and Development of Strategies for Resistance Breeding and Management*. Dissertation ETH No. 12777. Zürich: Swiss Federal Institute of Technology.

Thinlay, X., M. R. Finckh, A. C. Bordeos, and R. S. Zeigler. 2000a. Effects and possible causes of an unprecedented rice blast epidemic on the traditional farming system of Bhutan. *Agriculture, Ecosystems and Environment* 78:237–248.

Thinlay, X., R. S. Zeigler, and M. R. Finckh. 2000b. Pathogenic variability of *Pyricularia grisea* from the high- and mid-elevation zones of Bhutan. *Phytopathology* 90:621–628.

Thrall, P. H. and J. J. Burdon. 2003. Evolution of virulence in a plant host–pathogen metapopulation. *Science* 299:1735–1737.

Thurston, H. D. 1992. *Sustainable Practices for Plant Disease Management in Traditional Systems*. Boulder, CO: Westview Press.

Trutmann, P., J. Fairhead, and J. Voss. 1993. Management of common bean diseases by farmers in the Central African highlands. *International Journal of Pest Management* 39:334–342.

Trutmann, P. and M. M. Pyndji. 1994. Partial replacement of local common bean mixtures by high yielding angular leaf spot resistant varieties to conserve local genetic diversity while increasing yield. *Annals of Applied Biology* 125:45–52.

Trutmann, P., J. Voss, and J. Fairhead. 1996. Indigenous knowledge and farmer perception of common bean diseases in the central African highlands. *Agriculture and Human Values* 13:64–70.

Tushemereirwe, W. K. 1996. *Factors Influencing the Expression of Leaf Spot Diseases of Highland Bananas in Uganda*. PhD thesis, University of Reading, United Kingdom.

Ullstrup, A. J. 1972. The impacts of the southern corn leaf blight epidemics of 1970–1971. *Annual Review of Phytopathology* 10:37–50.

Voss, J. 1992. Conserving and increasing on-farm genetic diversity: Farmer management of varietal bean mixtures in Central Africa. In J. Lewinger Moock and R. E. Rhoades, eds., *Diversity, Farmer Knowledge, and Sustainability*, 34–51. Ithaca, NY: Cornell University Press.

Wolfe, M. S. 1985. The current status and prospects of multilane and variety mixtures. *Annual Review of Phytopathology* 23:251–273.

Wolfe, M. S. and M. R. Finckh. 1997. Diversity of host resistance within the crop: Effects on host, pathogen and disease. In H. Hartleb, R. Heitefuss, and H. H. Hoppe, eds., *Plant Resistance to Fungal Diseases,* 378–400. Jena, Germany: Fischer Verlag.

Zhu, Y., H. Chen, J. Fan, Y. Wang, Y. Li, J. Chen, J. Fan. S. Yang, L. Hu, H. Leung, T. W. Mew, P. S. Teng, Z. Wang, and C. C. Mundt. 2000. Genetic diversity and disease control in rice. *Nature* 406:718–722.

12 ⑫ Crop Variety Diversification for Disease Control

Y. Y. ZHU, Y. Y. WANG, AND J. H. ZHOU

Current modern agricultural practices with high input and high output have played a tremendously important role in enhancing rice productivity to meet the increasing food demands and have contributed significantly to food security in China (Lu 1996a, 1996b).

Yet this intensive cultivation, most commonly of a few improved high-yielding varieties on extensive rice farming land, and the long-term application of excessive amounts of chemical fertilizers and pesticides have severely deteriorated the rice ecological systems, rendering agricultural production environments vulnerable. As a result, the occurrence of diseases has become more common and the evolution of pathogens more rapid. The cycles of epidemics and outbreaks of diseases have also become more frequent (Shigehisa 1982; Bonman et al. 1992; Dai et al. 1997; Zhu et al. 2000a, 2000b). All of these factors have brought about a significant reduction in crop yield.

Rice blast, caused by *Pyricularia oryzae* Sacc. (teleomorph *Magnaporthe grisea* Barr.), is one of the epidemic diseases that has been a limiting factor to rice production in Yunnan Province, southwestern China. High inputs of chemical fertilizers and pesticides have not been effective in controlling the rice blast. On the contrary, they have led to the deterioration of rice ecosystems and limited further increases in rice productivity. This chapter discusses how biodiversity deployment was used to control rice blast, in which mixed planting of different rice varieties was adopted (box 12.1), and how the genetic diversity of blast fungi was studied (Shigehisa 1982; Staskawicz et al. 1995; Baker and Staskawicz 1997).

Box 12.1 Varietal Mixture Plantings in China

When traditional blast-susceptible glutinous varieties were interplanted with resistant modern hybrid rice, the result was up to 94% disease reduction of the susceptible varieties. The yield of glutinous rice per unit area on mixture farms was 84% higher than that on monoculture farms, and the yield of hybrid rice decreased by only 1% in mixtures (Zhu et al. 2000a), leading to increased incomes for farmers. The simplicity and effectiveness of this approach attracted active participation by farmers. The alternating rows of short and tall rice varieties has become a prominent feature of many rice fields of Yunnan and other provinces in China, with rapid expansion of the use of this diversification strategy, and the rural landscape has changed greatly. From 1998 to 2002, the area under mixtures expanded in China. As the interplanted area increased, so did the number of varieties used in mixtures. Farmers began interplanting modern rice varieties with other high-quality but blast-susceptible traditional rice varieties and obtained an average gain of 0.5 to 1.0 ton/ha. The rapid adoption of the diversification scheme can be attributed to a systematic extension campaign involving county and village officials, researchers, and extension workers. The extension network ensured that farmers were trained and that a sufficient amount of seed was available at planting time. The expansion was sustained by profitability (average income increase of US$150/ha per farmer) and farmers' preference for certain high-quality varieties for consumption. The production system was also relevant to on-farm genetic conservation of high-quality traditional varieties, which had not been grown for more than 40 years because of their susceptibility to disease but have been brought back into production.

The diversification concept has been extended to other major crops for pest and disease control in Yunnan, particularly for wheat (*Triticum aestivum*), barley (*Hordeum vulgare*), and broad bean (*Vicia faba*). As part of a rice–wheat cropping system, wheat and broad bean were planted in winter on more than 250,000 ha in Yunnan. Wheat stripe rust, caused by *Puccinia striiformis,* is a major disease, causing yield losses as high as 20%. Broad bean is an important cash crop planted in the same season as wheat, but its yield often declines because of serious leaf and stem damage caused by bean fly maggots (*Ophiomyia phaseoli*). Wheat and broad bean were intercropped, with the result that the incidence of wheat rust was reduced by 24% in five locations in Yunnan. Damage caused by bean fly maggots also decreased. The intercrop maintained the same yield of wheat as in monoculture, but broad bean yields increased. By the end of 2002, the area under crop species mixtures (wheat and broad bean, barley and broad bean, oil rape and broad bean, potato and maize, and maize and peanut) had expanded in Yunnan.

Box 12.1 continues to next page

Box 12.1 continued

Although the detailed mechanisms underlying the reduction of pests and diseases in mixed varietal plantings remain to be elucidated, our data indicate that it is possible to combine modern and traditional rice varieties to achieve high productivity and to provide good-quality food and income for the rural population. By reintroducing traditional varieties into a productive but sufficiently diverse ecosystem, in situ conservation can harmonize with intensive production systems. The diversification concept as a means to sustain productivity has spread to other rice-growing countries. In the Philippines, field trials showed that varietal mixtures could reduce the incidence of tungro, a serious viral disease in the tropics. Diversification experiments are also being planned to control blast in the Mekong Delta and central Vietnam, where commonly grown crops are no longer disease resistant. Positive results from these different ecosystems would further support diversification as an important strategy in modern agriculture.

Genetic Diversity of Rice Varieties in Mixture Planting

Yunnan Province, located in southwestern China, is known for rich biodiversity and has been recognized as a part of the origin center of cultivated rice (*Oryza sativa* L.) (Cheng 1976; Oka 1988; Shi et al. 1999). Yunnan has rich rice genetic resources, and new resistance genes have been identified among local rice varieties (Pan et al. 1998).

Sequence analysis of 30 cloned resistance (R) genes and the predicted amino acid sequence showed that the R genes are classified in five groups based on their common molecular features (Baker and Staskawicz 1997). Given the known conserved sequences of the genes, primers (or degenerated primers) to isolate DNA fragments can be designed with sequences corresponding to conserved motif of the R genes from different plant species. The resistance gene analogue (RGA) analysis provides an effective approach for evaluating genetic diversity and identifying candidate resistance genes. RGA markers have been used to characterize germplasm and breeding lines in rice (Chen et al. 1998).

A total of 137 rice varieties were collected from different rice ecological regions in Yunnan Province. These included traditional and hybrid varieties, *Indica* and *Japonica* types, glutinous and nonglutinous ones, and upland rice varieties. The study objectives were

- To evaluate the diversity of rice varieties in Yunnan Province using RGA polymerase chain reaction (PCR) analysis
- To look for DNA markers related to rice blast disease resistance
- To provide a molecular basis for rice disease resistance breeding and efficient use of local rice varieties

PCR Amplification for RGA Analysis

Three RGA primer pairs (S1/AS3, XLRR for/XLRR rev, and Pto-kin1/Pto-kin2) were used in this study (table 12.1). The PCR primer sequences were designed based on the conserved motifs of disease-resistance genes of the Xa21 gene (LRR) for XLRR for/XLRR rev and the DNA sequence encoding for protein kinase in the Pto gene for Pto-kin1/Pto-kin2 and N gene (NBS-LRR) for S1/AS3. Using these primer pairs, it is possible to scan for these three kinds of sequences in total genome DNA and find fragments of related resistance genes with NBS-LRR, LRR, and Pto. The procedures for PCR amplification, denatured polyacrylamide gel electrophoresis, and silver staining were adopted from Chen et al. (1998).

Cluster Analysis

To determine genetic relationships between rice cultivars, all the amplified bands were treated as dominant genetic marks. Cluster analysis was performed based on the binary data using unweighted pair-group

Table 12.1. Polymorphism of 137 rice cultivars based on resistance gene analogue primers.

Primer	Sequence 5'–3'	No. of Amplified Bands	Polymorphic Bands Number	%
S1	GGTGGGGTTGGGAAGACAACG	82	48	58.5
AS3	IAGIGCIAGIGGIAGICC			
XLRR for	CCGTTGGACAGGAAGGAG	41	23	56
XLRR rev	CCCATAGACCGGACTGTT			
Pto-kin1	GCATTGGAACAAGGTGAA	52	28	54
Pto-kin2	AGGGGGACCACCACGTAG			
Total		175	99	57

average for amalgamation (linkage) rule and percentage disagreement for distance measure in joining (tree clustering) method of STATISTICA (release 4.5).

DNA Polymorphisms Detected by RGA-PCR

RGA banding patterns produced by the three primer pairs revealed a high degree of intervarietal polymorphism. The total number of bands scored among these rice varieties generated by the three primer pairs ranged from 30 bp to 2 kb, and their mean polymorphisms are shown in table 12.1. A 350-bp band derived from XLRR for/XLRR rev was specific to *Japonica* cultivars. The *Indica–Japonica* differentiation between partial RGAs probably is the result of long interaction and coevolution of the rice pathogen in different environmental conditions.

Dissimilarity Analysis

The distribution and evolution of RGA in plant genomes partially reflect disease resistances of the plant species. The cluster analysis was performed based on the data of three primer pairs using unweighted pair-group average for amalgamation (linkage) rule and percentage disagreement for distance measures in joining (tree clustering) method. In general, abundant RGA polymorphism was observed among the varieties tested. The varieties were divided into three lineage groups on the basis of the RGA banding data at 96% dissimilarity. Varieties in the first group were *Japonica* and a few landraces. Most varieties in the second group were *Indica*, together with some *Japonica* varieties, such as Xunza 29, Xunza 36, Liming 251, Jingguo 92, and Huangkenuo. It is known that Xunza 29 and Xunza 36 are *Japonica* hybrid rice with the *Indica* pedigree. Varieties in the third group were *Indica*. Dissimilarity between traditional varieties with the same maternal and paternal parents varied from 8% to 70%. This differentiation probably is caused by directional selection and stabilizing selection during rice breeding.

RGA banding patterns and the dendrogram from the cluster analysis revealed a higher degree of polymorphism in *Indica* rice than between *Indica* and *Japonica* rice and in *Japonica* rice. This may be one of the reasons that the suppression of rice blast disease by mixed planting or rotations of *Indica* varieties or between *Indica* and *Japonica* varieties was more efficient than that among *Japonica* varieties (Zhu et al. 1999a).

Genetic Diversity of Rice Blast in Mixtures

A total of 251 isolates were collected from the fields of monocultures and mixtures in Shiping County in 1999 and 2000 and were tested for pathogenicity based on their DNA clustering data (24 isolates from mixture fields, 28 isolates from hybrid monoculture fields, and 10 isolates from glutinous monoculture fields). Two primers (pot2–1:5' CGGAAGCCCTAAAGCTGTTT3' and pot2–2:5' CCCTCATTCGTCACACGTTC3') were used in genetic diversity analysis.

Distinct banding patterns were generated by using the two primers in combination with PCR conditions that favored the amplification of long fragments. The amplified bands ranged in length from 400 bp to more than 23 kb, and there were 83.7% polymorphic bands. A dendrogram was constructed from the Pot2 repetitive element-based PCR fingerprint data.

The 113 isolates from 1999 were grouped into four genetic lineages (G1, G2, G3, and G4) at the 0.65 linkage distance, and the 138 isolates in 2000 were grouped into six genetic lineages at the 0.65 linkage distance (G1', G2', G3', G4', G5', and G6'). Each genetic lineage contained different cultivation pattern and isolates. G1 (G1') comprised 134 isolates, of which 95 isolates were from fields planted with hybrid monocultures and the other 39 were from those planted with mixtures. There were 11 isolates in G2 from fields planted with glutinous monocultures and 20 isolates from those planted with mixtures. G3 (G3') contained 25 isolates, of which 7 were from fields planted with glutinous monocultures and 18 were from those planted with mixtures. G4 (G4') comprised 57 isolates, of which 55 were from fields planted with glutinous monocultures and only 2 were from mixtures. There were 4 isolates belonging to G5' and G6', of which 2 in G5' were from fields planted with mixtures and 2 isolates in G6' were from glutinous monoculture ones.

There were fewer genetic lineages and more obvious dominant ones in fields with monocultures than with mixtures. G1 (G1') was the dominant lineage in fields with hybrid monocultures, and G4 (G4') was the dominant lineage in glutinous monocultures. There were no great changes in the composition of genetic lineage between 1999 and 2000. Rice variety diversity created a stabilizing environment for the pathogens.

Physiological Race Composition of Rice Blast Isolates from Monoculture and Mixture Fields

Sixty-two isolates in 2000 were divided into different physiological races based on their resistant or susceptible reaction with seven different varieties. There were 7 races belonging to 6 groups (ZB, ZC, ZD, ZE, ZF, and ZG) in fields of mixtures, 4 races belonging to 4 groups (ZC, ZD, ZE, and ZG) in fields of glutinous monocultures, and 10 races belonging to 3 groups (ZA, AB, and ZC) in fields of hybrid monocultures. There were more groups in fields of mixtures than in those of monocultures, which was evidence of pathogen-stabilizing selection. The frequency of the dominant race (ZB13) in fields of hybrid monocultures was 50.0%, and that of the dominant race (ZG1) in fields of glutinous monocultures was 70.0%, which resulted in directional selection on the virulent race. It can be concluded that rice variety diversity created an environment that reduced directional selection, limiting the ability of any single pathogen to become more virulent.

Effect of Relative Humidity and Rice Surface Area on Yield

The loss of rice yield caused by blast depended on variety, cultivation techniques, and climatic conditions. Many studies have been carried out on ecological and climatic factors affecting rice blast (Kong and Zhou 1989; Yu et al. 1994; He et al. 1998; Ding et al. 2002). These have shown that pathogen sporulation capacity and rice resistance were greatly affected by temperature, humidity, rainfall, fog, dewdrop, and light. When the temperature was above 20°C, with dewdrop and fog in the morning or the evening, the blast pathogen was found to sporulate most quickly (Dong et al. 2001). Blast conidia were not produced unless relative humidity was above 93%, and the higher the humidity, the greater and faster the conidia production. The development of spores depended on the presence of water droplets, with a critical relative humidity (RH) of more than 96%. In the absence of water droplets, when RH was 100%, only 1.5% of the conidia sprouted (Qiu 1975). Xu et al. (1979) reported that many fungal spores sprouted only when humidity was close to saturation, but they would sprout better under a water drop. Yang et al. (2000) inferred that pathogen sporulation capacity and the prevalence were distinctly affected by humidity and that fungal spores could sprout easily and become infective under saturated humidity.

In recent years, rice variety mixtures for controlling rice blast have been extended to more than 350,000 ha, which has resulted in economic, social, and ecological benefits in Yunnan, Sichuan, and Hunan provinces of China (Zhu et al. 2000b). To study the effect of key factors involved in the control of rice blast in crop mixtures, the field relative humidity and the surface area of a hill of rice were surveyed for moisture drops, which could then provide a theoretical basis for blast control through rice mixture planting.

One short-stem hybrid variety (Shanyou63) and two high-stem glutinous varieties (Huangkenuo and Zinuo) were used in this study (table 12.2). The two high-stem glutinous varieties had similar resistant gene fingerprints (similarity distance of 91%), but there was a big difference in resistant gene fingerprints between the glutinous varieties and the hybrid variety (similarity distance 59%) (Zhu et al. 1999b).

When high-stem glutinous varieties were grown alternately with the short-stem hybrid varieties, the surface areas covered with moisture droplets on a rice hill were much smaller. In 2000, for monocultures of the high-stem glutinous varieties, the average surface area of a rice hill covered by moisture drops was more than twice that in the mixtures. Similar results were obtained in 2001.

When high-stem glutinous varieties were mixed with the short-stem hybrid variety, relative humidity of the field microenvironment was substantially lower in both 2000 and 2001 (table 12.3).

The blast incidence and severity index of glutinous varieties declined in the mixture plots, with no significant differences between the blast control effects of monoculture and mixture for the hybrid rice, Shanyou63.

Table 12.2. Rice varieties and their agronomic traits.

Variety	Type	Resistance to Blast	Growth Period (d)	Plant Height (cm)	1,000-Gram Weight	Grains/ Panicle	Yield (kg/ha)
Shanyou63	*Indica*	Resistant	158	120	30.3	143	10,250
Huang-kenuo	Glutinous	Susceptible	168	160	30	205	3,975
Zinuo	Glutinous	Susceptible	165	155	28	198	3,675

Table 12.3. Relative humidity in mixture and monoculture.

Year	Type	Variety	100%	95%–100%	90%–95%	<90%
			Relative Humidity Range (days)			
2000	Monoculture	H	24	11	11	12
	Mixture	H/S	2	14	22	20
	Monoculture	Z	19	13	6	20
	Mixture	Z/S	6	17	12	23
2001	Monoculture	H	19	12	7	20
	Mixture	H/S	0	9	21	28
	Monoculture	Z	18	7	8	25
	Mixture	Z/S	1	12	16	29

H=Huangkenuo in monoculture; H/S=Huangkenuo with Shanyou63; Z=Zinuo in monoculture; Z/S=Zinuo with Shanyou63.

Plant Silica Content of Rice Varieties

Rice is a typical silicic acid plant. Silicon is the most abundant mineral element in rice, in both percentage content and total content (Chen et al. 1998; Chen 1990). Silica is important in plants because it makes the cells hard and difficult for pathogens to penetrate. When rice plants lack silicon, they are much more susceptible to diseases and insects, such as rice blast, rice brown spot, rice culm rot (*Sclerotium oryzae* and *S. oryzae* var. *irreyulare*), rice stem borders, and rice planthoppers. In addition, the underpart (bottom) leaves of rice easily flag, and this gradually expands to upper leaves; heading stage (period) is delayed two to three days; grains are easily infected by brown spot and neck blast; and stems are weaker and easily lodged (Shui et al. 1999; Hu et al. 2001; Chen et al. 2002). Silicon contributes to resistance to rice blast (Qin 1979), and silicon deposited in rice epidermal tissue forms siliceous cells and a cornified bisilica layer that acts as a mechanical barrier to prevent infection and extension of a pathogen (Yoshida and Kitagishi 1962; Nanda and Gangopadhyay 1984). The silica content in rice directly affects resistance to diseases, insects, and lodging; moreover, it can improve plant shape and increase yield (Hu et al. 2001). Many studies have been carried out on soil and crop silicon nutrient, including the effect of applying silica fertilizer on disease resistance and yield (Ye 1992; Hu et al. 2001).

Because it has been demonstrated that intrafield varietal diversification can be effective for controlling rice blast (Zhu et al. 2000a), this technology of large-scale mixture planting has been extended in 10 provinces of China, including Yunnan Province. The disease resistance of susceptible varieties with high quality is greatly enhanced in mixed rice fields, with an accompanying reduction in fungicide applications and lodging of tall varieties and an increase in grain yields of high-quality glutinous rice. At the same time, economic, social, and environmental benefits have been achieved (Zhu et al. 2000a).

To understand the mechanism of varietal diversification mixtures for blast and lodging management, two main traditional varieties for mixture extension were chosen to research their silica content in mixture and monoculture. The results could provide a scientific basis for varietal combinations that are effective in diversification experiments for blast and lodging management.

The two traditional varieties included a high-stem glutinous variety and a high-stem upland rice variety of high quality but susceptible to blast and easy lodging. One short-stem hybrid variety with high yield and resistance to blast was also included. A medium-rich fertility field was chosen for this experiment at Donghong Village, Mile County, Yunnan Province. Experimental treatment and plot setting were reported in Zhu et al. (2000b). All plots were managed by research staff and treated in the same manner as the surrounding mixed varieties without application of fungicide.

Samples were gathered for scanning electron microscope (SEM) analysis of form and numbers of siliceous cells; other samples were gathered for measuring silica content.

Rice neck blast was assessed seven days before harvest. Each sampled panicle was visually examined by experienced personnel to estimate the percentage of branches that were necrotic because of infection by *Magnaporthe grisea*, in which each panicle was given a rating from 0 to 5, where a disease severity of 0 would indicate no disease and 5 would indicate that 100% of panicle branches were necrotic. Disease severity was summarized within each plot as $\{[(N_1 \times 1)+(N_2 \times 2)+(N_3 \times 3)+(N_4 \times 4)+(N_5 \times 5)]/ \Sigma N_0 \ldots N_5\} \times 100$, where $N_0 \ldots N_5$ is the number of culms in each of the respective disease categories.

The cleaned rice stems were reduced to ash for silica content analysis. The average silica content of traditional varieties in mixture was higher than in monoculture (table 12.4). The difference of silica content between mixture and monoculture was significant except for Milexianggu in maturation stage.

For SEM analysis, observation samples were prepared (Revel et al. 1983), and then examined in the KYKY-1000B SEM (amplifying power 800×,

Table 12.4. Silica content of rice stems (%).

Variety	Growth Stage	Type	Replication	Replication 2	Replication 3	Mean	Increase Rate	T test ($t_{o.o5} = 2.78$)
Huangkenuo	Booting	Monoculture	8.11	7.57	7.22	7.63	14.68	3.89*
		Mixture	8.64	8.61	8.99	8.75		
	Maturation	Monoculture	7.52	7.51	8.05	7.69	11.83	3.14*
		Mixture	8.28	8.48	9.04	8.60		
Milexianggu	Booting	Monoculture	6.4	6.37	6.89	6.55	14.81	3.13*
		Mixture	7.03	7.91	7.63	7.52		
	Maturation	Monoculture	6.63	5.48	5.55	5.89	16.47	1.89
		Mixture	6.28	6.8	7.51	6.86		

*Significantly different at .05 level.

accelerating voltage 18 kV) to observe the form and numbers of siliceous cells.

There were prominent effects on the form and numbers of cuticular siliceous cell in mixture fields. The SEM (800×) showed that the form and numbers of siliceous cells of traditional varieties in mixture were very different from those in monoculture. The siliceous cells of traditional varieties in mixture were bigger, and their numbers were greater than those in monoculture.

Rice Blast, Lodging, and Yields in Mixture Fields

A mixture of traditional and modern varieties can be effective for controlling rice blast. The incidence, severity index for neck blast, and lodging ratio of traditional varieties were significantly lower (table 12.5). Survey results in 1998–2002 revealed that the disease incidence for traditional varieties in monoculture was 5.73–100% and the disease index was 0.011–0.804, whereas those of mixed plantings were only 1.14–58.79% and 0.0024–0328, respectively. For modern varieties, the disease incidence in monoculture was 1.3–81.9% and the disease index was 0.0026–0.486, whereas those of mixed plantings were 1.27–65.1% and 0.0045–0.297, decreases of 36.75% and 39.82%, respectively.

Because of ecological differences in the planting regions and different resistances of the varieties, large differences were found between different regions and different variety combinations in the effect on blast control. However, planting mixtures always had a positive effect on blast incidence when compared with monocultures.

In the mixed planting system, the total yield was 8,577.9 kg/ha, which included both the average yield of modern varieties and traditional varieties, 8,044 kg/ha and 533.9 kg/ha, respectively. In monocropping system, the average yield of the modern varieties was 8,060.5 kg/ha and that of traditional varieties was 3,663 kg/ha. Thus the total yield of mixed planting with both modern and traditional varieties was higher than in a monoculture of either the modern or traditional varieties.

Expanding the Scale of Mixed-Variety Planting

The expansion of mixed-variety planting to a large scale has been carried out since 1998, and more and more varieties have been selected to make

Table 12.5. Rice blast and lodging ratio.

Variety	Type	Lodging		Neck Blast			
		Ratio (%)	Resistance (%)	Incidence (%)	Decrease Rate (%)	Severity Index	Decrease Rate (%)
Huangkenuo	Monoculture	99.38	0.62	56.02**	77.26	43.61**	82.69
	Mixture	0	100	12.74**		7.55**	
Milexianggu	Monoculture	97.68	2.32	66.2**	80.80	42.7**	81.01
	Mixture	0	100	12.71**		8.11**	

**Significantly different at .01 level.

Table 12.6. Number of varieties used in planting mixtures in Yunnan Province.

Varieties	1998	1999	2000	2001	2002
Traditional	2	4	40	62	94
Modern	2	3	12	15	20
Combination	4	8	65	121	173

combinations (table 12.6). In 1998, expansion of planting of rice mixtures for blast control was carried out in Baxing, Maohe, Baoxiu, Yafangzi, and Taochun in Shiping County in an area of 812 ha. In 1999 the mixture planting was conducted in six counties, including Shiping, Jianshui, Honghe prefecture, in an area of 3,534 ha. It was increased to 34,740 ha over 40 counties in 2000, 84,467 ha in 2001, and 136,189 ha in 2002. The total area in 15 prefectures of Yunnan Province increased to 259,742 ha from 1998 to 2002.

Varietal Combinations

Varietal combination was based on a comprehensive analysis of the varieties' resistance background, agronomic character, economic value, local cultivation conditions, and the planting habits of farmers. The selection norm for resistance background was that genetic similarity had to be lower than 70%, using RGA analysis. Tall varieties were combined with short varieties based on the requirement that the difference in height had to be more than 30 cm and the difference in maturity period no more than 10 days. To boost the participation of farmers, mixed plantings had to provide a complementary economic effect and meet the demands for both high yield and high quality. Traditional varieties were selected for mixing with modern varieties based on local cultivation conditions such as irrigation, fertility, soil productivity, and elevation. At the same time, the varieties favored by farmers were chosen in mixed plantings so as to conform to local planting habits.

Growing Procedures

For convenient harvesting, it was required that the different varieties have the same harvest time. Therefore the sowing was adjusted according to the growth period of the different varieties. For example, the tall, high-quality traditional varieties Nuodao, Xiangdao, Zidao, and Ruanzhimi were sown 10 days earlier than the modern high-yielding hybrid varieties. Seedlings were transplanted to field from the nursery bed in April and May. Survey plots were randomly selected to record blast occurrence and rice yields in each area. The blast scoring system described in the Chinese national standard (Anonymous 1996) was used, and the yields were effective output.

Training

Because farmers were the major implementers of mixed plantings, it was important to raise awareness among them so that they would consciously adopt the relevant planting techniques.

At agricultural meetings held at different government levels, village leaders were trained in the social and ecological effects of planting mixtures in order to gain their support. At the same time, the methods, procedures, and essential points of the planting techniques and key measures of expansion were also introduced to enable them to become the technical supervisors. Agrotechnicians were the main agents of technique expansion, consequently becoming the main focus for technical training. Ninety-three training stations for agrotechnicians were established in 15 prefectures of Yunnan Province. These trained agrotechnicians then organized the farmers of a village for training through farmer schools at night or other free times. During the period of nursery and transplanting, the actual operation of the techniques was demonstrated in the field to village officers and representative farmers. Skilled farmers then took the lead in carrying out the work. Radio, television, and posters were widely used to introduce the technique. Thirty-three demonstration stations, three TV stations, and 29 video compact disc stations were established in Dali, Kunming, Dehong, Lijiang, Linchang, Simao, Zhaotong, Chuxiong, Luxi, Xiangyun, and Binchuan in 2002. Three hundred and eighty-two thousand copies of training files were printed, 836 issues of TV programs and PowerPoint files were made, and 5,871 training courses were organized in Yunnan Province. The total number of farmers trained was 929,000.

Demonstration fields were very important for the expansion process. Between 1998 and 2002, 64,133 ha of such fields were established in 90 counties of 15 prefectures. In Mile, Jianshui, and Tengchong counties, 6,667 ha of continuous demonstration fields in normal cultivation were built, and these boosted the expansion greatly.

Conclusion

From these results, we can conclude that crop variety diversification is an effective solution to the vulnerability of monoculture crops to disease. Both theory and observation indicate that genetic heterogeneity provides greater disease suppression when used over large areas. Our results support the view that crop variety diversification provides an ecological approach to disease control that can contribute to the sustainability of crop production while reducing fungicide applications. Our results also demonstrate that collective efforts from groups of scientists, institutions, and farmers are vital to the development and dissemination of an effective diversification technology. Wide adoption of a diversification technology depends on simplicity, effectiveness, and an ability to bring about obvious economic benefits to farmers.

On the other hand, crop variety diversification could not provide all the answers to the problems of controlling diseases and producing stable yields in modern agriculture. More research is needed to find the best packages for different purposes and to breed varieties with specific use in mixtures.

References

Anonymous. 1996. *Rules for Investigation and Forecast of the Rice Blast*. The State Standard of the People's Republic of China, No. Gb/t 15790–1995, 1–13. Beijing: China Standard Press.

Baker, J. and Z. J. Staskawicz. 1997. Signaling in plant–microbe interactions. *Science* 276:726–733.

Bonman, J. M., G. S. Khush, and R. J. Nelson. 1992. Breeding rice for resistance to pests. *Annual Review of Phytopathology* 30:507–528.

Chen, J., G. Mao, G. P. Zhang, and H. D. Guo. 2002. Effects of silicon on dry matter and nutrient accumulation and grain yield in modern Japonica rice (*Oryza sativa* L). *Journal of Zhejiang University (Agriculture & Life Sciences)* 28(1):22–26.

Chen, X. M., R. F. Line, and H. Leung. 1998. Genome scanning for resistance-gene analogs in rice, barley and wheat by high-resolution electrophoresis. *Theoretical and Applied Genetics* 97:345–355.

Chen Y. Q. 1990. Characteristics of silicon uptaking and accumulation in rice. *Journal of Guizhou Agricultural Sciences* 6:37–40.

Cheng, T. T. 1976. The origin, evolution, cultivation, dissemination, and diversification of Asian and African rice. *Euphytica* 5:425–441.

Dai, S. F., Z. H. Ye, Y. Z. Cao, and Y. Y. Guo. 1997. Disaster-causing characters and disaster-reducing strategies of crop pests in China. *Chinese Journal of Applied Ecology* 10:119–122. [in Chinese]

Ding, K., G. Tan, Z. Gao, and B. Ji. 2002. Effects of ecological factors on infection process of *Pyricularia grisea*. *Chinese Journal of Applied Ecology* 13(6): 698–700.

Dong, J., H. L. Li, J. M. Wang, A. Y. Ding, J. Chen, J. H. Zhu, W. Wang, B. D. Li, and Y. Q. He. 2001. *Agricultural Plant Pathology* (northern edition), 2–7. Beijing: China Agricultural Press.

He, M., D. Lu, and J. Mao. 1998. The effect of key ecological factors on rice blast disaster. *Journal of Southwest Agricultural University* 20(5):392–395.

Hu, R., S. Fang, and G. Q. Chen. 2001. Effects of silicon ion the physiological targets and yield of modern rice. *Journal of Hunan Agricultural University (Natural Sciences)* 27(5):335–338.

Kong, P. and R. Zhou. 1989. The multi-effect and modeling of dew temperature and time and nitrogen application on infection on *Pyricularia grisea*. *Acta Phytopathologica Sinica* 19(4):223–227.

Lu, L. S. 1996a. Agriculture and agricultural science and technology in the 21st century. *Science and Technology Review* 12:1–8. [in Chinese]

Lu, L. S. 1996b. The current status, perspectives and strategy of modern agriculture development. *Science and Technology Review* 2:41–44. [in Chinese]

Nanda, H. P. and S. Gangopadhyay. 1984. Role of silicated cells in rice leaves on brown spot disease. *International Journal of Tropical Plant Disease* 2:89–98.

Oka, H. I. 1988. *Origin of Cultivated Rice*. Tokyo: Japan Scientific Societies Press.

Pan, Q. H., L. Wang, T. Tanisaka, and H. Ikehashi. 1998. Allelism of rice blast resistance genes in two Chinese rice varieties and identification of two new resistance genes. *Plant Pathology* 47:165–170.

Qin, S. 1979. The analysis about the effects of rice resistance diseases and increasing yield using silicon fertilizer. *Zhejiang Agricultural Sciences* 5:12–15.

Qiu, W. 1975. *Agricultural Plant Pathology*, 1–11. Beijing: Agricultural Press.

Revel, J. P., T. Bernard, G. H. Haggis, and S. A Bhatt. 1983. Science of biological specimen preparation for microscopy and microanalysis. In *Proceedings of the 2nd Pfefferkorn Conference*. O'Hare, IL: SEM Inc.

Shi, Z. M., S. C. Qin, and S. H. Jiang. 1999. *Famous Flowers from Yunnan*. Kunming, Yunnan, China: Yunnan Science and Technology Press. [in Chinese]

Shigehisa, K. 1982. Genetics and epidemiological modeling of breakdown of plant disease resistance. *Annual Review of Phytopathology* 20:507–528.

Shui, M., D. Chen, S. C. Qin, and S. H. Jiang. 1999. The silicification of young tissues of rice and relationship with its resistance to blast of rice. *Plant Nutrition and Fertilizer Science* 5(4):352–358.

Staskawicz, B. J., F. M. Ausubel, J. Baker, J. G. Ellis, and J. D. G. Jones. 1995. Molecular genetics of plant disease resistance. *Science* 268:661–667.

Xu, Z. G., X. B. Zhen, H. F. Li, H. S. Shang, and W. Z. Liu. 1979. *Common Plant Pathology*, 2nd ed., 2–7. Beijing: China Agricultural Press.

Yang, X. M., G. Q Li, X. Li, J. G. Wang., L. Li. 2000. *Plant Ecological Phytopathology*, 51–52. Beijing: China Agricultural Science and Technology Press.

Ye, C. 1992. The relationship between soluble-silicon in soil, yield grain and rice physiology. *Journal of Agricultural Science Translation Series* (1):24–27.

Yoshida, S. and K. Kitagishi. 1962. Histochemistry of silicon in rice plant. *Soil Science and Plant Nutrition* 8(1):30–41.

Yu, L., J. Y. Zhang, and W. J. Fan. 1994. The effect and forecast of weather factors on rice blast. *Heilongjiang Weather* (2):35–36.

Zhu, Y. Y., J. X. Fan, Y. H. Wang, and S. F. Yu. 1999a. Demonstration trial of mixture variety culture for rice blast management. In S. Yu, ed., *Symposium of the Key Laboratory for Plant Pathology of Yunnan Province*, Vol. 2, 93–100. Kunming, Yunnan, China: Yunnan Science and Technology Press. [in Chinese]

Zhu, Y. Y., Y. Y. Wang, H. R. Chen, J. H. Fan, J. B. Chen, and Y. Li. 1999b. Exploiting crop genetic diversity for disease control: A large-scale field test. In *Articles Collection of Key Laboratory for Plant Pathology of Yunnan Province*, 75–80. Kunming, Yunnan, China: Yunnan Science and Technology Press.

Zhu, Y. Y., H. R. Chen, J. H. Fan, Y. Y. Wang, Y. Li, J. B. Chen, J. X. Fan, S. S. Yang, L. P. Hu, H. Leung, T. W. Mew, P. S. Teng, Z. H. Wang, and C. C. Mundt. 2000a. Genetic diversity and disease control in rice. *Nature* 406:718–722.

Zhu, Y. Y., H. R. Chen, J. H. Fan, Y. Y. Wang, Y. Li, J. B. Chen, Z. S. Li, J. Y. Zhou, J. X. Fan, S. S. Yang, M. G. Liang, L. P. Hu, C. C. Mundt, E. Borromeo, H. Leung, and T. W. Mew. 2000b. Current status and prospects of mixture planting for the control of rice blast in Yunnan [A]. In T. W. Mew, E. Borromeo, and B. Hardy, eds., *Impact Symposium on Exploiting Biodiversity for Sustainable Pest Management*, 21–23. Kunming, Yunnan, China: International Rice Research Institute.

13 ❦ Managing Biodiversity in Spatially and Temporally Complex Agricultural Landscapes

H. BROOKFIELD AND C. PADOCH

Farmers manage biodiversity. At one extreme, they may minimize it by planting thousands of hectares to a chemically enhanced and protected single crop or, at the other, create a diverse landscape of patches under multiple crops and trees interspersed with edges and woodlots. This chapter departs significantly from the subject matter of the preceding chapters. It is about biodiversity management at the scale of whole farms and farming regions, including not only agrobiodiversity but also natural and other managed biodiversity.

This chapter also views biodiversity in agricultural landscapes at a somewhat broader temporal scale. By rotating crops and modifying and managing natural regrowth after cropping, farmers ensure continued production of crops. Farmers take advantage of seasonal changes in water and soil conditions to introduce or encourage plant complexes that can survive and flourish in different seasons. Some cope with problems such as soil degradation, salinity, and waterlogging by changing management practices to compensate and thereby create mosaics of land use better adapted to environmental dynamics. All these modifications affect biodiversity at the landscape scale. The purpose of this chapter is to evaluate these wider changes and discuss some of the scientific efforts made to understand and measure them.

Agricultural Landscape

Much recent work on biodiversity has focused on small plots and detailed analyses. On the other hand, reconnaissance work for conservation purposes has often been carried out in large areas that are thought to be of special value and over which protective regimes have been proposed or applied. In the more specialized study of agricultural biodiversity (agrobiodiversity), farmer selection, deliberate or inadvertent, is an important element. Thus farms and their fields, orchards, gardens, fallows, and pastures become significant units for sampling and investigation. Farmers, too, are a diverse group of people.

Biodiversity comes together in patches, fields come together in farms, and farms come together in rural communities. If we are concerned with the maintenance of biodiversity on farm, we need to look at areas within which metapopulations, interconnected by gene flow and subject to change and replacement, have meaning. Everything comes together, then, at a level somewhere between the patch, or field, and the large region. This is where the structure of diversity is expressed, where its generating processes operate, and where interrelations can be observed and understood. This is the landscape, but we need to try to define this in more positive terms before we can begin.

As a scientific entity, as opposed to its qualitative meaning of a view as seen from a particular viewpoint, *landscape* is not easily defined. It came into Anglophone science from German geography of the late 19th century, in which the *Naturlandschaft* and *Kulturlandschaft* of specific regions were analyzed, sometimes in an integrated manner. Analysis depended on maps, and nowadays on remote sensing, but the definition remains linked to what is visible at ground level, and thus the units of landscape are defined within the topographic range of scales. These have become significant in ecology since the 1970s through the evolution of the notion of patches and mosaics of patches, and Forman (1995:13) has usefully defined landscapes as areas in which a mix of local ecosystems and land uses is repeated in similar form over a wide area. By the empirical evidence of writings about ecology and land cover, landscape areas may range from a few square kilometers to several hundreds, even more in sparsely peopled areas with poorly described landscape history. Even the smaller areas contain micro-environmental diversity, often dynamic. Various systems of management, adapted to this diversity, create the pattern of land uses.

Pure science apart, the most common purpose of biodiversity analysis at landscape level is to measure or estimate change resulting from human use and change in the conditions of that use. This has become of particular importance because of the great changes that have taken place since the 19th century and especially since 1950. Population growth has been a basic driving force of change, with global totals increasing from 1.25 billion to more than 6 billion since 1850. Huge changes in agricultural technology have taken place since 1950 alone, with great success in terms of production but with serious ecological consequences. It is common wisdom that a great loss of species and genetic diversity has taken place, in areas both with and without modern agricultural technology.

It took less than 30 years of what is already called conventional agricultural technology before consequences in terms of pollution, soil loss and deterioration, deforestation and landscape homogenization, genetic erosion, and the impoverishment of areas unsuited to mechanization and chemicalization became matters of serious concern among policymakers and among a minority of farmers. In the regions most changed by the new technologies, these concerns have overtaken the earlier and still widespread concerns simply because of intensification of human use.

In Europe, where only about 3% of the landscape carries what can still be described as a natural vegetation and where 44% is managed in farms, land degradation and other changes became matters of public concern as early as 1980. By the 1990s, these concerns had led to initiation of what are now becoming major changes in the common agricultural policy of the European Union. These involve new basic standards of environmental management, which will be applied to all farms receiving subsidies, and specifically funded agro-environmental programs, which are now in use in all member countries, though with very different levels of participation (Piorr 2003). About one farm in seven is involved, and 17% of farmland in the pre-2004 European Union is subject to some type of agro-environmental program (Bureau 2003). With almost the entire European area subject to anthropogenic land use, solutions must be found through land use management. Whereas some agro-environmental programs involve no more than reducing livestock densities, others are more constructive, and some seek to create or recreate hedgerows and copses to link remaining areas of woody species and break up the wholly cleared areas that have been greatly enlarged since 1950. The aim is to restore a measure of diversity in a mosaic of suitable habitat patches at landscape level.

Characterizing Landscape-Level Biodiversity: Europe and the Developing Countries

Although the fact that diversity is disappearing is rarely disputed, monitoring change precisely continues to challenge researchers. Europeans are prepared to have part of their taxes spent on restoring their agricultural environment, and farmers participating in agro-environmental programs are paid to do so. This creates a need for monitoring, and for several years there has been a rising effort to find ways to characterize and monitor changes in biodiversity at landscape level. Although Europe is very different from the developing countries that are the main focus of this volume, the fairly intricate mosaic of land uses that still characterizes a large part of the continent makes it more similar to the latter than are the wide landscapes of, for example, North America. It is therefore worthwhile to examine some of this work, most of it in Germany.

A range of methods has been explored. Some have concentrated on inventories of the plant biodiversity on land that has come to be used in different ways; one such study in an area where farming has been given up in stages since the 1950s found, unsurprisingly, that biodiversity increased with the number of years since cultivation ceased (Waldhardt and Otte 2003). In order to avoid the large input of time and money that such standard inventories entail, a large amount of effort has been put into the search for indicator species that can be readily identified and used to monitor change. There has been particular focus on insect fauna, such as beetles, that can be trapped quickly (Duelli 1997; Büchs 2003). Sampling is major problem, and some approaches have focused specifically on the subclassification of landscape into habitat type areas. Landscape structure, involving the nature and scale of the mosaic, can itself be a valuable surrogate indicator, taking account of the influence of the matrix surrounding managed sites on species richness (Dauber et al. 2003).

One study used a combination of Landsat imagery and a detailed biotope mapping, carried out some years earlier, to develop a stratified sample (Osinski 2003). An ecological area sampling project used satellite-generated land cover data to develop an initial classification of 28 land classes for Germany, within which samples of $1km^2$ were drawn (of agricultural land only) for detailed analysis of their biotope content (Hoffmann-Kroll et al. 2003). This work was carried out in the mid-1990s, at about the same time as the large-scale British countryside survey, which used a similar approach to the search for country-wide information based on representative sites (Haines-Young et al. 2000). Opperman (2003) proposed an even more indirect but thoroughly participatory method, evaluating the ecological management of specific farms by presence of a few indicator species—both flora and fauna—but principally by physical characteristics of the farm space and its management.

In a comparative review of recent work mainly in Germany and Switzerland, Waldhardt (2003) and Waldhardt et al. (2003) concentrate on the value of combining organismic and landscape indicators, which may well be the road forward. However, the search for indicators raises many problems, and the methods of sampling and assessment proposed all have high costs. Both the species groups and the reference areas considered in most of this work are small, and the search for indicator species that can be used widely to monitor progress in agroenvironmental work still has a long way to go. The whole European effort, despite its regionally dense level of scientific input, is still at an early stage, although an enormous amount of valuable information has been gathered. The long-term aim of developing a set of indicators for agricultural landscapes that have international validity, as proposed by the OECD (1997), remains almost as far from achievement as when it was first proposed.

Surrogate indicators can barely be envisaged in the developing countries in view of the great range of agricultural systems, climates, and biotic and abiotic conditions. Although habitat diversity and pattern are potentially important, their interpretation from remote sensing and ground-truthing demands skills and resources that are available in only a few of these countries. Sample area surveys on the ground still have to provide most of the information. Despite their limited power of explanation, the 50 or more quantitative measures of biodiversity to be found in the literature, most developed several years ago, remain the only tools available for classifying biological diversity, whether in agriculturally used or natural areas (Whittaker 1972; Magurran 1988).

The 12-country People, Land Management and Environmental Change (PLEC) project set out with the hypothesis that agricultural management using diversity strategies can sustain and even enhance biodiversity. This view has gained support in Europe, where 1,000 years of agriculture, until the development of modern technology in the 1950s, had the effect of creating a dynamic mosaic of habitat or ecotope patches that enhanced not only species diversity but also structural and functional diversity, and probably genetic diversity as well, among plants and animals (Waldhardt et al. 2003). For PLEC, which was mandated to prepare biodiversity inventories, it was necessary to record diversity in all its demonstration site areas, and a sampling scheme was set up to do this in 1999 (Zarin et al. 2002), followed by a database design (Coffey 2000) and detailed guidelines on calculation of the most relevant indices of α (and its area summation γ) and β diversity[1] (Coffey 2002). PLEC was concerned only with the diversity of vascular plants, not with fauna at any level.

Full stratified random sampling was, even more than in Europe, logistically infeasible, so our sampling procedure was more purposive than random. It went through three stages. In each of 12 countries, one to seven landscape areas (the demonstration site areas) were chosen to represent the territories of particular villages or groups of farmers with whom contact had been established and where the project was invited to work. They ranged from less than 10 km^2 to a notional maximum (never achieved) of 100 km^2 but often lay within transect bands in which reconnaissance work had been done before final selection. Within these landscapes, broad land use classes distinguished by a superficially common groundcover were first identified. Because we were working largely in areas where land rotational practices were or had recently been present, and to stress the impermanence of land cover, we called them land use stages. In 12 countries, 27 such stages were identified, reducible for comparative purposes into seven main categories, including edges (Pinedo-Vasquez et al. 2003a).

Within these larger classes, we sought characteristic types or assemblages of habitats or biotopes. Because of an emphasis on defining these by farmers' management practices, we called them field types, although they also included different stages of managed or unmanaged fallow and of forest. Actual sample areas were then selected within these field types, in a biased manner with emphasis on greatest apparent diversity or on the land worked by particular households on which other information was collected (Guo et al. 2002). Within these, sample quadrats were marked for enumeration of species. Details on management practices in the whole sampled field around the biodiversity enumeration quadrat were collected at the same time (Brookfield et al. 2002). Home gardens and edges between fields were separately sampled and treated in different ways (Zarin et al. 2002).

PLEC's biodiversity assessment was done on the ground, only sometimes with partial aid from air photographs and remote sensing images. The system was designed to make the best use of limited human and financial resources. PLEC's purpose was to study farmers' management and its effects. Such work has to be done in close collaboration with the farmers. In a small area of a few square kilometers on the upper slopes of Mt. Meru in Tanzania, Kaihura et al. (2002) found the order of detail that is summarized in box 13.1, noting that because planting takes place three times a year in this area, the crop composition of fields may change every few months. The crop composition was one important criterion for distinction of field types, and a great deal of other information was also recorded, including land ownership, age and wealth of the farmer, slope, fertility rating, evidence of nitrogen, phosphorus, and potassium deficiency, type of tillage and tillage tools used, livestock raised, methods used to control pests and weeds, and methods used to manage erosion, soil moisture, and drainage. Further inquiry on one farm, with 12 field types encountered in 1999, revealed 10 different food and cash crops, 6 types of trees, more than 10 medicinal plants used in curing more than 30 diseases, 17 types of nursery seedlings for propagation and sale, 6 vegetable crops, 18 fruit tree s, and 7 ornamental plants (Kaihura 2002:136). Thus does management diversity give context to agrobiodiversity.

Scale is an important consideration. Habitat types or field types can be hard to distinguish from land use stages when they are repeated over large areas. Though also distinguishable by different floristic composition, they are always determined by differences in farmers' management. Over much of southeast Asia the irrigated pond fields, alternately cultivated and fallowed dry fields, planted and managed agroforests or woodlots, and intensively managed home gardens constitute just four main classes of field type, each constituting a land use stage, but each can be subdivided in terms of crop content or management. Similarly, on the Fouta Djallon of Guinée, West Africa, all land except small areas of forest and uncultivable waste can be classified into three land use stages: the intensively cultivated infields that are cultivated all year and every year, the more extensive outfields and the associated fallow land, and small areas of planted and managed agroforest. At the level of a single Fouta Djallon village, these could be subdivided into a larger number of field types, together with the edges between them. Both levels of classification are valid, and both relate to the whole landscape. Which is chosen depends on the purposes of characterization.

BOX TABLE 13.1.

Land Use Stage	Field Types	Field Type Description
Natural forest	Least disturbed	Upper foothills of Mount Meru; inaccessible because of steepness and deep incised valleys. Slopes from 85% to 50%; humid tropical climate; some wild animals; area gazetted.
	Slightly disturbed	Upper footslopes of Mount Meru; used for timber, firewood, and medicinal plants; distance from village and steepness limits use. Slopes from 15% to 35%; humid tropical climate with few wild animals; area gazetted.
	Highly disturbed	Cone-shaped hilltops some times used for recreation; used for timber, firewood, and medicinal plants. Treeharvesting controlled by village, butmost economic trees and shrubsalready harvested.
Planted forest	*Pinus* with temporary cropping	*Pinus* trees planted after clearing natural forest; maize and beans commonly in rotation with cabbage and potatoes; crop combinations and sequences differ between farmers and seasons. Slopes from 10% to 20%.
	Cypress with temporary cropping	Cypress trees planted; cropping system similar to *Pinus* plantations.
	Eucalyptus plantation	Natural forest cleared and planted with eucalyptus only.

Box 13.1 continues to next page

Box 13.1 continued

Land Use Stage	Field Types	Field Type Description
Agroforestry	Crops and trees	Complex mixes of crops and trees depending on farm size, season, and farmer preference; coffee, banana, and trees, with maize and beans most typical. Varying slopes.
	Maize and beans with trees	Maize and beans as intercrops, with trees as hedges on contours and boundaries; the most economic crops occupy the largest area.
	Potatoes in rotation with vegetables	Commercial potatoes in first season, followed by cabbage and fallow in the third season of the year.
	Maize	Maize planted as monocrop.
	Potatoes	Potatoes as a commercial monocrop.
	Farm boundaries	Boundary fences and partitioning structures with trees, shrubs, and climbers. Species have diverse uses, but most have thorns to limit trespass.
	Plot boundaries	Structures separating field types within farm, including crop residue and weed piles along boundaries, creepers, and shrubs of economic value. These may be destroyed and spread for soil fertility improvement.
	House gardens	Near the house with local and introduced vegetables. Mostly on flat areas or gentle slopes with irrigation.
Water source	Microcatchment	Delineated patches less than 30 m^2 protecting water seepage points; planted with perennial trees and bananas. No tree harvesting; trespass limited to fetching water; owned communally.

Box 13.1 continued

Land Use Stage	Field Types	Field Type Description
Fallows	Regenerating fallows	Communal or individual plots temporarily left uncultivated for fertility recovery. Steep to moderately steep slopes.
	Pastures, recreation, or fallows	Lands left fallow or family recreation places; goats may graze.
	Tethering and cut-and-carry fields	Pastures where cows are tethered for grazing or servicing (in case of bulls); grass may also be cut for fodder.

Source: Kaihura et al. (2002:155).

Farmers and Other Users of Biodiversity

Whether the landscape is a big region or the territory of only a single community, farms are the units through which most of its diversity is managed for production. Farmers rarely manage only one field type or even one land use stage; they often include areas of forest, planted woodland, and water bodies as well as arable and pasture land and the edges between these types. Fallow land may or may not be managed, and it very often provides resources that are harvested. The measurement and recording of diversity at landscape level must have not only the agreement of the landholder or user but also his or her active cooperation. Even on tracts of common property, there is much to be learned from those who use the resources.

In PLEC, we made extensive use of the concept of agrodiversity, first proposed by Brookfield and Padoch (1994), going beyond the natural versus cultural division of most landscape study to interrelate agrobiodiversity, management diversity, and biophysical diversity and put them into the context of a fourth dimension, which we called organizational diversity (Brookfield 2001; Brookfield et al. 2002). The latter term needs explanation. Whether or not it sets out to make money, a farm is a working enterprise with a distinctive set of relationships with parallel enterprises and the higher levels of the community, the authorities, and the regional, national, and global economies. Like any other enterprise, it is both a social and an economic system nested within larger social and economic systems. The operators of the farm are land managers in the sense used by Blaikie and Brookfield (1987). Even if they have to work within a system that determines what crops and livestock are produced, the farmers or farming households have to make the yearly, monthly, and daily rounds of decisions needed to obtain that production. Farms differ greatly from one another, and the resources and skills of farm operators also differ greatly.

This is a central part of diversity. It includes diversity in the manner in which farms are owned or rented and operated and in the use made of resource endowments and the farm workforce. Elements include labor, household size, the differing resource endowments of households, and reliance on off-farm employment. Also included are age group and gender relations in farm work, dependence on the farm rather than external sources of support, the spatial distribution of the farm, the amount of mutual aid that is practiced between farms, and differences between farmers in access to land. Tenure of resources, the conditions of access to them, and what Leach et al. (1999) describe as environmental entitlements are fundamentally important. Organizational diversity is involved in all management of resources, including land, crops, labor, capital, and other inputs.

Whatever the conditions of tenure, the skills needed in simple organization of the workforce at periods of peak demand are much undervalued in the general literature on agricultural development. The shift from single to double or even triple cropping made possible by Green Revolution innovations was enormously demanding of these skills. Yet farmers received little guidance and instruction on how to manage their resources and workforces at such times. They learned this by themselves. Organizational diversity is highly dynamic. Farmers change their organization of labor and resources according to circumstances, sometimes in a very short time, and are quick to respond to signals that call for new ways of combining the factors of production.

The expert farmers who do this best are not often political or social leaders in their communities. PLEC in China found a remarkable example of an innovating expert, Mr. Li Dayi, a former shifting cultivator and hunter. In the 1980s he became interested in experimenting with domestication of a rare but valuable timber species found in the forest, *Phoebe puwenensis*. Although no botanically established means existed, he succeeded in two years in growing viable seedlings. He then converted 0.13 ha of maize-growing land allocated to him in the privatization of collective land that took place in 1983 into a mixed-tree plantation. With support from PLEC, he has extended his technology to 95 village farmers (Dao et al. 2003).

In the past few years farmers in a remote Papua New Guinea village have modified their subsistence farming system to incorporate cash crops. A number of them are planting cacao or coffee seedlings in the garden during the first and second year of yam cultivation. Until around 1990 the only cash crop in the area was robusta coffee, introduced by the extension service in the 1960s, and it was grown on small plots averaging 150 trees per plot, surrounded by secondary forest and shaded by *Leuceana*. Very little additional coffee was planted after the initial enthusiasm of the 1960s, when all families planted at least one plot and often two. However, between 1990 and 2001, more than 70,000 cacao trees were planted. In this case, the shifting cultivation system is being modified in response to new conditions. In the old final stage of a three-year cropping life, plots became dominated by weeds, but these have been controlled in modern times by the introduction of ground-mantling sweet potatoes. Fallow tree species and tall grasses now are weeded out, and *Gliricidia* is planted to shade the cacao. Thus the food garden is transformed into a cash crop garden. Farmers argue that in 20 years, they will clear the cacao and plant food again. They know that land cleared from cacao and *Gliricidia* or *Leuceana* grows food crops as well as a 20-year forest fallow. So the consequences of this practice will not be a reduction in food production. Rather it will be, over 20 years, a significant loss of natural successional fallow species, many of which have uses for the people who gather them. Farmers recognize this problem but believe the loss will not be serious because not every site cleared for food crops will be converted into cacao or coffee. They will not have the labor to harvest and process this amount of cacao or coffee (Sowei and Allen 2003).

Many other examples of this kind could be cited. The most famous case in modern history is the creation of a major export industry in southern Ghana, West Africa, by enterprising migrant farmers who established big areas of cacao among secondary forest in that country between 1890 and 1920 and developed new land tenure systems in order to facilitate their colonization of land purchased from others (Hill 1963). Later in this chapter we describe how farmers in the Brazilian Amazon

Box 13.2 Agricultural Biodiversity and the Livelihood Strategies of the Very Poor in Rural Bangladesh

The fact that the poor people depend on uncultivated foods for their survival and livelihoods is well known in the villages of rural Bangladesh. But what is the nature of this dependence? Our study explores the use by the very poor of the food and plants they collect from the lands, water bodies, and forests where they live. When we asked villagers, "Where are the poor?" the answer was "Chak," meaning in the cultivated fields of others or out on the roadsides. From the months of Bhadra to Kartik they are busy in the sugarcane fields harvesting for farmers. In the months of Agarhayan, Poush, and Magh they are busy harvesting potatoes and preparing seedlings for the paddy fields of farmers. They may receive some money for this labor, which they will use for oil, salt, school expenses, and debt repayment. But they will also take potatoes as partial payment and collect the straw that is no longer needed to cover the ground in the potato field and bring it home for fuel. They will pick the jute leaves in the farmers' field for food and collect the uncultivated leafy greens along the side of the rice field, some of which they will sell. They will sell eggs from their free-range chickens to buy rice and collect small fish in the water bodies for the daily meal. This is their livelihood.

What is an appropriate response to the challenges of ensuring their access to these food sources? Agricultural development based on a few crops cannot adequately compensate the very poor for the losses in access to uncultivated food sources caused by farming practices such as the extensive use of pesticides and monocropping. Nor can they compensate for the erosion of the common property regimes and social rules that enable people to use these food sources. Analysis of the contributions of uncultivated foods to food security in Bangladesh suggests that the appropriate level for enhancing access to these food sources is the community landscape, not the individual plant species, farm, or backyard. Simply by promoting biodiversity-based farming systems and protecting village lands from pesticides and enclosure of common lands, an enormous resource of uncultivated foods is also ensured. Such a strategy can be called cultivating the landscape, in contrast to more limited definitions of agriculture based on cultivated plants in cultivated fields. Improvements in agriculture should be pursued in the context of a broader strategy to increase the capacity of communities to create and maintain the conditions needed for biodiverse food systems. Ultimately, biodiversity is not cultivated but rather nurtured in biodiverse agroecosystems.

Source: Farida Ahkter, the Centre for Policy Research for Development Alternatives, Bangladesh.

are responding to price signals by converting a field crop system into an agroforestry system. Activities of this type transform the biodiversity of whole landscapes.

There are other users of biodiversity apart from farmers. Transhumant pastoralists may use different landscapes at different times of year, and some of them enter into contractual arrangements with farmers to pasture their livestock on fallow land. This is a widespread practice in the savanna regions of western Africa. Toulmin (1992) showed how the households of one Bambara village in northern Mali dug wells by hand in the 1970s and 1980s in order to attract migrating Fulani herders, whose livestock would then be corralled on the fields of the well owners to provide manure and thus permit expansion of their cropping. But the villagers make sure that the herders remain their clients and that if Fulani settle in the vicinity they do not acquire land on which they might dig their own wells. Maintaining this security of resource access takes a lot of organization among the Bambara.

Within any resident population, some have very little land or are landless. They may depend on the resources of common lands or almost anywhere in the landscape. The foods they use may not be cultivated at all. Box 13.2, derived from an abstract of a paper presented in Montreal by Farida Ahkter of the Centre for Policy Research for Development Alternatives in Bangladesh, describes graphically how the poorest of the poor depend on landscape diversity.

Temporal Dimension

Biodiversity, whether found in agroecosystems or outside them, is always in a state of flux. The seasonal variations of temperate zones, their orderly crop rotations, and short-term farm and fallow sequences may be familiar types of temporal variation in northern agriculture. But the temporal complexities of smallholder systems in the tropics often are unfamiliar, poorly understood by scientists, and often ignored or condemned by governments.

Among the most commonly studied smallholder patterns, which typically involve both complex temporal changes in management and high levels of biological diversity, are swidden or shifting cultivation systems. These pan-tropical—and perhaps near-global—forms of smallholder agriculture are highly varied, but they generally feature clearing of fields by cutting and burning and alternations of a brief intensive cropping phase with years of forest or bush fallowing. Until recently, the phase of swiddening systems characterized by less intensive management of crops—or the "fallow" phase—usually was understood to be a temporary abandonment. It was assumed that during this part of the cycle, when planting, weeding, and harvesting of most crops were done, all active management of plants and animals ceased, and direct economic benefits derived from the plot became negligible. Indeed, fields under "fallow" often appear to have reverted to completely natural vegetation.

Research carried out over the last several decades, particularly in South America and Southeast Asia, has shown otherwise. An increasing number of studies have demonstrated that many shifting cultivation systems are more accurately described as cyclic agroforestry and that although management of swiddens may change dramatically over time, many plots are never really abandoned. Even when a substantial part of the vegetation seems to be wild or spontaneous, active and skilled, though subtle, management may be going on, shaping the species and frequencies of plants and animals on the site. Which among the plants in a particular fallow plot has actually been cultivated or which has not often is difficult or impossible to determine. And although the species sampled may not change, the wild/cultivated ratio may shift as natural regeneration and volunteers join or replace cultivated plants over the months or years in which a fallow is subtly managed.

Economic pressures are rising and rural populations increasing throughout tropical Asia and in other parts of the world where swiddening has been a common way to make a living. The management of swidden–fallows is undergoing dramatic changes in response to these shifts. Management of all phases of the cycle is becoming more intensive and visible, with market-oriented species increasingly featured. The forms these intensified swidden–fallow systems are now taking remain varied, with agroforests often dominated by rubber, fruit, or fast-growing timbers. Other, more intensive—but still complex and cyclic systems—include fallows dominated by economically valued shrubs and even by herbaceous legumes (Cairns 2006).

How can we accurately measure the biodiversity in systems such as swidden–fallow agroforestry systems that change continuously? Taking the broader landscape as a unit of research and including fields and "fallows" of various management levels and ages helps the researcher capture a good amount of the richness and complexity of such agricultural patterns. Resampling over time is desirable to catch seasonal and other variations. The PLEC project found that in each region researchers must be flexible in their methods, varying them to suit local conditions. And they must understand the limitations of their data when long-term research and resampling are not possible.

Temporal complexity takes many forms and therefore presents many difficulties to the researcher. In the floodplain of the Amazon River, where PLEC has several research sites, plots of land annually pass through both terrestrial and aquatic phases. Farmers' plots disappear under river waters in the annual flood, normally about 10 meters high in the upper section of the river in Peru. The fields that appear after floodwaters recede several months later are changed not only in vegetative cover but often in size, soil type, and other qualities that determine both present and future agricultural uses. Complicating the issue for the researcher is the fact that while a plot is under water and its biodiversity changes drastically, it is often not unproductive but is merely passing through a different, aquatic phase. Many floodplain farmers in the PLEC site of Muyuy in Peru, for example, manage streamside and lakeside vegetation, including fruit-bearing trees, not only to produce fruits for human consumers in the terrestrial phase but also to attract fish during the flood season (Pinedo-Vasquez et al. 2003b). The multipurpose aquatic and terrestrial phase management developed by Amazonian farmers is difficult for agricultural researchers to see, much less appreciate; its biodiversity components are certainly difficult to measure.

Multifunctionality and simultaneous management of agricultural, agroforestry, and forestry resources in a single field are common to smallholder enterprises throughout the tropics. Despite their pervasiveness, these approaches are rarely mentioned in the literature and appear to be invisible to most researchers. Many farmers manage annual crops in their fields for harvest in a few months while also tending interspersed tree seedlings that will be cut in 30 years or so. The tree seedlings may be spontaneous volunteers or be deliberately planted or transplanted from neighboring forests or gardens. While the crops are planted, weeded, and harvested, the slower-growing trees may receive little more attention than a cursory cleaning and an occasional pruning. The continued non-mechanized nature of much smallholder farming in the tropics makes such diversity possible. The knowledge local farmers have of the growth characteristics of many organisms and their combinations, as well as of the specific capabilities and limitations of each corner of their fields, make such complex management profitable. Box 13.3 demonstrates the end result of such a process of management in the Amazon floodplain of Brazil, beginning in the swidden stage with planting or nuturing the seedlings of valued trees, continuing through the fallow stage to final incorporation in the developed forest.

Dynamism can be easy to misinterpret, particularly in systems that are rich in diversity and managed by farmers or communities that are politically marginal or culturally distinct. The work of anthropologists Fairhead and Leach (1996) in Guinea illustrates this point in what has now become a famous case. They describe a situation in which there is general agreement that the fields, forests, and grasslands of Kissidougou Province are now, and have long been, in flux. However, official and local understandings of the direction of those changes, and of the essential nature of human–forest interactions in Kissidougou, are at complete odds. If the local contention that most of the diverse and large forest islands that dot the landscape are largely human creations is accepted, measures of regional agrobiodiversity and notions of human manipulation of local landscapes are essentially opposite to those suggested if the competing scenario of advancing deforestation is affirmed. Making effective use of the modern evidence of air photography and remote sensing, as well as earlier descriptions, Fairhead and Leach confirmed the local interpretations.

In another part of the world, Yin (2001) has shown how the remaining swidden farmers in Yunnan, China, have greatly modified their systems, some of them recently, some a long time ago. Crops and cultivation methods have been changed, rotations have been shortened and means have been found to sustain fertility within these shortened rotations, cash crops have been introduced or cash-earning uses have been found for plants formerly used only for subsistence, and even terracing has been incorporated. Yet despite the highly skilled adaptations that have been made and continue to be made, many officials and some scientists continue to regard all that the swidden farmers do as primitive, to be replaced rapidly. Only with modern appreciation of the ecological advantages of agroforestry has a new understanding of traditional skills begun, belatedly, to arise.

Change in smallholder agroecosystems often occurs in incremental and seemingly disjointed steps (Doolittle 1984; Padoch et al. 1998) that again add a measure of complexity and ambiguity. Limitations in human

Box 13.3 Biodiversity in the Forest Stage at Two Sites on the Lower Amazon in Brazil

In two sites we found that the forest areas that are part of the landholdings of smallholders are the results of successive management operations that began in the field stage and continued into the fallow and forest stages. Inventories conducted in a sample of 10 ha (5 ha in Mazagão and 5 ha in Ipixuna) show a great diversity of species (box table 13.3).

In both sites the forests contain high levels of species richness and evenness. However, the average number of species found in the Mazagão forests (51) is slightly higher than the average found in Ipixuna (36). In contrast, the sampled forests of Ipixuna have more trees (average 1,117) than the ones sampled in Mazagão (average 1,041). These results reflect the histories of management and resource extraction practiced by smallholders in both sites. In Mazagão people are more dedicated to forest activities, and they tend to continually enrich their forest with desirable timber, medicinal, and fruit species. Farmers in Ipixuna are more dedicated to agroforestry and the collection of fruits and medicinal products than to timber extraction.

Despite the differences in forest uses and management practiced by the inhabitants of Mazagão and Ipixuna, forests in both sites show very high diversity or Shannon's index. Based on the estimated diversity indices, forests in Mazagão have higher values (average $H' = 2.59$) than forests in Ipixuna (average $H' = 1.77$). These results are very similar to the reported estimated Shannon's index for forest areas in other regions of the estuarine *várzea* floodplain (Anderson and Ioris 1992).

Although forests in Mazagão are richer in species than those in Ipixuna, the two most commercially valued species (*Euterpe oleraceae* and *Calycophyllum spruceanum*) are some of most dominant and abundant species in both sites. This indicates that people are encouraging the establishment and growth of

BOX TABLE 13.3. Diversity in forest samples comparing the number of species, number of individuals, and Shannon index (H)

Mazagão				Ipixuna			
Sample Plot	Number of Species	Number of Individuals	H	Sample Plot	Number of Species	Number of Individuals	H
1	48	892	2.96	6	26	623	1.66
2	55	1,096	2.66	7	41	1,032	1.91
3	54	1,118	2.43	8	38	1,610	1.68
4	45	778	2.66	9	43	1,696	1.80
5	55	1,322	2.26	10	34	923	1.80

Box 13.3 continues to next page

these and other valuable species in their forests. Similarly, the presence of a high number of timber, fruit, and medicinal species suggests the intensity and frequency of management by local people in both sites. The inventory data also show that people maintain a low number of individuals of several noncommercial species. Among these species are some pioneer species, such as *Cecropia palmata* and *Croton* sp., that play an important role in attracting game animals.

The estimated importance value index shows that 8 of the 10 most important species found in the forests of Mazagão and Ipixuna produce commercial products. As in the case of managing fallows, people also are adapting and developing new management technologies that correspond to specific environmental and economic conditions. The abundance and dominance of economically important species are maintained by smallholders through the application of management operations that promote the regeneration of species under different light and environmental conditions. For instance, the majority of farmers conduct preharvest operations to avoid excessive damage to the forests, thus optimizing production. Among the most recent and innovative preharvest operations is broadcasting seeds or planting seedlings of valuable species before cutting timber. Most seedlings are collected from other parts of the forests; however, the seedlings of andiroba (*Carapa guianensis*) are produced mainly in home gardens.

Source: Pinedo-Vasquez et al. (2003c:69–71).

labor and other resources available to rural households to produce substantial environmental alterations make long-term, farm-as-you-go approaches necessary. In what are agriculturally marginal environments, the very creation of arable land may take years and many labor-intensive operations. For instance, the conversion of peat swamps to productive coconut orchards or rambutan gardens on the coasts of Indonesia's West Kalimantan Province takes years of ditch-digging, drainage, and creation and destruction of several forms of rice planting before a profitable orchard is established. A typical landscape in such a region may comprise a multitude of patches of varying use, management, and diversity. All of the components of this landscape mosaic are parts of a diverse, complex, and dynamic system of smallholder tree cropping.

Inland from Kalimantan's peat swamps, change from one system to another is taking place in what is also a discontinuous, incremental, and visually confusing process. Padoch and others have documented how Dayak villagers are switching from swidden farming of upland rice to irrigated rice cropping (Padoch et al. 1998). The process of change creates multiple intermediate stages, many of them diverse, and all of them differing in productivity and appearance. The study of agrobiodiversity in such dynamic systems poses challenges to scientists who would accurately represent their richness.

Conclusion

Biodiversity at landscape scale presents a number of challenges in both measurement and interpretation. Because comprehensive survey is logistically impossible at this scale, the nature of the sampling frame is centrally important, and it creates a number of difficulties, especially in the definition of the nested samples. A strong purposive element is almost always introduced. In PLEC, we selected fields within field types and comprehensively sampled quadrats within these, as well as entire home gardens. In the European work, efforts have been made to find indicators to overcome the scale problem of landscape-wide surveying, but none have been found that are universally applicable. Because the purpose is to evaluate the success of management improvements, a combination of selected biotic indicators with structural aspects of the farming matrix and even specific management characteristics seems a likely way to proceed. Modified, such an approach could be applicable in developing countries.

However difficult it may be to measure in a scientifically defensible manner, appreciating agricultural biodiversity at the landscape scale is necessary for understanding many of the strengths of smallholder farming, particularly in developing countries. A large part of the managed—if not directly planted—diversity of these systems is found in the margins of fields, along the paths, between the houses, and along the watercourses. These patches of vegetation are harvested regularly and their fruits are eaten, sold, and used to fill a hundred economic needs. When farmers are deprived of these invisible resources, their diets and incomes often decline, and their ability to deal with climatic or economic perturbations often is lost.

Larger questions are raised by the temporal element of biodiversity management by farmers. The alternation of aquatic and terrestrial phases in annually inundated floodplains has been discussed in some detail, but the larger issue is the purposive management of land at one land use stage to create a modified biodiversity in a later land use stage or, put another way, the different biodiversity that results from deliberate changes in land use. These sequences are central to the understanding of landscape-level management, and they reveal how large an effect management can have on biodiversity. This chapter has discussed the modification of biodiversity through management, both deliberately to affect biodiversity and indirectly through changes determined only by the needs of production. The constant flux of biodiversity emerges as a central conclusion, and it is one that questions all notions of conserving "static" conditions in plants, plant assemblages, and managed landscapes.

Note

1. Alpha diversity is the diversity within a site or quadrat (i.e., local diversity), beta diversity is the change in species composition from site to site (i.e., species turnover), and gamma diversity is the diversity of a landscape or of all sites combined (i.e., regional diversity).

References

Anderson, A. and E. Ioris. 1992. The logic of extraction: Resource management and resource generation by extractive producers in the estuary. In K. Redford and C. Padoch, eds., *Conservation of Neotropical Forests*, 175–199. New York: Columbia University Press.

Blaikie, P. M. and H. Brookfield. 1987. *Land Degradation and Society*. London: Routledge.

Brookfield, H. 2001. *Exploring Agrodiversity*. New York: Columbia University Press.

Brookfield, H. and C. Padoch. 1994. Appreciating agrodiversity: A look at the dynamism and diversity of indigenous farming practices. *Environment* 36(5):6–11, 37–45.

Brookfield, H., M. Stocking, and M. Brookfield. 2002. Guidelines on agrodiversity assessment. In H. Brookfield, C. Padoch, H. Parsons, and M. Stocking, eds., *Cultivating Biodiversity: Understanding, Analysing and Using Agricultural Diversity*, 41–56. London: ITDG Publishing.

Büchs, W. 2003. Biotic indicators for biodiversity and sustainable agriculture: Introduction and background. *Agriculture, Ecosystems and Environment* 98:1–16.

Bureau, J.-C. 2003. *Enlargement and Reform of the EU Common Agriculture Policy: Background Paper* [mimeo]. Washington, DC: Inter-American Development Bank.

Cairns, M. 2006. *Voices from the Forest: Farmer Solutions Towards Improved Fallow Husbandry in Southeast Asia*. Washington, DC: Resources for the Future Press.

Coffey, K. 2000. *PLEC Agrodiversity Database Manual.* New York: New York Botanical Garden for the United Nations University.

Coffey, K. 2002. Quantitative methods for the analysis of agrodiversity. In H. Brookfield, C. Padoch, H. Parsons, and M. Stocking, eds., *Cultivating Biodiversity: Understanding, Analysing and Using Agricultural Diversity,* 78–95. London: ITDG Publishing.

Dao, Z., H. Guo, A Chen, and Y. Fu. 2003. China. In H. Brookfield, H. Parsons, and M. Brookfield, eds., *Agrodiversity: Learning from Farmers Across the World,* 195–211. Tokyo: United Nations University Press.

Dauber, J., M. Hirsch, D. Simmering, R. Waldhardt, A. Otte, and V. Wolters. 2003. Landscape structure as an indicator of biodiversity: Matrix effects on species richness. *Agriculture, Ecosystems and Environment* 98:321–329.

Doolittle, W. E. 1984. Agricultural change as an incremental process. *Annals of the Association of American Geographers* 82:369–385.

Duelli, P. 1997. Biodiversity evaluation in agricultural landscapes: An approach at two different scales. *Agriculture, Ecosystems and Environment* 62:81–91.

Fairhead, J. and M. Leach. 1996. *Misreading the African Landscape: Society and Ecology in a Forest–Savanna Mosaic.* Oxford, UK: Clarendon Press.

Forman, R. T. 1995. *Land Mosaics: The Ecology of Landscapes and Regions.* Cambridge, UK: Cambridge University Press.

Guo, H., C. Padoch, Y. Fu, Z. Dao, and K. Coffey. 2002. Household-level biodiversity assessment. In H. Brookfield, C. Padoch, H. Parsons, and M. Stocking, eds., *Cultivating Biodiversity: Understanding, Analysing and Using Agricultural Diversity,* 70–125. London: ITDG Publishing.

Haines-Young, R. H., C. Barr, H. Black, D. Briggs, R. Bunce, R. Clarke, A. Cooper, F. Dawson, L. Firbank, R. Fuller, M. Furse, M. Gillespie, R. Hill, M. Hornung, D. Howard, T. McCann, M. Morecroft, S. Petit, A. Sier, S. Smart, G. Smith, A. S. Stott, R. Stuart, and J. Watkins. 2000. *Accounting for Nature: Assessing Habitats in the U.K. Countryside.* London: Department of Environment, Transport and Regions.

Hill, P. 1963. *The Migrant Cocoa-Farmers of Southern Ghana: A Study in Rural Capitalism.* Cambridge, UK: Cambridge University Press.

Hoffmann-Kroll, R., D. Schäfer, and S. Seibel. 2003. Landscape indicators from ecological area sampling in Germany. *Agriculture, Ecosystems and Environment* 98:363–370.

Kaihura, F. B. S. 2002. Working with expert farmers is not simple: The case of PLEC Tanzania. In H. Brookfield, C. Padoch, H. Parsons, and M. Stocking, eds., *Cultivating Biodiversity: Understanding, Analysing and Using Agricultural Diversity,* 132–144. London: ITDG Publishing.

Kaihura, F. B. S., P. Ndondi, and E. Kemikimba. 2002. Agrodiversity assessment and analysis in diverse and dynamic small-scale farms in Arumeru, Arusha, Tanzania. In H. Brookfield, C. Padoch, H. Parsons, and M. Stocking, eds., *Cultivating Biodiversity: Understanding, Analysing and Using Agricultural Diversity,* 153–166. London: ITDG Publishing.

Leach, M., R. Mearns, and I. Scoones. 1999. Environmental entitlements: Dynamics and institutions in community-based natural resource management. *World Development* 27:225–247.

Magurran, A. E. 1988. *Ecological Diversity and Its Measurement.* Princeton, NJ: Princeton University Press.

OECD. 1997. *Environmental Indicators for Agriculture,* Vol. 1, *Concepts and Framework.* Paris: Organisation for Economic Cooperation and Development Publications Service.

Opperman, R. 2003. Nature balance for farms: Evaluation of the ecological situation. *Agriculture, Ecosystems and Environment* 98:463–475.

Osinski, E. 2003. Operationalisation of a landscape-oriented indicator. *Agriculture, Ecosystems and Environment* 98:371–386.

Padoch, C., E. Harwell, and A. Susanto. 1998. Swidden, sawah and in-between: Agricultural transformation in Borneo. *Human Ecology* 26:3–20.

Pinedo-Vasquez, M., K. Coffey, L. Enu-Kwesi, and E. Gyasi. 2003a. Synthesizing and evaluating PLEC work on biodiversity. *PLEC News and Views NS* 2:3–8.

Pinedo-Vasquez, M., P. P. del Aguila, R. Romero, M. Rios, and M. Pinedo-Panduro. 2003b. Peru. In H. Brookfield, H. Parsons, and M. Brookfield, eds., *Agrodiversity: Learning from Farmers Across the World,* 232–248. Tokyo: United Nations University Press.

Pinedo-Vasquez, M., D. G. McGrath, and T. Ximenes. 2003c. Brazil (Amazonia). In H. Brookfield, H. Parsons, and M. Brookfield, eds., *Agrodiversity: Learning from Farmers Across the World,* 43–78. Tokyo: United Nations University Press.

Piorr, H.-P. 2003. Environmental policy, agri-environmental indicators and landscape indicators. *Agriculture, Ecosystems and Environment* 98:17–33.

Sowei, J. and B. Allen. 2003. Papua New Guinea. In H. Brookfield, H. Parsons, and M. Brookfield, eds., *Agrodiversity: Learning from Farmers Across the World,* 212–231. Tokyo: United Nations University Press.

Toulmin, C. 1992. *Cattle, Women and Wells: Managing Household Survival in the Sahel.* Oxford, UK: Clarendon Press.

Waldhardt, R. 2003. Biodiversity and landscape: Summary, conclusions and perspectives. *Agriculture, Ecosystems and Environment* 98:305–309.

Waldhardt, R. and A. Otte. 2003. Indicators of plant species and community diversity in grasslands. *Agriculture, Ecosystems and Environment* 98:339–351.

Waldhardt, R., D. Swimmering, and H. Albrecht. 2003. Floristic diversity at the habitat scale in agricultural landscapes of central Europe: Summary, conclusions and perspectives. *Agriculture, Ecosystems and Environment* 98:79–85.

Whittaker, R. H. 1972. Evolution and measurement of species diversity. *Taxon* 21:213–251.

Yin, S. 2001. *People and Forests: Yunnan Swidden Agriculture in Human-Ecological Perspective*. Kunming, Yunnan, China: Yunnan Education Publishing House.

Zarin, D. J., H. Guo, and L. Enu-Kwesi. 2002. Guidelines for the assessment of plant species diversity in agricultural landscapes. In H. Brookfield, C. Padoch, H. Parsons, and M. Stocking, eds., *Cultivating Biodiversity: Understanding, Analysing and Using Agricultural Diversity*, 57–69. London: ITDG Publishing.

14 🎗 Diversity and Innovation in Smallholder Systems in Response to Environmental and Economic Changes

K. RERKASEM AND M. PINEDO-VASQUEZ

In a world where most agricultural products are industrially produced, smallholder farmers in the tropics are among the few groups that are still planting and managing a great diversity of crops and other biological resources in their landholdings. Experts have identified the most rare species and varieties of crops in small farmers' fields, and they have called for their preservation. The dangers posed to world agriculture by the loss of the diversity of crops and the flexibility they offer are well known, and programs to conserve these priceless resources in situ and ex situ have been established. However, based on the experience of the People, Land Management and Environmental Change (PLEC) project, we argue that the important contributions of smallholders are not summed up by the array of rare crops and varieties their farms feature. Smallholder farmers have also created and continue to develop a great abundance of complex and diverse resource use systems that are also important and threatened resources. The meaningful biodiversity of these production systems reflects and integrates both biological and technological resources. The conservation of agrobiodiversity will be greatly advanced if the resource use systems that produce the valuable products are documented, tested, improved, and promoted as diligently as are their products.

In this chapter we look at these resource use systems and at how small farmers constantly generate and change them. We focus particularly on smallholder production and management technologies that fall outside the strict categories of indigenous or nonindigenous and traditional or modern. Field experiences show that a broad spectrum of smallholders maintain

362

systems of great biodiversity value, and it is often impossible or unproductive to separate smallholder technologies into such categories. The repertoire of complex and diverse resource use technologies and conservation practices that we discuss creates and maintains environments that are home to a great diversity of crops and other cultivated or managed organisms. We know that the high levels of agrobiodiversity and other forms of biological diversity managed by smallholders are among the most important economic assets of the world's small farmers. We suggest that smallholder agrodiversity also reflects and supports smallholders' capacity to respond technologically and economically and to adjust to the large number of ecological and social processes that characterize their rapidly changing environments and societies.

To support our generalizations we present, analyze, and discuss some empirical data on dynamic agriculture, agroforestry, and forest management technologies—as well as on conservation practices—from PLEC demonstration sites throughout the tropics. The PLEC project is being carried out in the landholdings of small farmers in China, Papua New Guinea, Tanzania, Kenya, Uganda, Ghana, Guinea, Mexico, Jamaica, Peru, and Brazil (see also chapter 13). The PLEC approach focuses on the diversity and dynamism of farmers' strategies and the resulting agrobiodiversity and treats these not only as traditional products of long histories of community experience but also as results of individual creativity and individual decisions made in the face of unrelenting environmental, sociopolitical, and economic changes. The changes we discuss here stem from a great variety of causes—often a complex of causes—and vary in duration, severity, and configuration. We focus on local-level changes and responses; that is, we do not review responses to great disasters that have received global attention, nor do we generalize about the reaction of smallholders to a generality such as "globalization," although several of our examples feature what might be classified as a "disaster," that is "sudden calamitous event[s] bringing great damage, loss, or destruction" (Merriam-Webster Online Dictionary), and most describe shifts in prices and markets with global ties. The collapse of several villages into swirling Amazon waters seems a disaster, as is the arrival of a new disease that wipes out a community's main source of cash or plummeting prices for an important coffee harvest. However, the case studies we discuss that do feature abrupt and catastrophic events are observed on a local scale and described with attention not only to the general but also to the specific. The "invisibility" of the calamities we include reflects not only their small scale and rural locations but also the essential flexibility and diversity that the smallholders have displayed in successfully coping with such events.

Smallholder systems may be changeable, idiosyncratic, and site specific. However, we emphasize that such systems usually integrate rather than separate many types of resource use technologies and conservation practices. The variety of technologies that aim not only to exploit but also to conserve resources among smallholders continues to be underappreciated. We conclude by making a few general observations and recommendations concerning research and development strategies. We firmly believe that if we aim to help poor countries use their biological diversity sustainably and to ameliorate the plight of some of their poorest farmers, we need to value and incorporate the technological resources and conservation practices that underlie the high levels of agrobiodiversity found in the production landscapes engineered and managed by smallholders for centuries and into the present.

Agrodiversity: Processes and Products

An emphasis on products rather than processes undoubtedly has helped to obscure the dynamism of systems developed by smallholders. Likewise, the stress placed by many researchers—ironically including some who most value smallholders—on long-term persistence and adaptation implies that these systems never change. We certainly recognize the importance of the long-term view of many villagers and the foundation of long-term experience that characterizes smallholder production. However, we also assert that the capacity of small farmers to generate innovative technologies and strategies may be the most important asset they have to address the problems and opportunities generated by environmental and economic changes. Diversity, both biological and technological, often is the key to that adaptability. The many integrated "ways in which farmers use the natural diversity of the environment for production, including their choice of crops, and management of land, water and biota as a whole," or "agrodiversity" (Brookfield and Padoch 1994:8), is essential in allowing smallholders to deal with changes, both positive and negative, in their social and natural landscapes.

Researchers have long studied technological adaptation and innovation in production systems in smallholder societies as a response to change

(Feder and O'Mara 1981). For instance, perhaps the most influential work in the field of the past four decades, Boserup's *Conditions of Agricultural Growth,* established a relationship between land use intensification and growth of populations (Brookfield 2001). However, with a few notable exceptions (Richards 1993; Scoones and Thompson 1994) most researchers have taken technological change among smallholders to mean long-term, evolutionary change or, alternatively, abrupt forced abandonment, degradation, or collapse of the system. We focus instead on change and adaptation that has been and must be a constant in production by small farmers if they hope to persist, and on the agrodiversity that has been the adaptable farmer's essential tool.

Rich agrodiversity data collected over several years by PLEC researchers show that farmers are continually innovating and adapting technologies to take advantage of opportunities and to solve problems (see Brookfield et al. 2002, 2003 for a broader sample of these data). We will outline several examples to illustrate the great variety of systems used and of difficulties overcome. For instance, farmers in Amazonian Brazil have developed an agroforestry system called *banana emcapoeirada* that allows them to produce bananas despite infestation with a bacterial disease that has wiped out bananas planted in monocultural plantation systems (Pinedo-Vasquez et al. 2002b). An increase in market demand for wild vegetables and changing land policies have motivated poor farmers in northern Thailand to develop an edge farming system (Rerkasem et al. 2002). Decline in coffee prices in Kenya has led producers to develop a cluster system to plant several species and varieties of annual and semiperennial crops while maintaining coffee stands of suitable sizes (Kang'ara et al. 2003). The potential disaster of riverbank erosion in Peru has been dealt with by farmers who have developed a "tradition of change" to produce crops and manage their dynamic economic and biophysical surroundings (Pinedo-Vasquez et al. 2002a).

Agrodiversity and Variation in Farmers' Responses to Changes

Approximately two decades ago, bush fires in an El Niño year severely damaged cacao and other fruit species planted in the fields of small Ghanaian farmers. This event took place in an economic landscape where cacao was already a failing crop: The price of cacao beans was low, and most Ghanaian farmers were already experiencing a rapid transition from a boom to a bust period in the cacao economy. In the early 1990s PLEC scientists began assessments of selected Ghanaian villages. They found that a majority of farmers had preserved the diversity necessary for them to switch their production system to rotations featuring corn and cassava (Gyasi et al. 2003). Large swaths of rural Ghana changed from a landscape dominated by cacao plantations to a mosaic of small farms where a great diversity of plants was planted. To this production landscape farmers continued adding products, including chickens and even snails. A great diversity of yams was also brought back; they had nearly disappeared as a result of the boom in cacao farming (box 14.1).

The Ghanaian farmers' change from monoculture to "agrodiversity" proved important economically for many when prices of cacao beans fell and the series of fires in the El Niño year destroyed cacao stands (Gyasi and Uitto 1997). However, the resultant diverse production landscape had other important functions. It helped farmers reduce the risk of fires and reestablished and maintained vegetative cover, which in turn restored a variety of ecosystem services that had been severely affected by the extensive cacao plantations.

Changes in products and production systems as a strategy to overcome the loss of markets for cash crops and forest products by Hmong communities in the highlands of Thailand is yet another example of how farmers rely on agrodiversity to take advantage of changes. Hmong farmers grow cabbage commercially using a complex crop rotation system (Rerkasem et al. 2002). The core function of such a rotation system allows Hmong farmers to maintain high levels of agrobiodiversity by rearranging their crops on both temporal and spatial scales to optimize production and labor efficiency and to minimize the adverse effects of fertilizers and pesticides in their fields and surrounding environments. The multiple products of the rotational system help farmers deal with fluctuations in cabbage prices and to realize profits even during declines in markets (Rerkasem et al. 2002).

Identifying and documenting variation in the ways smallholders respond to changes in PLEC has yielded a rich repertoire of technologies and practices for producing, managing, and conserving crops and other forms of biological diversity at the farm, household, and landscape levels. We found that despite great differences in the technological responses of small farmers to changes, they tend to be consistently different from the solutions proposed by experts. A multilayer and multiple-use system developed by small

Box 14.1 Ghana Crop Rotation Systems

Crop rotation systems are helping farmers plant more species and varieties of crops in their agriculture and agroforestry fields (box table 14.1). Currently farmers are planting and conserving an average of 13 varieties of cassava and

BOX TABLE 14.1. Rotational cropping systems and their benefits.

Practices or Regime	Major Characteristics and Advantages
Bush–fallow land rotation, using fire to clear land	A means of regenerating soil fertility and conserving plants in the wild.
Minimal tillage and controlled use of fire for vegetation clearance	Minimal disturbance of soil and biota.
Mixed cropping, crop rotation, and mixed farming	Maximizes soil nutrient usage; maintains crop biodiversity; spreads risk of complete crop loss; enhances a diversity of food types and nutrition; favors soil regeneration.
Mixed agroforestry: cultivating crops among trees left in situ	Conserves trees; regenerates soil fertility through biomass litter. Some trees add to productive capacity of soil by nitrogen fixation.
Oprowka, a no-burn farming practice that involves mulching by leaving slashed vegetation to decompose in situ	Maintains soil fertility by conserving and stimulating microbes and by humus addition through the decomposing vegetation; conserves plant propagules, including those in the soil by the avoidance of fire.
Use of household refuse and manure in home gardens	Sustains soil productivity.
Use of *nyabatso* (*Neubouldia laevis*) as live stake for yams	The basically vertical rooting system of *nyabatso* favors expansion of yam tubers, and the canopy provides shade and the leaf litter mulch and humus. It also is suspected that *nyabatso* fixes nitrogen.
Staggered harvesting of crops	Ensures food availability over the long term.

Box 14.1 continues to next page

Box 14.1 *continued*

Practices or Regime	Major Characteristics and Advantages
Storage of crops, notably some yam species, in the soil for future harvesting	Enhances food security and secures seed stock.
Conservation of forest in the backyard	Conserves forest species; source of medicinal plants at short notice; favors apiculture, snail farming, and shade-loving crops such as yams.

Source: Gyasi et al. (2003).

140 yam cultivars in their landholdings. The complex rotation systems are also helping farmers to interplant several varieties of nitrogen-fixing cowpea, maize, chili pepper, vegetables, and legumes with fruit species such as mango, avocado, citrus, coconut, and adesaa (*Chrysophyllum albidum*).

farmers in China to reforest slopes and other deforested areas (box 14.2) provides an example of this contrast (Guo et al. 2003).

Smallholders in southwestern China are participating in state-sanctioned reforestation programs. These programs promote the planting of two fast-growing species that bring few benefits to farmers. Small farmers have instead incorporated several native tree species as part of a multilayered and multiuse system. The addition of native tree species initially was observed as peculiar because the rotation time for harvesting these species is three times longer than that of the ones recommended by the foresters. Through field work, PLEC researchers found that farmers planting the local species do not need to wait until the end of the rotation of the local trees to reap benefits (Dao et al. 2001). The native species create habitats for insects and herbaceous vegetation that favor the growth of mushrooms and wild vegetables and even the raising of chickens. In contrast, areas reforested with only the species recommended by foresters are very low in insects and do not provide habitats for the growth of mushrooms and wild vegetables. This example from China is one of many cases documented by PLEC members who observed and recorded technologies locally developed in response

Box 14.2 Multilayer and Multiuse Reforestation System in Baihualing, Baoshan, China

The multilayer and multiuse reforestation system developed by Chinese farmers is an improvement on their traditional agroforestry systems. Some of the most common tree species that are either planted or protected include the following:

- *Phoebe puwenensis*, a naturally regenerated fast-growing timber species
- *Alnus nepalensis*, a species that is commonly planted in agroforestry fields, which provides cover for the natural regeneration of tree species in plots that are reforested
- *Toona sianensis*, a timber and wild vegetable tree species
- *Toona ciliate*, a slow-growing hardwood timber species
- *Cunninghamia lanceolata*, a timber and firewood species
- *Punica granatum*, a timber and firewood species
- *Pinus armandii*, a fast-growing timber species
- *Lindera communis*, a fast-growing timber species
- *Trachycarpus fortunei*, a tree that produces fiber for weaving and an edible flower
- *Crateva unilocularis*, a fast-growing timber species that also produces an edible flower
- *Paris* sp., a timber species used medicinally

When these species are established, farmers enrich the stands by planting walnuts, chestnuts, several varieties of pears, and a number of medicinal plants such as *Dendrobium candidum,* a medicinal orchid species. In the process of enriching their forests they create small clusters (usually 1. 5×3 m long) where they plant or protect wild vegetables. In addition, the branches of the timber species are stored at the base of the largest trees for the production of mushrooms. The reforested plots using the multilayer and multiuse system contain 73 species, and 52 (71%) of them are economically important.

Sources: Dao et al. (2001, 2003).

to imposed changes. The dissemination of the techniques used by the Chinese farmers in reforesting their land with native species has greatly facilitated the incorporation of other small farmers in reforestation programs. This case also illustrates our contention that focusing on farmer-developed practices is not backward-looking and is not limited to traditional practices. Many poor farmers are dynamic and forward-looking.

We found that poor farmers are continually creating and providing society with new technologies that yield both short- and long-term benefits, and they do this without radically altering important ecological processes, as do most modern production systems. By keeping patches of vegetation at different stages of succession and interconnected with agriculture fields, poor farmers often maintain important ecosystem services, conserving soils and water as well as biodiversity. These locally developed systems also offer readily accessible and practical solutions to seemingly devastating problems of disease and rapid environmental change. Two examples from PLEC's work in Amazonia show how farmers take advantage of agrodiversity and of their knowledge of ecosystem functions and services to overcome what appear to be imminent disasters.

Many villages in the tidal floodplain of the Amazon estuary were major exporters of bananas until recently. Not only did the smallholder farmers of the area supply the urban markets of the state of Amapá with bananas, they also exported to the major Amazonian city of Belém. In the last several years, however, banana production in the region has been almost completely wiped out by Mokko disease, known locally as *febre de banana*. The disease, which is common in many banana-producing areas, can be controlled by a concerted and very expensive campaign of destruction of infected plants, repeated disinfection of all tools, and constant inspection of all plantings (Stiver and Simmonds 1987). These control measures are not economically feasible for local banana producers. The local villagers or *ribeirinhos* have developed an agroforestry system, locally known as the *banana emcapoeirada* system, by which farmers manage the disease, although they do not eliminate it.

The *emcapoeirada* agroforestry system is a new adaptation of a system that combines agroforestry techniques and forest management practices. Many discussions of Amazonian agroforestry systems, including how Amazonians adapt traditional patterns to modern needs and opportunities, are available (Padoch et al. 1985; Padoch and de Jong 1987, 1989, 1995; Denevan and Padoch 1988; Irvine 1989; Posey 1992). However, these studies do not examine the agroforestry systems as resources for controlling plant diseases such as the *febre de banana*.

In Amapá farmers now plant bananas and encourage the natural regeneration and growth of sororoca and pariri, two native wild species of the Musaceae family. Villagers report that these two species do not compete (*brigar*, in local terms) with bananas. Rather, they protect banana stands from dispersal of the disease. Apart from these two species, *banana emcapoeirada* stands include a large number of other plants and feature a forest-like appearance. Since the system was adopted, yields of banana per hectare have increased by 500% (table 14.1).

At the other end of the Amazon River, Sector Muyuy is part of the highly dynamic floodplain in Peru (Kalliola et al. 1993). The region comprises heterogeneous landscapes that include a great diversity of human settlements, land formations, water bodies, and vegetation cover. Based on examination of three sets of Landsat images, major changes in the direction of the river and the location of structural features took place from 1987 to 2000 (Pinedo-Vasquez et al. 2002a). During this period a new lake was formed, a secondary channel was sedimented, and the Amazon River changed course significantly, reducing the size of the large and populated island of Padre. Two of the five villages located on the island moved to the other side of the river, and two new villages were founded. Many farmers completely lost their fields, and the changes in area and number of landforms and streams caused multiple other important economic shifts.

Table 14.1. Increase in yield and variety of bananas used in the *banana emcapoeirada* agroforestry system.

Year	Yield (bunches/ha)	Banana Varieties Planted
1997	63	3 (local shade-tolerant varieties)
1998	165	3 (local shade-tolerant varieties)
1999	247	5 (3 local and 2 brought from the Santarém region)
2000	284	6 (3 local, 2 from San tarém, and 1 from Obidos)
2001	332	9 (3 local, 2 Santarém, 1 Obidos, and 3 EMBRAPA)
2002	378	9 (3 local, 2 Santarém, 1 Obidos, and 3 E MBRAPA)

The banana production system, known locally as *banana emcapoeirada*, is used concurrently with the management of timber species.

These included changes in fish populations and increased silt bars available during the dry season and high levees.

Villagers have responded to these extreme changes not only by moving villages but also by varying their broad agricultural repertoire. They have planted rice and a host of other annual crops in the dry lake beds and on the increased expanse of low levees. They used the emergent high levees to plant fruit trees and other agroforestry crops. The richness of both agricultural biodiversity (e.g., 18 varieties of beans that can be planted in a great variety of growing conditions) and of technologies (reportedly at least 12 different agricultural systems [Padoch and de Jong 1989]) for exploiting the many landforms are the villagers' most important resources in this dynamic environment.

In addition to changes that affect agriculture, a strong current is removing the vegetation along the new stream, forming a number of small new stream channels that facilitate access to forest resources at the interior of the large island. Increased access to this area has led to an increase in the extraction of catahua (*Hura crepitans*) and other timber species. This in turn has prompted the villagers to control access to the resources and their extraction. For instance, community rules were established to prohibit outsiders from extracting timber within the community territory, and residents limited their extraction to four adult trees (diameter at breast height greater than 55 cm) per year. Such shifts in village control illustrate how the *ribereños,* as Amazon River dwellers are known in Peru, effectively respond to changes not only by varying exploitation techniques but also by transforming social rules to fit new opportunities for exploitation.

Appreciating Change and Hybridity

There are several reasons why it is difficult for researchers and technicians to appreciate poor farmers' production systems even when they are profitable and biodiversity-rich. A common problem is that many farmer-developed technologies in the tropics confound the categories and concepts that are familiar to those who study, develop, and disseminate agricultural innovations. Even some common terms used to describe how farmers organize their crops (e.g., monocropping, polycropping, or intercropping) reveal little about the diversity found in many smallholder farms (Scoones and Thompson 1994). Insights into locally developed responses to change are difficult to achieve without observing the technical diversity of seemingly standard tasks such as clearing, hoeing, plowing, planting, weeding, protecting, harvesting, and fallowing fields (Agrawal 1997). These categories of farm activity can conceal as much as they reveal.

Many smallholder systems that PLEC has promoted seemed to be invisible or incomprehensible to agricultural experts for several reasons. Agrodiversity itself can be visually and conceptually ambiguous. Biologically diverse systems are difficult to understand, especially to those trained to look for order, simplicity, and uniformity in agricultural undertakings. Many of the management systems PLEC promotes initially were difficult to even identify as managed systems. Dynamism also can be easy to misinterpret. Among environmentally concerned development personnel who now emphasize the desirability of sustainable production, dynamism can be easy to confuse with degradation. Smallholder technologies also regularly combine pieces of the local and the borrowed into unexpected hybrid wholes (Gupta 1998). These production technologies therefore are often ignored or disparaged by the defenders of the autochthonous and indigenous and by those who champion modern technologies.

Combination of the new and old, of the local and foreign, is characteristic of all knowledge and all resource management practices, of course. In our world of "production packages" and of modernist (and conservationist) ideologies, however, such obviously mixed systems sometimes are not appreciated. More often they are just not seen. These technologies continue to reflect profound locally developed knowledge of specific soils, watercourses, climates, and biodiversity and management of ecological processes (box 14.3), but they have been updated or hybridized with knowledge and practices learned by farmers outside their villages (Fairhead 1993).

Hybrid systems commonly introduce much higher levels of biodiversity into what are monocrop systems. One recently recognized example includes the systems dubbed "jungle rubber" by International Center for Research in Agroforestry researchers in Southeast Asia. These production systems combine the planting of rubber trees with management of many other species in stands that mimic natural forests and provide many of the ecosystem services of standing forests. Jungle rubber is the product of probably two of the most common patterns of formation of hybrid systems: integrating knowledge gained while engaged in wage labor away from smallholder farms and adapting modern production systems and techniques brought by the development agencies.

The objective of this study was to understand the influence of ecosystem components on sociocultural factors that affect the decision making of farmers in the daily management of their plant genetic resources. Six cultivated species were the focus of study: sorghum (*Sorghum bicolor*), millet (*Pennisetum glaucum*), groundnut (*Arachis hypogeae*), cowpea (*Vigna unguiculata*), okra (*Abelmoschus esculentus*), and fabirama or fra fra potato (*Solenostemon rotundifolius*).

The study was carried out in three different agroecological zones of Burkina Faso: north zone, less than 500 mm annual rainfall, Ouahigouya village; center-north zone, less than 400 to less than 600 mm, Tougouri village; and southeast zone, more than 1,000 mm, Thiougou village.

A multidisciplinary team administered questionnaires to farmers at three sites. One site in each agroecological zone was chosen according to the following criteria: the importance of agriculture for the local populations, the degree of genetic erosion already in progress, the role of the species studied in the lives of the local populations, the influence of the environment on the conservation of agricultural biological diversity, and the importance of the interdependence of sociological phenomena on the development of agriculture. The data collected at the three sites were farmers' observations of the development of the plants (phenology), the behavior of animals (ethology), the movement of the stars, and the changes in the atmosphere relating to their predictions about the nature of the rainy and agricultural seasons to come.

Two groups of signs of the season to come were considered by farmers: the different stages of plant development and signs related to animal behavior, such as the appearance of or the duration of the song of certain birds, their manner of building a nest and its position, the relocation of reptiles, and the movements and the cawing of batrachians (box table 14.3). These signs are used to predict the rainy season to come. The next most used groups are the movement of certain stars such as "la petite ours," the direction and orientation of the winds, consultation of the traditional lunar calendar, and the prediction of the usual person in charge (called Tengsoba) from the Gnignonsés ethnic group.

An understanding of the role of ecosystem components as a basis for prediction is vital for the Burkinabè farmer. Because the agroecosystem components are very variable, farmers understand the need to dispose of and maintain diversity according to their interpretation of the signs of nature, by which they then recommend the varieties to be sown, choose the cultivation methods to use, determine sowing dates, and predict whether the crop will be good or bad. These practices play a big role in control of the agroecological environment through knowledge of climatic conditions, soils, and biotic factors. This shows that indigenous systems of environmental management that guarantee the maintenance

Box 14.3 continued

BOX TABLE 14.3. Environmental signs used by farmers to predict the rainy season in Ghana.

	Plants	Animals	Stars and Weather	Rites
Signs that indicate a good rainy season	Good flowering of kapok tree (*Ceiba pentandra*). Good emergence of seeds. Even fruiting of *Xemenia americana, Ficus platyphylla, Diospiros mespiliformis, Vitelaria paradoxa, Sclerocarya birrea, Heeria insignis* (lebnoré), and *Lannea microcarpa*. Heavy rains on Bêga day (traditional fair). Good leafing of rônier (Borassus palm, *Borassus aethiopum* Mart.).	Prolonged singing by Frouko and Falaogo. From its nest, the Waali (stork) has its head turned westward. The Taba builds nests at the tops of the trees. Only one stork drones back and forth in the village. Appearance of the Taba. The buffalo toad looks toward the west. Good hunting for wild guinea-fowl and varanus. Red caterpillars arrive in the village. Abundance of Wonnonwondo. Termites construct many territaries. The crickets dig and reclose their holes from January to February. Appearance of three types of Yantyaaki.	The first rains fall late in the night. The winds blow strong from south to north. March is very warm. From December, it is very cold for 3 months. Appearance of water mirages during the day. It rains everywhere right from the first rains. The waters seem to run from west to east (in village of Kougny).	According to the predictions of the rainmakers. Forecasts of the land chiefs.

Box 14.3 continues to next page

Box 14.3 continued

Box 14.3 continued

Plants	Animals	Stars and Weather	Rites
Market garden plants grow poorly.	From its nest, the Waali looks toward the east.	The cold season (winter) lasts less than 3 months.	According to the predictions of the rainmakers.
Withering of the first leaves of rônier and *Heeria insignis* (lebnoré).	The songs of Frouko are broken. The Falaogo sings less.	Winds come from all directions (wind-madness).	Consultations of the lunar calendar.
Weak and nonuniform production of karité, raisinier, kuna(s), and nobga (*Sclerocarya birrea*).	The turtledoves sitting on their eggs look toward the east. Many storks are in the village. Birds' nests are oriented toward the west. Good hunting for partridge.	The monsoon winds blow from north to south. The waters of the flood-plain flow from east to west.	
Good fruiting of tamarind and fig, while keeping their leaves.	The tortoises look toward the east. The buffalo toad looks toward the east.	Excessive thunder with the first rains.	
Premature fall of raisinier and karité fruit. Late second leafing of karité.	Many tortoises appear in the fields. Early appearance of locusts. Crickets do not close their holes.	Many winds laden with red dust at the beginning of the rainy season.	
Poor flowering of kapok tree.	Termites move into the trees. Repeated passings of the Mimimanas.		

Signs that indicate a bad rainy season

Box 14.3 continued

of agricultural biological diversity exist. Farmers have their own management criteria, which they use variably according to area, ethnic group, environment, cultures and rites, and agricultural activities. These methods ensure the maintenance of the genetic variability of the crop plants in an evolutionary way, which also guarantees gene flow between local cultivars and their wild relatives. It is therefore vital for on-farm conservation of plant genetic resources that the components of the ecosystem such as the trees and the animals are also preserved. The ecosystem is thus inseparable from the lives of farmers.

Source: Sawadogo et al. (2005).

Responding to Conservation Priorities

Smallholders have shown an ability to respond quickly and appropriately to new problems and opportunities, whether environmental, economic, or political (Agrawal 1997). Their responses tend to include varying the mix of resource uses, the location of their economic activities, their residence patterns, and the organization of their work, all elements of agrodiversity. In some areas where biodiversity conservation priorities have taken the form of protected areas, agrodiversity has also helped smallholders adapt. The multiplicity of economic options that local farmers identify and create has allowed them to respond effectively to the changes that conservation programs often impose, including zonation of their territories, restrictions on their use of resources, and increases in wildlife populations.

The importance of smallholder production systems to biodiversity preservation takes many forms. However, attempts to interpret and understand the conservation value of small farmers' technologies have often overvalued or oversimplified them. For instance, many experts base their conclusions about the conservation value of a production system solely on the levels of crop diversity that are incorporated (Brush 2000; Hamlin and Salick 2003). Data collected on agrodiversity and field observations show that in many instances the conservation value of agrodiversity cannot be measured simply by counting crops, especially because these may change frequently. The previous example from China clearly illustrates this point.

The reforestation systems created by farmers have higher values both economically and in terms of biodiversity than do those proposed by state foresters. Both the added number of timber species planted and the indirect increase in associated plant and animal biodiversity should be taken into account. The same complexity holds true when income or economic value is measured in these systems; often these values are both difficult to measure accurately and difficult to identify.

The conservation value of agrodiversity in smallholder societies can also be measured in the creation, maintenance, and management of niches in the boundaries of fields, fallows, forests, and streams. Numerous agricultural systems rich in agrobiodiversity where crops are planted on the bunds and along the boundaries between irrigated fields were observed and documented in China, Tanzania, Kenya, and other PLEC sites. The economic and ecological importance of these systems is appreciated by local communities but largely unknown to outsiders. In some cases the importance of these edge-cropping systems was overlooked because these systems may not have been significant sources of income in years of plenty. However, they may demonstrate their usefulness in the years when agricultural harvests are poor. Perhaps if analysis of these systems included their social and ecological value in addition to their economic worth, their importance would be readily evident.

Conclusion

We have sketched out only a few of the most important reasons why focusing on the biodiversity-maintaining processes, and not just the products, of smallholder agriculture has been a priority for the more than 200 researchers involved in the PLEC project. We have focused on both the diversity and the dynamism that characterize these systems. Small farmers and their communities commonly maintain patchy production landscapes where forest plots may be as important as agricultural fields for livelihoods and for the conservation of crops and other forms of biological diversity. We have emphasized the role of diverse technologies and products in maintaining smallholder households' ability to respond effectively to new opportunities and problems.

Our emphasis on the importance of patchiness of production landscapes, system hybridity, and change may appear to contradict some of the recent concerns of many in the biodiversity community. Although we do not dispute that ecosystem fragmentation, unsustainable production, and destruction of traditional or indigenous systems and species are pressing problems in some contexts, these concepts should not be misapplied or misunderstood. Many of the systems PLEC scientists have described and promoted are too modern for those concerned with cultural preservation, too traditional for those concerned with modernization, and too production oriented for conservationists. They defy the categories of scientists and elude the understanding of researchers. Yet in their seeming lack of order these systems preserve the biodiversity that even the most modern agriculture still depends on, serve the needs of a billion people, and allow them to adapt to a changing environment. Our policies and programs must strive at the very least not to destroy this most important resource.

Acknowledgments

We would like to thank Christine Padoch for providing comments and critical observations. Much of the information reported and analyzed in this chapter was provided by our colleagues in PLEC. We are very grateful to them and hope that we have represented their work well. We would also like to thank the United Nations University for providing financial and administrative support. The field work was made possible by funds from the Global Environment Facility. We would like to thank the reviewers and the editor for their valuable comments.

References

Agrawal, A.1997. *Community in Conservation: Beyond Enchantment and Disenchantment*. Gainesville, FL: Conservation & Development Forum.

Brookfield, H. 2001. *Exploring Agrodiversity*. New York: Columbia University Press.

Brookfield, H. and C. Padoch. 1994. Appreciating agrodiversity: A look at the dynamism and diversity of indigenous farming practices. *Environment* 36(5):6–11, 37–45.

Brookfield, H., C. Padoch, H. Parsons, and M. Stocking. 2002. *Cultivating Biodiversity: The Understanding, Analysis and Use of Agrodiversity*. London: ITDG Publications.

Brookfield, H., H. Parson, and M. Brookfield. 2003. *Agrodiversity: Learning from Farmers Across the World*. Tokyo: United Nations University Press.

Brush, S. 2000. *Genes in the Field: On-Farm Conservation of Crop Diversity.* Rome: IPGRI; Ottawa: IDRC; Boca Raton, FL: Lewis Publishers.

Dao, Z., X. H. Du, H. Guo, L. Liang, and Y. Li. 2001. Promoting sustainable agriculture: The case of Baihualing, Yunnan, China. *PLEC News and Views* 18:34–40.

Dao, Z., H. Guo, A. Chen, and Y. Fu. 2003. China. In H. Brookfield, H. Parson, and M. Brookfield, eds., *Agrodiversity: Learning from Farmers Across the World,* 195–211. Tokyo: United Nations University Press.

Denevan, W. M. and C. Padoch. 1988. Swidden–fallow agroforestry in the Peruvian Amazon. *Advances in Economic Botany* 5.

Fairhead, J. 1993. Representing knowledge: The "new farmer" in research. In J. Pottier, ed., *Practicing Development: Social Science Perspectives,* 187–204. London: Routledge.

Feder, G. and G. T. O'Mara. 1981. Farm size and the adoption of green revolution technology. *Economic Development and Cultural Change* 30:59–76.

Guo, H., Z. Dao, X. H. Du, L. Liang, and L. Yingguang. 2003. China. In H. Brookfield, H. Parson, and M. Brookfield, eds., *Agrodiversity: Learning from Farmers Across the World,* 195–211. Tokyo: United Nations University Press.

Gupta, A. 1998. *Postcolonial Developments: Agriculture in the Making of Modern India.* Raleigh, NC: Duke University Press.

Gyasi, E. A., W. Oduro, L. Enu-Kwesi, G. T. Agyepong, and J. S. Nabila. 2003. Ghana sub-cluster final report. In H. Brookfield, H. Parson, and M. Brookfield, eds., *Agrodiversity: Learning from Farmers Across the World,* 79–109. Tokyo: United Nations University Press.

Gyasi, E. A. and J. I. Uitto, eds. 1997. *Environment, Biodiversity and Agricultural Change in West Africa.* Tokyo: United Nations University Press.

Hamlin, C. C. and J. Salick. 2003. Yanesha agriculture in the Upper Peruvian Amazon: Persistence and change fifteen years down the "road." *Economic Botany* 57:163–180.

Irvine, D. 1989. Succession management and resource distribution in an Amazonian rain forest. *Advances in Economic Botany* 7:223–237.

Kalliola, R., M. Puhakka, and W. Danjoy. 1993. *Amazonía Peruana: Vegetación húmeda tropical en el llano subandino.* Jyvaskyla, Finland: Gummers Press.

Kang'ara, J. N., E. H. Ngoroi, C. M. Rimui, K. Kaburu, and B. Okoba. 2003. Kenya. In H. Brookfield, H. Parson, and M. Brookfield, eds., *Agrodiversity: Learning from Farmers Across the World,* 154–168. Tokyo: United Nations University Press.

Padoch, C., J. Chota Inuma, W. de Jong, and J. Unruh. 1985. Amazonian agroforestry: A market-oriented system in Peru. *Agroforestry Systems* 3:47–58.

Padoch, C. and W. de Jong. 1987. Traditional agroforestry practices of native and ribereño farmers in the lowland Peruvian Amazon. In H. L. Gholz, ed., *Agroforestry:*

Realities, Possibilities and Potentials, 179–194. Dordrecht, The Netherlands: Martinus Nijhoff/Dr. W. Junk Publishers.

Padoch, C. and W. de Jong. 1989. Production and profit in agroforestry: An example from the Peruvian Amazon. In J. Browder, ed., *Fragile Lands of Latin America,* 102–113. Boulder, CO: Westview Press.

Padoch, C., and W. de Jong. 1995. Subsistence- and market-oriented agroforestry in the Peruvian Amazon. In T. Nishizawa and J. I. Uitto, eds., *The Fragile Tropics of Latin America: Sustainable Management of Changing Environments,* 226–237. New York: United Nations University Press.

Pinedo-Vasquez, M., J. Barletti Pasqualle, D. Del Castillo Torres, and K. Coffey. 2002a. A tradition of change: The dynamic relationship between biodiversity and society in Sector Muyuy, Peru. *Environmental Science & Policy* 5:43–53.

Pinedo-Vasquez, M., C. Padoch, D. McGrath, and T. Ximenes. 2002b. Biodiversity as a product of smallholders' strategies for overcoming changes in their natural and social landscapes. In H. Brookfield, C. Padoch, H. Parsons, and M. Stocking, eds., *Cultivating Biodiversity: The Understanding, Analysis and Use of Agrodiversity,* 167–178. London: ITDG Publications.

Posey, D. 1992. Interpreting and applying the "reality" of indigenous concepts: What is necessary to learn from the natives? In K. Redford and C. Padoch, eds., *Conservation of Neotropical Forests: Working from Traditional Resource Use,* 21–34. New York: Columbia University Press.

Rerkasem K., N. Yimyam, C. Korsamphan, and B. Rerkasem. 2002. Agrodiversity lessons in mountain land management. *Mountain Research and Development* 22:4–9.

Richards, P. 1993. Cultivation: Knowledge or performance? In M. Hobart, ed., *An Anthropological Critique of Development: The Growth of Ignorance,* 61–78. London: Routledge.

Sawadogo, M., J. Ouedrago, M. Belem, D. Balma, B. Dossou, and D. Jarvis. 2005. Influence of ecosystem components on cultural practices affecting the in situ conservation of agricultural biodiversity. *Plant Genetic Resources Newsletter* 141: 19–25.

Scoones, I. and J. Thompson. 1994. Knowledge, power and agriculture: Towards a theoretical understanding. In I. Scoones and J. Thompson, eds., *Beyond Farmers First: Rural People's Knowledge, Agriculture Research and Extension Practice,* 16–32. London: Intermediate Technology Publications Ltd.

Stiver, R. H. and N. W. Simmonds. 1987. *Bananas.* Tropical Agriculture Series. Rome: FAO.

15 🍃 Agrobiodiversity, Diet, and Human Health

T. JOHNS

Plant biodiversity is essential to human health. Plants provide both nutrients and medicinal agents, form components of robust ecosystems, and contribute to sociocultural well-being. Traditional values and scientific conceptions concur on the necessity of dietary diversity, particularly of fruits and vegetables, for health. In the face of economic and environmental changes, increased simplification of the diets of large numbers of people to a limited number of high-energy foods presents unprecedented obstacles to health. Cultural knowledge of the properties of plants erodes at the same time. Conservation of biodiversity and the knowledge of its use therefore preserves the adaptive lessons of the past and provides the necessary resources for present and future health.

Nutrition and health considerations forge a strong connection between the imperatives to ensure human well-being and conserve biodiversity. Accordingly, a nutritional perspective informs ways of thinking about plant genetic resources (PGRs), and nutrition can assume a prominent place in the effort to conserve and use PGRs. Although the links between agrobiodiversity, dietary diversity, and health appear logical in principle, empirical data on the validity of food-based approaches to health needed to convince decision makers on how the relationships work in practice are inadequate. Such data also are essential for the implementation of strategies that promote the conservation of plant genetic resources by enhancing their use and value to producers and consumers in developing countries. Most importantly, empirical and participatory research on the links between dietary diversity, health, and biodiversity can provide the foundation for the design of programs that enable developing countries to respond effectively to current problems and future changes in food systems, environment, and patterns of disease.

Dietary Diversity and Health

A handful of epidemiological studies uphold the conventional wisdom embodied in dietary guidelines concerning the benefits of a varied diet (Johns 2003; Johns and Sthapit 2004). For example, American women (mean age, 61 years) who consumed a greater number of recommended foods had a lower risk of mortality (Kant et al. 2000). Women in the highest quartile (median variety scores of 15) had an odds ratio of dying in a 5.5-year period of 0.69 in comparison to the lowest quartile (variety score of 7). The association of dietary diversity with longevity and reduced rates of chronic degenerative diseases such as cardiovascular disease, diabetes, and cancer for men and women was shown in previous work of Kant et al. (1995).

In an Italian study, dietary diversity, most strongly in vegetables and fruits, was associated with a lower incidence of stomach cancer (La Vecchia et al. 1997). This coincides with the recognized relationship of the benefits of Mediterranean diets in reducing risk for chronic degenerative diseases with fruit and vegetable consumption (Trichopoulou and Vasilopoulou 2000). Similarly, Drewnowski et al. (1996) show that although French diets are higher in fats than those in the United States and therefore lower on indices of dietary quality, overall diversity probably accounts for their recognized benefits.

Fewer data exist to support the contribution of dietary diversity to health in developing countries (Johns 2003). However, dietary diversity has been linked to improved growth in children one to three years old in Kenya (Onyango et al. 1998). In Mali, Harløy et al. (1998) demonstrated a strong correlation of diversity of fruits and vegetables with overall nutrient adequacy and with specific nutrients such as vitamins A and C.

Among different studies, inconsistent measurement of diversity by indices of the number of individual foods and numbers of high-quality foods makes comparisons and general conclusions difficult. Nonetheless, data from different approaches consistently support the assumption that diversity in fruits and vegetables contributes to nutrition and health.

Food Functionality in Relation to Dietary Diversity

Dietary quality as it contributes to the health benefits of dietary variety can be attributed partly, but by no means exclusively, to nutrient content. Nutritional quality of the diet does improve with consumption of greater food diversity (Hatløy et al. 1998; Johns 2003). However, content of vitamins, minerals, protein, and energy alone does not explain the benefits associated with Mediterranean, Korean (Kim et al. 2000), or other diets. Various nonnutrients such as phytochemicals and fiber, as well as quality of energy sources, also play an important role (Trichopoulou and Vasilopoulou 2000).

Such scientific insights have stimulated attention to so-called functional foods (Johns and Romeo 1997; Hasler 1998; Milner 2000), more so in developed countries, where both consumer demand and entrepreneurial initiative drive interest. The licensing of Foods of Specific Health Use in Japan starting in 1991 (Arai 2000; Arai et al. 2001) and the U.S. Food and Drug Administration's acceptance of health claims under the Nutrition Labeling and Education Act of 1990 (www.cfsan.fda.gov/~dms/lab-hlth.html; Ross 2000) provide sanction and impetus for recognition of the ability of foods to promote health beyond their contribution to basic nutrition. In parallel, many dietary supplements and natural health products sold in dosage form, such as garlic and grape seed, are targeted to diseases and conditions associated with diet (Blumenthal et al. 2000). Such products derive from both conventional foods and herbal sources and together contribute to the growing profusion of plant diversity ingested by consumers in developed countries.

Many of these products have longstanding traditional reputations and use in various parts of the world. Relevant food- and beverage-derived examples include major commodities such as soy, tea, flax, and tomato, as well as locally produced species such as *Vaccinium* (cranberry, bilberry, and blueberry), maca (*Lepidium meyenii*) (Johns 1981; Quiros and Aliaga-Cardenas 1997), rooibos (*Aspalanthus* sp.) (Standley et al. 2001), prickly pear leaf (*Opuntia ficus-indica*), fish oil, and other marine foods.

Table 15.1 lists examples of commercially important functional foods. In many cases functional activities can be attributed to specific chemical constituents (Johns and Romeo 1997). For example, many phenolics (e.g., flavonoids), carotenoids, and other phytochemicals are antioxidants with important roles in lipid metabolism and as antimutagenic agents. These

Table 15.1. Examples of functional foods, beverages, and condiments from plant sources.

Plant	Physiologic Effect	Known Active Constituents	References
Bitter melon	Hypoglycemic		Marles and Farnsworth (1995)
Blueberry, bilberry	Antioxidants	Polyphenols	Wang et al. (1999)
Broccoli and other cruciferous vegetables	Anticancer	Indole-3 carbinol, isothiocyanates	Hasler (1998)
Citrus fruits	Anticancer	Limonoids	Montanari et al. (1997)
Cranberry	Urinary tract infections	Proanthocyanidins	Howell et al. (1998)
Fenugreek	Hypoglycemic		Marles and Farnsworth (1995)
Flaxseed	Anticancer, estrogenic, hypocholesterolemic	Lignan precursors, α-linolenic acid	Hasler (1998), Blumenthal et al. (2000)
Garlic	Hypolipidemic; antihypertensive; antibacterial	Thiosulfinates	Hasler (1998), Blumenthal et al. (2000)
Oat	Hypocholesterolemic	β-glucans	Hasler (1998)
Olive oil	CVD risk reduction	Oleic acid, polyphenols	Visioli and Galli (1998)
Prickly pear leaf	Antidiabetic	Soluble fiber?	Trejo-Gonzalez et al. (1996)
Psyllium	CVD risk reduction	Soluble fiber	Hasler et al. (2000)
Rooibos	Antimutagenic, antioxidant	Polyphenols	Standley et al. (2001)
Soy	Hypocholesterolemic, CVD risk reduction, anticancer, estrogenic	Soy protein, isoflavones	Hasler (1998)
Tea	Antioxidant	Polyphenols	Mukhtar and Ahmad (2000)

Table 15.1. continued

Plant	Physiologic Effect	Known Active Constituents	References
Thyme	Antioxidant, bronchospasmolytic	Phenols: thymol, carvacrol	Nakatani (1997); Blumenthal et al. (2000)
Tomato	Antioxidant, anticancer	Lycopene	Hasler (1998)
Wine and grapes	Antioxidant, CVD risk reduction	Polyphenols	Hasler (1998)

CVD = cardiovascular disease.

activities positively mediate the risk of cardiovascular disease, cancer, and other diseases. Other anticancer agents include indoles and isothiocyanates from cruciferous vegetables and phytoestrogens such as isoflavonoids and lignans from soy and flax, respectively. Phytosterols, which are ubiquitous in plants, and thiosulfinates (allicin and its derivatives) from species of the genus *Allium* appear to reduce risk of coronary heart disease by lowering blood lipids. In addition, the nature of the lipid composition of foods, specifically the chain length and unsaturation of fatty acids, can mediate health affects. Oleic acid, a monounsaturated fatty acid of 18 carbons, contributes to the health benefits of olive oil. Long-chain polyunsaturated fatty acids that are preformed in animal foods have a number of important metabolic and health-related roles that are mediated by their metabolism to eicosanoids (Simopoulos 1994). Dietary fiber, particularly soluble fiber, contributes to the role of grain products such as oats and psyllium and many fruits and vegetables in reducing the risk of coronary heart disease and cancer. Soluble and insoluble fiber, digestibility-inhibiting phytochemicals in food, and the nature of particular carbohydrates improve glycemic control and reduce hyperlipidemia in patients with diabetes (Johns and Chapman 1995; McIntosh and Miller 2001).

Thousands of phytochemicals comprising diverse structural types and individual compounds vary both within and between plants. Few foods have been fully characterized for the range and functionality of their chemical constituents, but as data increase, so should the list of functional foods and associated compounds. Focused research on the diverse traditional plants ingested as food and medicine in developing countries, such as leafy vegetables (Chweya and Eyzaguirre 1999; Trichopoulou et al. 2000), apart from expanding the list of more widely known species in table 15.1, would greatly enhance understanding of their importance in traditional subsistence systems (Johns and Sthapit 2004).

Global Change, Diet, and Health

Rapid processes of change in both industrialized and developing countries that profoundly alter relationships between humans and the ecosystems in which they live have dietary implications. Traditional subsistence systems often represent finely tuned and unique human resource interactions that ensure that nutritional and health needs are met (Johns 1996; Kuhnlein and Receveur 1996). Disruption in environmental integrity in turn affects patterns of human health, disease, and nutritional status (Johns and Eyzaguirre 2000). Dietary inadequacy, including loss of diversity, represents a key outcome of change as it directly precipitates challenges to human health. Degradation of diet coupled with environmental stresses challenges the health of human communities in unprecedented ways, including through malnutrition, immunity and infection, environmental toxins, and oxidative stress.

Malnutrition

Overpopulation and factors leading to ecosystem and biodiversity destruction that undermine the capacity to produce food result in inadequate intake or consumption of nutritionally poor foods, and perhaps consequently protein energy malnutrition. Micronutrient malnutrition may reflect a disruption of traditional patterns of subsistence, resulting in reduced access to and intake of specific biological resources.

Traditional subsistence patterns couple energy expenditure for food procurement and other activities with intake of foods with low energy density. In addition to energy overconsumption, increased reliance on processed foods may affect health by reducing intake of nutrients and functional compounds that protect health more subtly (Johns 1999).

Data on dietary patterns for most populations are inadequate to establish shifts in diversity of individual fruit and vegetable intake over time. However, national consumption trends in many cases are sufficiently profound to underscore emerging disease phenomena such as diabetes and coronary heart disease (Popkin et al. 2001b). The United Nations Food and Agriculture Organization (FAO) Food Balance Sheets (apps.fao.org/) for several East African countries over the past 35 years, most dramatically Kenya, show a large reduction in per capita consumption of legumes that parallels an increase in the energy intake from edible oils (figure 15.1). Considering the recognized nutritional (iron, fiber, protein) and functional (Milner 2000) benefits of beans and pulses, the emerging public health impact in terms of micronutrient and possibly protein deficiencies and of diseases of energy overconsumption could be significant, the possible mediating effects of fruit and vegetable diversity notwithstanding.

Conventional interventions to nutritional deficiencies, including clinical and dietary supplementation and fortification of processed food (Allen and Gillespie 2001), though effective where warranted and adequately monitored, under normal circumstances offer imperfect solutions for people in developing countries for economic, technical, and cultural reasons. Moreover, single-nutrient responses to identifiable deficiencies, though immediately appropriate, may occur at the expense of addressing

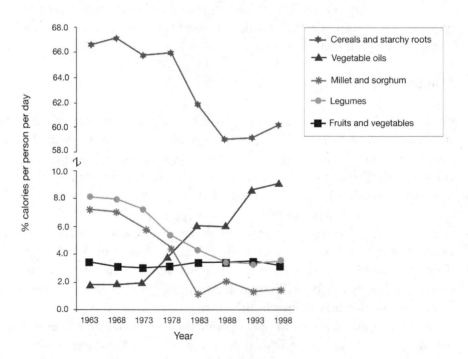

FIGURE 15.1. A comparison of Kenyan food available for consumption (1963–1998). Three-year averages (year ± 1) are presented (from FAO Food Balance Sheets, apps.fao.org/default.jsp).

multiple, usually more cryptic deficiencies and fail to provide the balance necessary for long-term health.

Food-based solutions—in addition to increasing the availability and intake of vitamins A and C, folate, iron, and other micronutrients—are reasonably likely to be sustained (Allen and Gillespie 2001), particularly if they are ecologically and culturally appropriate. However, evaluation of food-based intervention programs has been inadequate and should be a priority for any efforts to use PGRs for this objective.

Immunity and Infection

Disease factors of environmental origin compromise nutritional status, which in turn plays a critical role in the severity and prevalence of illnesses. Disruption of natural ecosystems can elevate rates of infectious disease by increasing exposure to vector-borne disease such as malaria, leishmaniasis, and dengue (Spielman and James 1990) or through impacts on density-related factors such as sanitation and direct person-to-person transmission. Major public health problems of global importance such as tuberculosis, gastrointestinal diseases, measles, and respiratory disease all reflect the interaction of nutritional and environmental factors (Platt 1996). Malnutrition may result in deficiencies in micronutrients such as vitamin A and iron that affect the immune system and compound these and other diseases (Tomkins 2000), such as HIV and AIDS. The impact of potential functional (such as immunostimulatory or antioxidant) properties in traditional diets and medicines is much less understood.

Oxidative Status

Oxidative status plays an important role in many disease states, including chronic diseases such as diabetes, cardiovascular disease, and cancer, as both a causal factor and an adverse outcome. Environmental contamination from industrial and agricultural chemicals such as heavy metals, organochlorines, and radionuclides compromises nutritional status (Kuhnlein and Chan 2000) and has both local and global impacts on diet and health, including as serious contributors to oxidative stress. Exogenous antioxidants, particularly dietary vitamins and nonnutrients, form a key component of the normal defense against oxidative stress. Reduction in intake of plant dietary diversity therefore has further negative consequences, whereas increased use offers positive solutions.

Urbanization and the Nutrition Transition

Urban populations make increasing impacts on the environment through market demands, by settling in natural and agricultural areas, and through pollution associated with industrial growth and urban waste. In this situation the urban poor are doubly affected by deficiencies in diet and by the negative consequences of living in unhealthful conditions.

While making greater numbers of people secure in terms of energy, the high-input, high-yield agriculture and long-distance transport that increase the availability and affordability of refined carbohydrates (wheat, rice, sugar) and edible oils (WHO 2003) also underpin the nutrition transition (Popkin et al. 2001a; Chopra et al. 2002; Popkin 2002). In addition, the globalization of culture and commerce fosters a Westernization of developing country food systems and diets. Urban populations depend to a greater degree on purchased food than people in rural areas while having less access to diverse wild and locally cultivated foods. Choice of purchased food is determined by availability and affordability, and so in cities poverty probably becomes a greater limit to dietary diversity than it does in traditional subsistence systems. In parallel, as rural producers become more linked to urban markets for their livelihoods, a lack of demand for products most consumers cannot afford can further reduces market volumes and makes production less economically viable.

Nutrition Transition

Consumption of a diet derived from high-energy plant and animal source foods coincides with low energy expenditure. The greater diversity, including fruits and vegetables, generally available to urban populations does not necessarily translate into consumption (Popkin et al. 2001b), particularly for the poor. Processed foods available for purchase through contemporary market systems, though potentially variable in brand and formulation, may comprise limited actual biological diversity, often related to the use of imported replacements for local foods.

This nutrition transition is leading to emerging epidemics of type 2 diabetes mellitus, cardiovascular disease, obesity, cancer, and other chronic noncommunicable diseases, even in poor countries (Popkin et al. 2001a; Chopra et al. 2002; Popkin 2002; WHO 2003). The consequences of a high-carbohydrate, high-fat diet are further complicated and compounded among the disadvantaged in developing countries, where dietary changes in combination with poverty, high rates of infectious disease, and under-nutrition create a double burden (Popkin et al. 2001a; Popkin 2002). Cheap food energy combined with low diversity and nutritional quality produces a pattern of obesity, particularly of women, in combination with household undernutrition (Doak et al. 2000). Early childhood mal-nutrition (fetal programming) probably increases susceptibility to diabe-tes and other conditions in later life (Popkin et al. 2001a). Epidemics of chronic noncommunicable diseases can be expected to further accelerate in countries with aging populations. The WHO Global Strategy on Diet, Physical Activity and Health (WHO 2003) affirms the centrality of food-based approaches to combating noncommunicable diseases.

What few dietary studies have been conducted in African cities point to similar trends to those already well recognized in Asia and Latin America: decreased energy expenditure couples with increased dependence on deep-fried foods derived from starchy sources such as cassava, wheat, and pota-toes and decreased intake of fresh fruits and vegetables (Mennen et al. 2000). For the poorest and most vulnerable segments of the population these products often take the form of street foods (Van t' Riet et al. 2001) of low nutrient density. As a result, in Africa for large segments of the pop-ulation the conditions of energy overconsumption probably will coexist with classic nutrient deficiencies and with infectious disease (Bourne et al. 2002; Johns 2003).

Senegal shows an even more dramatic increase in edible oil consumption than in Kenya (figure 15.1), with available calories from edible oil and fat having increased between 1963 and 1998 from 8% to 20% (figure 15.2). Although increased fats are beneficial in impoverished diets to increase en-ergy and facilitate vitamin A availability, the increase of total available cal-ories from fat in the Senegalese diet from 18% to 29%, in other words to levels equal to recommendations in developed countries, suggests that a sig-nificant number of people consume much more than recommended. At the same time there appears to be a drop by half (and by an even higher pro-portion in Kenya) in consumption of traditional cereals of millets and sor-ghum, foods that, though suboptimal in their nutrient content, digestibility, and palatability, offer potential antioxidant (Sripriya et al. 1996) and hypo-glycemic benefits relative to exotic cereals of wheat, rice, and maize.

With urbanization in sub-Saharan Africa projected by the United Na-tions Human Settlements Programme (www.unchs.org/unchs/english/stats/table2.htm) at 4% per annum, to approach 50% of the population of the region in the next 15 years, solutions to forestall the nutrition and

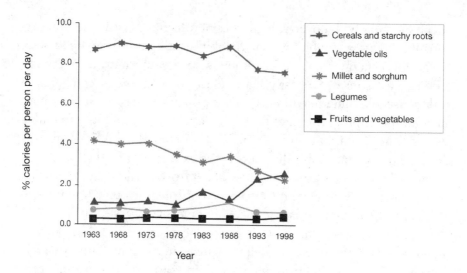

FIGURE 15.2. A comparison of Senegalese food available for consumption (1963–1998). Three-year averages (year ± 1) are presented (from FAO Food Balance Sheets, apps.fao.org/default.jsp).

health impact of this trend are acutely needed. Here and globally a greater use of plant biodiversity based on scientific evaluation of plant properties, cultural support programs, dietary education, innovative processing, and marketing provide possible avenues for mediating the impacts of change.

Importance of Crop Diversity and Neglected and Underutilized Species

Erosion of Food Crop Diversity

Whereas more than 7,000 plant species traditionally have been used for food, three species—rice, wheat, and maize—account for 60% of the total caloric intake in the human diet today (Eyzaguirre et al. 1999). Global modern agriculture typically focuses on yields of a few crops, and years of breeding research have brought about high-yielding, pest- and drought-resistant varieties of a small number of distinct food species. The sheer magnitude of agricultural effort applied to the three principal crops has led to a decline in the consumption of more diverse grains. There has been an accompanying decrease in the variety of vegetable and fruit species consumed. Cultural change and urbanization compound this trend (Chweya and Eyzaguirre 1999). Additionally, many traditional foods are now associated with being poor or backward. The result is disruption of dietary patterns and loss of dietary diversity. Little is known about the impact of these dietary changes on human nutrition and health.

Neglected and Underutilized Species

Given the well-supported principles of dietary diversity, a variety of foods undoubtedly contribute to a balanced diet in local communities (Padulosi 1999). In Africa, for example, neglected and underutilized species (NUSs) of local dietary importance include cereal grain crops such as fonio (*Digitaria exilis*), roots and tubers such as yams, pulses, and oil seeds such as bambara groundnut (*Vigna subterranea*) (Heller et al. 1997), leafy vegetables (Chweya and Eyzaguire 1999), and tropical fruits such as African plum (*Dacryodes edulis*) or the bush mango (*Irvingia gabonensis*).

Although the importance of diversity and the wisdom inherent in the traditional systems that incorporate NUSs can be appreciated even without knowledge of the specific nutrient constituents of the individual components of the diet, existing data on a few species provide useful insights into the ways they contribute to health. For example, baobab (*Adansonia digitata*), of which the young leaves and the fruits are eaten, has local dietary importance in several African countries (Diouf et al. 1999). The fruit pulp and dried leaves, which are added to porridge, made into sauces, and added directly to cooked dishes (Diouf et al. 1999; Maundu et al. 1999), are good sources of calcium (West et al. 1988; Glew et al. 1997; Boukari et al. 2001). In addition, they combine iron and vitamin C in amounts that can interact to increase iron absorption and prevent anemia. Even without comprehensive studies we know that leafy vegetables in general make important contributions in provitamin A, vitamin C, folate, iron, calcium, fiber, and protein (West et al. 1988; Uiso and Johns 1996; Chweya and Eyzaguirre 1999), recent controversies regarding the bioavailability of provitamin A notwithstanding (Solomons and Bulux 1997; de Pee et al. 1998).

Nutritional Value of Traditional Edible Species and Varieties

Although wild and cultivated biodiversity in most developing regions is ignored in dietary surveys, compositional analyses, FAO Food Balance

Sheets, and policy- and decision-making (Johns 2003), such resources unquestionably make essential contributions to dietary adequacy (Chweya and Eyzaguirre 1999; Burlingame 2000; Johns 2003; Kuhnlein and Johns 2003). Studies on home gardens have established links between diversity and nutritional status (Marsh 1998; Johns 2003). In some cases the contributions of gathered species for specific nutrients are clearly demonstrated (Ogle et al. 2001a, 2001b), and many indigenous species have exceptional nutritional properties (Rodriquez-Amaya 1999; Johns 2003).

Documentation of the contribution of intraspecific diversity to nutrition and health has received little attention and few analytical resources. Farmer-based research demonstrates the wealth of traditional knowledge and beliefs concerning the health, sensory, and culinary properties of local crop varieties (FAO 2001). Screening of major crops (Fassil et al. 2000; Graham and Rosser 2000; FAO 2001; Johns 2003; Johns and Sthapit 2004), though incomplete, clearly documents wide variation in nutritional and functional properties that undoubtedly has implications for nutritional status of populations and individual consumers (in addition to its usefulness to plant breeders). The potential genetic variation for nutrient composition within NUSS (cf. Calderon et al. 1991; Chweya and Eyzaguirre 1999; Burlingame 2000) has been even less documented.

Intraspecific Diversity

From the perspective of PGR conservation and use, intraspecific variation in the nutrient and nonnutrient composition of crop plants is of particular interest. Although few data have been compiled systematically, variations in the composition of β-carotene in sweet potatoes (Huang et al. 1999; Ssebuliba et al. 2001) and of carotenoids in maize (Kurilich and Juvik 1999) provide examples of the likely range of functional diversity that exists within species.

Within traditional agrofood systems, potatoes present an intriguing case. Andeans maintain large numbers of distinct genotypes on farm and in the diet, varying most strikingly in pigments characterized by polyphenols and xanthophylls (lutein and zeaxanthin), carotenoids (Brown et al. 1993) with known functional properties and, by extension, in implications for health.

Traditional concepts of diet often include associations with health that, generally speaking, refer not to nutrients but rather to specific functional properties. Some traditional concepts such as tonics or strengtheners may be understandable in nutritional terms. Other food attributes relate to physiological and pharmacological properties and can be supported by scientific investigations in these areas.

Many of the benefits of nonnutrients may exceed those attributable to nutrients. For example, vegetable diets that make modest contributions to improving vitamin A status result in significant increases in serum levels of lutein (de Pee et al. 1998), an antioxidant for which protective benefits in relation to ocular disease (Sommerburg et al. 1998; Brown et al. 1999; Gale et al. 2001), as well as cardiovascular disease and cancer, are increasingly recognized as significant to health in developed countries. Such insights have potential significance in tropical countries such as those of Africa, where cataracts represent a major cause of blindness (Lewallen and Courtright 2001). Although compilations of data on xanthophylls (Holden 1999; O'Neill et al. 2001) point to the richness of leafy vegetables in these carotenoids, extension of these analyses to indigenous plants foods is called for. In light of this important functional activity, the single-minded attention to the limitations of leafy vegetables and other plant foods as sources of provitamin A (Solomons and Bulux 1997; de Pee et al. 1998) seems myopic.

Potential health-related functions of indigenous dietary plants include antibiosis, immunostimulation, nervous system action, detoxification, and anti-inflammatory, antigout, antioxidant, glycemic, and hypolipidemic properties. Ethnobotanical and analytical work at the Centre for Indigenous Peoples' Nutrition and Environment and McGill University, among many other groups, has addressed a number of these functional health benefits of traditional dietary plants.

For example, the Luo of western Kenya and Tanzania attribute action against gastrointestinal disturbances to the leafy vegetables that are an important component of their traditional diet. Among these, *Solanum nigrum*, in particular, has strong activity against the protozoan parasite *Giardia lamblia* (Johns et al. 1995). Additionally, we have reported on the antioxidant activity of phenolics (Lindhorst 1998) and cholesterol-binding activity of saponins (Chapman et al. 1997; Johns et al. 1999) in roots and barks that the pastoral Masai add to fatty soups and milk, the potential hypolipidemic activity of gums chewed by the Masai (Johns et al. 2000), the antioxidant activity of Tibetan treatments of heart disease (Owen and Johns 2002), and the antidiabetic remedies of indigenous peoples of the boreal forests of eastern North America (McCune and Johns 2002). We have also identified xanthine oxidase activity in traditional remedies for gout and related symptoms from the latter region (Owen and Johns 1999) and in dietary additives of the Masai (unpublished results).

Because such functional effects on human health can be attributable to phytochemical constituents of these plants, diversity of function and chemical composition add further dimensions to the diversity inherent in the food and medicinal plants used around the world.

Despite the potential for income generation that comes with the commercialization of some traditional foods and medicinal products, functionality generally has a different significance in addressing the needs of the majority of the populations in developing countries than it does in Europe, North America, or Japan. Whether for rural subsistence or in the diets of urban populations, the function of culturally significant species has immediate biological and social importance to the present and future health of people in developing countries that warrants research and program support appropriate to this context.

Dietary Adaptation and Optimization

Rational use of dietary resources and application of knowledge concerning their value can define a course for optimal adaptation to the changes facing populations around the world (Johns and Eyzaguirre 2002). Considering the magnitude and unprecedented nature of the shifts occurring in lifestyle, scientific insights into the relationships between environment, diet, and health and the adverse consequence of current change and scientific evaluation of the properties of plant and animal foods seem essential tools for achieving novel solutions for contemporary problems. In this process of adaptation, however, the lessons of the past represented by the wealth of indigenous knowledge of biological resources and ecosystems, as well as the diversity of resources themselves, are essential. In this regard documentation and study of the world's biocultural diversity should take high priority.

Developing Links Between Nutrition, Health, and PGR Conservation

Nutrition and health offer several potential entry points into PGR programs and activities, and links between nutrition and PGRs can proceed simultaneously on several fronts. Given the imperative to ensure human well-being while conserving biodiversity, conservation and use of biodiversity and local and global human needs provide distinct but complementary approaches. In the former, nutrition-related activities can be defined within both ex situ and in situ strategies of PGR conservation and use. In turn, the nutritional and health needs of farmers and consumers as well as scientific and public health issues of global concern can guide PGR activities.

With greater awareness of nutrition and health priorities in agriculture and environmental sciences and of the role of plant biodiversity in the international health community, scientists and institutions engaged in agriculture, environmental conservation, and health can better address contemporary problems by creating and taking advantage of opportunities to collaborate.

Defining Priorities for Nutrition and Health Research

In health-related fields, PGRs offer useful perspectives on a number of issues of contemporary scientific and public health importance, including micronutrient deficiency and food-based strategies for addressing multiple concurrent deficiencies; bioavailability of provitamin A, iron, and other nutrients from fruits and vegetables; nutrition and disease; the nutrition transition; and medicinal plants as physiological mediators of health.

Within the scope of these health priorities, research activities linking nutrition and PGRs (Johns 1999, 2002) that might emerge might include laboratory analysis identifying crop varieties and minor crops with selective nutritional assets (cf. Booth et al. 1994); databases on composition with emphasis on intraspecific diversity; on-farm and community-based activities focusing on indigenous knowledge of health-related properties of plant resources; formulation and compilation of criteria and indicators for evaluating consumer quality (e.g., sensory, nutrition, culinary, toxicological, and medicinal properties) for both in situ and ex situ approaches to PGR conservation and use; dietary diversity indices to further establish the importance of PGRs and serve as simple, low-cost indicators of nutritional status in developing country context (Hatløy et al. 1998); and public health research (WHO 2003).

Because poverty is the single most important determinant of malnutrition and disease, a better understanding of the synergies between biodiversity conservation, economics, and nutrition is essential (Johns and Sthapit 2004).

Conclusion

Diversity in plant resources plays an essential role in enabling human populations to meet their nutritional, health, and sociocultural needs. Biodiversity equates with dietary diversity, which equates with health. In the contemporary world, where global change affects traditional ecology in ways that threaten biodiversity and at the same time undermine human subsistence, health is a vital rationale for managing biodiversity and conserving PGRs.

Plant resources coupled with the biocultural wisdom inherent in traditional systems can help address the serious problems of food insecurity and undernutrition facing developing countries. At the same time, plant biodiversity is an essential resource as societies adapt to changes, particularly those associated with urbanization. In this regard, rural–urban links are crucially important. The diverse nutrition and health functions that plants serve in traditional culture, and the indigenous knowledge of plant diversity, offer potentially valuable solutions that enable biodiversity to address the unique problems facing contemporary society.

Major international initiatives in nutrition, food security, and agriculture typically focus on single characteristics of food or on a few species and genotypes. Though understandable given the severity of problems of micronutrient deficiency and food insecurity for large population segments, such targeted approaches overlook the complex nature of human–environment relationships and the multifactorial nature of human diseases and health. Diversity of diet is a direct measure of dietary quality. Therefore, overemphasis on quantity of yield rather than quality and on single nutrients or a limited number of foods in programs of biofortification or dietary modification can be very short sighted. Immediate positive outcomes may fail or produce adverse consequences in the long run as they limit the complexity and functional diversity of diets and possibly precipitate diseases states.

Furthermore, short-term success in addressing nutritional needs may have negative consequences in eroding biodiversity, indigenous knowledge of its use, and the sociocultural values that support its maintenance. In turn, loss of diversity contributes to major health problems such as diabetes. Within the limited economic and technological options in a developing country, the consequences of a shift to dietary simplicity are likely to be magnified as they limit people's capacity to adapt to changing circumstances. Emphasis on technological solutions creates a dependence on technology that is likely not to be available to address consequential problems. Without poverty reduction or attention to economic factors limiting the availability of diverse diets, the benefits in traditional food systems cannot be sustained. Only a holistic approach to dietary diversity supporting the wide availability of diverse crops and edible plants can raise people's nutritional and health status in a sustainable way.

PGRs in human diet and medicine and the knowledge imbedded in culture as an integral component of the complexity in human ecological systems offer a time-honored buffer to destructive change. Proximate approaches provide neither optimal nor ultimate solutions. The health desired for all the world's people is much more than simply the absence of disease and infirmity (WHO 1946). As human health is recognized as a state of complete physical, mental, and social well-being, it intrinsically connects with the health of the ecosystems in which we live. Toward this end, plant genetic resources are of both profound utility and inherent value.

Acknowledgments

The Natural Sciences and Engineering Research Council of Canada, the International Plant Genetic Resources Institute, the Fond Québécois de la Recherche sur la Nature et les Technologies, and the Fonds de la Recherche en Santé du Québec provided financial support for this work. Pablo Eyzaguirre, Mikkel Grum, and others at the International Plant Genetic Resources Institute contributed to the development of the ideas explored in this chapter.

References

Allen, L. H. and S. R. Gillespie. 2001. *What Works? A Review of the Efficacy and Effectiveness of Nutrition Interventions.* Geneva: ACC/SCN.

Arai, S. 2000. Functional food science in Japan: State of the art. *Biofactors* 12:13–16.

Arai, S., T. Osawa, H. Ohigashi, M. Yoshikawa, S. Kaminogawa, M. Watanabe, T. Ogawa, K. Okubo, S. Watanabe, H. Nishino, K. Shinohara, T. Esashi, and T. Hirahara. 2001. A mainstay of functional food science in Japan: History, present status, and future outlook. *Bioscience, Biotechnology and Biochemistry* 65:1–13.

Blumenthal, M., A. Goldberg, and J. Brinckmann. 2000. *Herbal Medicine: Expanded Commission E Monographs.* New York: American Botanical Council.

Booth, S., T. Johns, J. A. Sadowski, and N. W. Solomons. 1994. Phylloquinone as a biochemical marker of the dietary intake of green leafy vegetables of the K'ekchi people of Alta Verapaz, Guatemala. *Ecology of Food and Nutrition* 31:201–209.

Boukari, I., N. W. Shier, X. E. Fernandez R., J. Frisch, B. A. Watkins, L. Pawloski, and A. D. Fly. 2001. Calcium analysis of selected western African foods. *Journal of Food Composition and Analysis* 14:37–42.

Bourne, L. T., E. V. Lambert, and K. Steyn. 2002. Where does the black population of South Africa stand on the nutrition transition? *Public Health Nutrition* 5(1A):157–162.

Brown, C. R., C. G. Edwards, C. P. Yang, and B. B. Dean. 1993. Orange flesh trait in potato: Inheritance and carotenoid content. *Journal of the American Society for Horticultural Science* 118:145–150.

Brown, L., E. B. Rimm, J. M. Seddon, E. L. Giovannucci, L. Chasan-Taber, D. Spiegelman, W. C. Willet, and S. E. Hankinson. 1999. A prospective study of carotenoid intake and risk of cataract extraction in US men. *American Journal of Clinical Nutrition* 70:517–524.

Burlingame, B. 2000. Wild nutrition. *Journal of Food Composition and Analysis* 13:99–100.

Calderon, E., J. M. Gonzalez, and R. Bressani. 1991. Caracteristicas agronomicas, fisicas, quimicas y nutricias de quince variedades de amaranto. *Turrialba* 41:458–464.

Chapman, L., T. Johns, and R. L. A. Mahunnah. 1997. Saponin-like in vitro characteristics of extracts from selected non-nutrient wild plant food additives used by Maasai in meat and milk based soups. *Ecology of Food and Nutrition* 36:1–22.

Chopra, M., S. Galbraith, and I. Darnton-Hill. 2002. A global response to a global problem: The epidemic of overnutrition. *Bulletin of the World Health Organization* 80:952–958.

Chweya, J. A. and P. B. Eyzaguirre, eds. 1999. *The Biodiversity of Traditional Leafy Vegetables*. Rome: IPGRI.

de Pee, S. C., W. West, D. Permaesih, S. Martuti, and J. G. A. J. Hautvast. 1998. Orange fruit is more effective than are dark-green, leafy vegetables in increasing serum concentrations of retinol and beta-carotene in schoolchildren in Indonesia. *American Journal of Clinical Nutrition* 68:1058–1067.

Diouf, M., M. Diop, C. Lo, K. A. Drame, E. Sene, C. O. Ba, M. Gueye, and B. Faye. 1999. Sénégal. In J. A. Chweya and P. B. Eyzaguirre, eds., *The Biodiversity of Traditional Leafy Vegetables*, 111–154. Rome: IPGRI.

Doak, C. M., L. S. Adair, C. Monteiro, and B. M. Popkin. 2000. Overweight and underweight coexist within households in Brazil, China and Russia. *Journal of Nutrition* 130:2965–2971.

Drewnowski, A., S. A. Henderson, A. B. Shore, C. Fischler, P. Preziosi, and S. Hercberg. 1996. Diet quality and dietary diversity in France: Implications for the French paradox. *Journal of the American Dietetics Association* 96:663–669.

Eyzaguirre, P. B., S. Padulosi, and T. Hodgkin. 1999. IPGRI's strategy for neglected and underutilized species and the human dimension of agrobiodiversity. In S. Padulosi, ed., *Priority-Setting for Underutilized and Neglected Plant Species of the Mediterranean Region*, 1–19. Rome: IPGRI.

FAO (Food and Agriculture Organization of the United Nations). 2001. *Specialty Rices of the World: Breeding, Production and Marketing*. Rome: FAO.

Fassil, H., L. Guarino, S. Sharrock, Bhag Mal, T. Hodgkin, and M. Iwanaga. 2000. Diversity for food security: Improving human nutrition through better evaluation, management, and use of plant genetic resources. *Food Nutrition Bulletin* 21:497–502.

Gale, C. R., N. F. Hall, D. I. Phillips, and C. N. Martyn. 2001. Plasma antioxidant vitamins and carotenoids and age-related cataract. *Ophthalmology* 108:1992–1998.

Glew, R. H., J. VanderJagt, C. Lockett, L. E. Grivetti, G. C. Smith, A. Pastuszyn, and M. Millson. 1997. Amino acid, fatty acid, and mineral composition of 24 indigenous plants of Burkina Faso. *Journal of Food Composition and Analysis* 10:205–217.

Graham, R. D. and J. M. Rosser. 2000. Carotenoids in staple foods: Their potential to improve human nutrition. *Food Nutrition Bulletin* 21:405–409.

Hasler, C. M. 1998. Functional foods: Their role in disease prevention and health promotion. *Food Technology* 52:63–70.

Hasler, C. M., S. Kundrat, and D. Wool. 2000. Functional foods and cardiovascular disease. *Current Atherosclerosis Reports* 2:467–475.

Hatløy, A., L. E. Torheim, and A. Oshaug. 1998. Food variety: A good indicator of nutritional adequacy of the diet? A case study from an urban area in Mali, West Africa. *European Journal of Clinical Nutrition* 52:891–898.

Heller, J., F. Begemann, and J. Mushonga, eds. 1997. *Bambara Groundnut* Vigna subterranea *(L.) Verdc.* Rome: IPGRI.

Holden, J. M. 1999. Carotenoid content of U.S. foods: An update of the database. *Journal of Food Composition and Analysis* 12:169–196.

Howell, A. B., N. Vorsa, A. Der Marderosian, and L. Y. Foo. 1998. Inhibition of the adherence of p-fimbriated *Escherichia coli* to uroepithelial-cell surfaces by proanthocyanidin extracts from cranberries. *New England Journal of Medicine* 339:1085–1086.

Huang, A. S., L. Tanudjaja, and D. Lum. 1999. Content of alpha-, beta-carotene, and dietary fiber in 18 sweetpotato varieties grown in Hawaii. *Journal of Food Composition and Analysis* 12:147–151.

Johns, T. 1981. The añu and the maca. *Journal of Ethnobiology* 1:208–212.

Johns, T. 1996. Phytochemicals as evolutionary mediators of human nutritional physiology. *International Journal of Pharmacognosy* 34:327–334.

Johns, T. 1999. The chemical ecology of human ingestive behaviors. *Annual Review of Anthropology* 28:27–50.

Johns, T. 2002. Plant genetic diversity and malnutrition: Practical steps for developing and implementing a global strategy linking plant genetic resource conservation and nutrition. *African Journal of Food and Nutritional Sciences* 2(2):98–100.

Johns, T. 2003. Plant biodiversity and malnutrition: Simple solutions to complex problems. *African Journal of Food, Agriculture, Nutrition and Development* 3:45–52.

Johns, T. and L. Chapman. 1995. Phytochemicals ingested in traditional diets and medicines as modulators of energy metabolism. In J. T. Arnason and R. Mata, eds., *Phytochemistry of Medicinal Plants,* Recent Advances in Phytochemistry 29, 161–188. New York: Plenum.

Johns, T. and P. B. Eyzaguirre. 2000. Nutrition for sustainable environments. *SCN News* 21:24–29.

Johns, T. and P. B. Eyzaguirre. 2002. Nutrition and the environment. In *Nutrition: A Foundation for Development.* Geneva: ACC/SCN.

Johns, T., G. M. Faubert, J. O. Kokwaro, R. L. A. Mahunnah, and E. K. Kimanani. 1995. Anti-giardial activity of gastrointestinal remedies of the Luo of East Africa. *Journal of Ethnopharmacology* 46:17–23.

Johns, T., R. L. A. Mahunnah, P. Sanaya, L. Chapman, and T. Ticktin. 1999. Saponins and phenolic content of plant dietary additives of a traditional subsistence community, the Batemi of Ngorongoro District, Tanzania. *Journal of Ethnopharmacology* 66:1–10.

Johns, T., M. Nagarajan, M. L. Parkipuny, and P. J. H. Jones. 2000. Maasai gummivory: Implications for Paleolithic diets and contemporary health. *Current Anthropology* 41:453–459.

Johns, T. and J. T. Romeo, eds. 1997. *Functionality of Food Phytochemicals*, Recent Advances in Phytochemistry 31. New York: Plenum.

Johns, T. and B. R. Sthapit. 2004. Biocultural diversity in the sustainability of developing country food systems. *Food and Nutrition Bulletin* 25:143–155.

Kant, A. K., A. Schatzkin, B. I. Graubard, and C. Schairer. 2000. A prospective study of diet quality and mortality in women. *JAMA* 283:2109–2115.

Kant, A. K., A. Schatzkin, and R. G. Ziegler. 1995. Dietary diversity and subsequent cause-specific mortality in the NHANES I epidemiologic follow-up study. *Journal of the American College of Nutrition* 14:233–238.

Kim, S., S. Moon, and B. M. Popkin. 2000. The nutrition transition in South Korea. *American Journal of Clinical Nutrition* 71:44–53.

Kuhnlein, H. V. and H. M. Chan. 2000. Environment and contaminants in traditional food systems of northern indigenous peoples. *Annual Review of Nutrition* 20:595–626.

Kuhnlein, H. V. and T. Johns. 2003. Northwest African and Middle Eastern food and dietary change of indigenous peoples. *Asia Pacific Journal of Clinical Nutrition* 12:344–349.

Kuhnlein, H. V. and O. Receveur. 1996. Dietary change and traditional food systems of indigenous peoples. *Annual Review of Nutrition* 16:417–442.

Kurilich, A. C. and J. A. Juvik 1999. Quantification of carotenoid and tocopherol antioxidants in *Zea mays*. *Journal of Agricultural and Food Chemistry* 47:1948–1955.

La Vecchia, C., S. E. Munoz, C. Braga, E. Fernandez, and A. Decarli. 1997. Diet diversity and gastric cancer. *International Journal of Cancer* 72:255–257.

Lewallen, S. and P. Courtright. 2001. Blindness in Africa: Present situation and future needs. *British Journal of Ophthalmology* 85:897–903.

Lindhorst, K. 1998. *Antioxidant Activity of Phenolic Fraction of Plant Products Ingested by the Maasai*. MSc thesis, McGill University, Montreal, Canada.

Marles, R. J. and N. R. Farnsworth. 1995. Antidiabetic plants and their active constituents. *Phytomedicine* 2:137–189.

Marsh, R. 1998. Building traditional gardening to improve household food security. *Food, Nutrition and Agriculture* 22:4–14.

Maundu, P. M., G. W. Ngugi, and C. H. S. Kabuye. 1999. *Traditional Food Plants of Kenya*. Nairobi: National Museums of Kenya.

McCune, L. M. and T. Johns. 2002. Antioxidant activity in medicinal plants associated with the symptoms of diabetes mellitus used by the indigenous peoples of the North American boreal forest. *Journal of Ethnopharmacology* 82:197–205.

McIntosh, M. and C. Miller. 2001. A diet containing food rich in soluble and insoluble fiber improves glycemic control and reduces hyperlipidemia among patients with type 2 diabetes mellitus. *Nutrition Reviews* 59:52–55.

Mennen, L. I., J. C. Mbanya, J. Cade, B. Balkau, S. Sharma, S. Chungong, and J. K. Cruickshank. 2000. The habitual diet in rural and urban Cameroon. *European Journal of Clinical Nutrition* 54:150–154.

Milner, J. A. 2000. Functional foods: The US perspective. *American Journal of Clinical Nutrition* 71:1654S–1659S.

Montanari, A., W. Widmer, and S. Nagy. 1997. Health promoting phytochemicals in citrus fruit and juice products. In T. Johns and J. T. Romeo, eds., *Functionality of Food Phytochemicals*, Recent Advances in Phytochemistry 31, 31–52. New York: Plenum.

Mukhtar, H. and N. Ahmad. 2000. Tea polyphenols: Prevention of cancer and optimizing health. *American Journal of Clinical Nutrition* 71:1698S–1702S.

Nakatani, N. 1997. Antioxidants from spices and herbs. In F. Shahidi, ed., *Natural Antioxidants: Chemistry, Health Effects, and Applications*, 64–75. Champaign, IL: AOCS Press.

Ogle, B. M., N. N. X. Dung, T. T. Do, and L. Hambraeus. 2001a. The contribution of wild vegetables to micronutrient intakes among women: An example from the Mekong Delta, Vietnam. *Ecology of Food and Nutrition* 40:159–184.

Ogle, B. M., P. H. Hung, and H. T. Tuyet. 2001b. Significance of wild vegetables in micronutrient intakes of women in Vietnam: An analysis of food variety. *Asia Pacific Journal of Clinical Nutrition* 10:21–30.

O'Neill, M. E., Y. Carroll, B. Corridan, B. Olmedilla, F. Granado, I. Blanco, H. Van den Berg, I. Hininger, A.-M. Rousell, M. Chopra, S. Southon, and D. I. Thurnham. 2001. A European carotenoid database to assess carotenoid intakes and its use in a five-country comparative study. *British Journal of Nutrition* 85:499–507.

Onyango, A., K. Koski, and K. Tucker. 1998. Food diversity versus breastfeeding choice in determining anthropometric status in rural Kenyan toddlers. *International Journal of Epidemiology* 27:484–489.

Owen, P. and T. Johns. 1999. Xanthine oxidase inhibitory activity of northeastern North American plant remedies for gout. *Journal of Ethnopharmacology* 64:149–160.

Owen, P. and T. Johns. 2002. Antioxidants in medicines and spices as cardioprotective agents in Tibetan highlanders. *Pharmaceutical Biology* 40:346–357.

Padulosi, S., ed. 1999. *Priority-Setting for Underutilized and Neglected Plant Species of the Mediterranean Region.* Rome: IPGRI.

Platt, A. E. 1996. *Infecting Ourselves: How Environmental and Social Disruptions Trigger Disease.* Washington, DC: Worldwatch Institute.

Popkin, B. M. 2002. An overview of the nutrition transition and its health implications: The Bellagio meeting. *Public Health and Nutrition* 5:93–103.

Popkin, B. M., S. Horton, and S. Kim. 2001a. The nutrition transition and prevention of diet-related diseases in Asia and the Pacific. *Food and Nutrition Bulletin* 22:S1–58.

Popkin, B. M., S. Horton, S. Kim, A. Mahal, and J. Shuigao. 2001b. Trends in diet, nutritional status, and diet-related non-communicable diseases in China and India: The economic costs of the nutrition transition. *Nutrition Reviews* 59:379–390.

Quiros, C. F. and R. Aliaga-Cardenas. 1997. Maca. *Lepidium meyenii* Walp. In M. Hermann and J. Heller, eds., *Andean Roots and Tubers: Ahipa, Arracacha, Maca and Yacon,* 173–197. Rome: IPGRI.

Rodriquez-Amaya, D. B. 1999. Latin American food sources of carotenoids. *Archivos Latinoamericanos de Nutricion* 49:74S–84S.

Ross, S. 2000. Functional foods: The Food and Drug Administration perspective. *American Journal of Clinical Nutrition* 71:1735S–1738S.

Simopoulos, A. P. 1994. Fatty acids. In I. Goldberg, ed., *Functional Foods: Designer Foods, Pharmafoods, Nutraceuticals,* 355–392. New York: Chapman and Hall.

Solomons, N. W. and J. Bulux. 1997. Identification and production of local carotene-rich foods to combat vitamin A malnutrition. *European Journal of Clinical Nutrition* 51:S39–S45.

Sommerburg, O. E., J. E. Keunen, A. C. Bird, and F. J. van Kuijk. 1998. Fruits and vegetables that are sources for lutein and zeaxanthin: The macular pigment in human eyes. *British Journal of Ophthalmology* 82:907–910.

Spielman, A. and A. A. James. 1990. Transmission of vector-borne disease. In K. S. Warren and A. A. F. Mahmoud, eds., *Tropical and Geographical Medicine.* New York: McGraw-Hill Information Services Company.

Sripriya, G., K. Chandrasekharan, V. S. Murty, and T. S. Chandra. 1996. ESR spectroscopic studies on free radical quenching action of finger millet (*Eleusine coracana*). *Food Chemistry* 57:537–540.

Ssebuliba, J. M., E. N. B. Nsubuga, and J. H. Muyonga. 2001. Potential of orange and yellow fleshed sweetpotato cultivars for improving vitamin A nutrition in central Uganda. *African Crop Science Journal* 9:309–316.

Standley, L., P. Winterton, J. L. Marnewick, W. C. A. Gelderblom, E. Joubert, and T. J. Britz. 2001. Influence of processing stages on antimutagenic and antioxidant potentials of rooibos tea. *Journal of Agricultural and Food Chemistry* 49:114–117.

Tomkins, A. 2000. Malnutrition, morbidity and mortality in children and their mothers. *Proceedings of the Nutrition Society* 59:135–146.

Trejo-Gonzalez, A., G. Gabriel-Ortiz, A. M. Puebla-Perez, M. D. Huizar-Contreras, M. del R. Munguia-Mazariegos, S. Mejia-Arreguin, and E. Calva. 1996. A purified extract from prickly pear cactus (*Opuntia fuliginosa*) controls experimentally induced diabetes in rats. *Journal of Ethnopharmacology* 55:27–33.

Trichopoulou, A. and E. Vasilopoulou. 2000. Mediterranean diet and longevity. *British Journal of Nutrition* 84:S205–S209.

Trichopoulou, A., E. Vasilopoulou, P. Hollman, C. Chamalides, E. Foufa, T. Kaloudis, D. Kromhout, P. Miskaki, I. Petrochilou, E. Poulima, K. Stafilakis, and D. Theophilou. 2000. Nutritional composition and flavonoid content of edible wild greens and green pies: A potential rich source of antioxidant nutrients in the Mediterranean diet. *Food Chemistry* 70:319–323.

Uiso, F. C. and T. Johns. 1996. Consumption patterns and nutritional contribution of *Crotalaria brevidens* in Tarime District, Tanzania. *Ecology of Food and Nutrition* 35:59–69.

Van t' Riet, H., A. P. den Hartog, A. M. Mwangi, R. K. N. Mwadime, D. W. J. Foeken, and W. A. van Staveren. 2001. The role of street foods in the dietary pattern of two low-income groups in Nairobi. *European Journal of Clinical Nutrition* 55:562–570.

Visioli, F. and C. Galli. 1998. The effect of minor constituents of olive oil on cardiovascular disease: New findings. *Nutrition Reviews* 56:142–147.

Wang, M., J. Li, Y. Shao, T. C. Huang, M. T. Huang, C. K. Chin, R. T. Rosen, and C. T. Ho. 1999. Antioxidative and cytotoxic components of highbush blueberry (*Vaccinium corymbosum* L.). In F. Shahidi and C.-T. Ho, eds., *Phytochemicals and Phytopharmaceuticals*, 271–277. Champaign, IL: AOCS Press.

West, C. E., F. Pepping, and C. R. Temalilwa. 1988. *The Composition of Foods Commonly Eaten in East Africa*. Wageningen, The Netherlands: Wageningen Agricultural University.

WHO. 1946. Preamble to the Constitution of the World Health Organization as adopted by the International Health Conference, New York, June 19–22, 1946.

WHO. 2003. *WHO Global Strategy on Diet, Physical Activity and Health*. Geneva: WHO.

16 ❦ Comparing the Choices of Farmers and Breeders

The Value of Rice Landraces in Nepal

D. GAUCHAN AND M. SMALE

Nepal is an important center of diversity for *Oryza sativa* ("Asian" rice). Asian rice probably was first cultivated in the geographically and culturally diverse region extending from Nepal to northern Vietnam (Vaughan and Chang 1992). Farmers' rice varieties (referred to here as landraces) still occupy more than 30% of the total cultivated rice area in Nepal (APSD 2001). These are typically more heterogeneous than the modern varieties that are bred for uniformity in stature and selected on the basis of particular performance criteria, and they are often adapted to specific local human needs and environmental niches (Simmonds 1979). An estimated 2,000 rice landraces are maintained by farmers in different parts of Nepal in association with their wild and weedy relatives (Shrestha and Vaughan 1989; Upadhyay and Gupta 2000). These landraces have evolved in response to wide variations in edaphic (soil), topographic, and climatic conditions, coupled with farmers' careful seed selection and management practices. In some locations, isolation from markets has contributed to the need of farmers to rely on their own seed sources and harvests to meet food needs, reinforcing this process.

On-farm conservation involves farmers' decisions to continue cultivating and managing landraces in the agroecosystems and communities where they have evolved, such as those of Nepal. Farmers choose to maintain the landraces they value by planting the seed, selecting the seed from the harvest or exchanging it with other farmers, and replanting (see chapter 4). Their choices also determine whether genetic resources of social value for crop improvement continue to be grown in situ. Farmers may choose to cease growing landraces if changes in the production or marketing environment cause them to lose their relative value.

Professional plant breeders also make decisions that affect the conservation of crop biodiversity on farms. Plant breeders select and cross materials in order to develop new varieties. The choices they make shape the range of genetic resources supplied to farmers as new varieties released by commercial seed systems. Breeders can expand farmers' options by introducing new or recombined genetic materials to better meet their needs or complement those already grown. Both genetic resources stored ex situ and those grown in situ are important for the crop improvement process that generates social value through enhanced productivity and lower food prices.

Not all landraces can be conserved on farms, and not all farmers can conserve them because of the costs involved, including direct program costs and costs in terms of forgone opportunities. Nepal is one of the lowest-income countries of the world in terms of gross national product (World Bank 2003). The challenge for the government of Nepal is to create incentives for maintaining the rice biodiversity that benefits farmers today as well as future society. Although future needs cannot be predicted with certainty, the expert assessments of rice breeders provide us with reasonable guesses, but rice breeders, like farmers, have differing points of view.

This chapter uses detailed sample survey data from research in Nepal to investigate the relationship between farmers' and breeders' choices for "in situ" conservation of rice biodiversity. Several breeders' criteria for choosing which materials to conserve on farms are advanced. A conceptual approach drawn from a microeconomic model of farmer decision making relates the likelihood that farmers continue to grow the choice sets defined by these criteria to explanatory factors that may be influenced by public investments and policies. The relationship is then estimated econometrically. If the effects of explanatory factors are the same regardless of the choice set, we can conclude that they are neutral to the choice criteria. If they differ, enhancing the prospects for conservation of one choice set may diminish prospects for another, implying policy trade-offs.

Some empirical studies have investigated trade-offs in one type of diversity compared with another when policies promote changes in an explanatory variable, such as investments in education and infrastructure (Van Dusen 2000; Benin et al. 2003; Smale et al. 2003). These analyses were based on indices that did not capture possible differences in social value between varieties. In the analysis presented here, we relate explicitly the preferences of rice breeders and conservationists to the preferences of farmers. The choices of rice breeders and conservationists reflect their views about the potential value to society of the landraces still grown by farmers. The choices of farmers reveal their preferences in the face of numerous economic and physical constraints, indicating the private value of the varieties. Reference to findings from studies in which similar methods have been applied to study other crops and economic contexts is also presented.

The next section describes the study sites and the methods used to collect data. The conceptual approach and econometric methods are then summarized, followed by presentation of descriptive statistics and results. Conclusions are drawn in the final section.

Study Sites

This research focuses on two of the three ecological sites (ecosites) of the project titled "*In Situ* Conservation of Agricultural Biodiversity On-farm," in Nepal. The ecosite includes a watershed area that includes a cluster of communities or villages. Criteria used to select ecosites included the significance of rice and other crop genetic diversity targeted for on-farm conservation, agroecological features, and market infrastructure. Kaski ecosite represents the hill physiographic region of the country, with an intermediate level of market infrastructure. Bara ecosite is found in the *terai* (lowlands) and has a more developed market infrastructure. In both ecosites, rice is the major crop in the food economy, and it is cultivated across a range of microecological conditions; upland, lowland, and swamp environments often are found on the same farm. Farmers typically plant several varieties to match land types, soils, moisture conditions, and cropping sequences. At the ecosite level, sample farmers maintain a total of 50 and 23 rice cultivars in the hill and lowland ecosites, respectively (table 16.1).

As expected, the highest number of rice landraces (39) and percentage of area allocated to landraces (72.5) was found in the hill ecosite. Although modern varieties and landraces coexist in both ecosites, almost all of the area in the lowlands is allocated to modern varieties (96%). Sample farmers in the lowland ecosite also cultivate a higher number of

Table 16.1. Farmers' cultivation of rice diversity in Bara and Kaski ecosites, Nepal.

Cultivation Pattern at Ecosite Level	Bara (Lowlands)	Kaski (Hills)
Total number of cultivars	23	50
Total number of landraces	5	39
Total number of modern varieties	18	11
Area share in landraces (%)	4	72.5
Area share in modern varieties (%)	96	27.5

modern varieties (18) than those in the hills (11). The total number of rice varieties in the hills is more than twice as high as that found in the lowlands.

Data Source

Sample Survey of Rice-Growing Households

The sample survey research and analysis reported here build on several years of intensive, participatory research with farmers as part of the Nepal national in situ conservation project. Initially the survey team listed all 1,856 households in both sites. Through local contacts, the team learned that some of the households were no longer engaged in farming, some were no longer located in the original settlement, and a few did not grow rice. A random sample representing 17.25% of actively farming, rice-growing households was drawn, numbering 159 in Kaski and 148 in Bara, for a total sample size of 307.

The survey instrument was a structured questionnaire administered in personal interviews. Questions covered social, demographic, and economic characteristics of farmers and their households, physical characteristics of their farms, economic aspects of rice production, and market access. The principal researcher coordinated the survey with the support of experienced, local staff. Both men and women involved in rice production and consumption decisions were interviewed. To improve data quality and uniformity, peer review of the questionnaires was undertaken at regular intervals to check for measurement errors, ambiguities, and missing information. Households were revisited immediately for missing information and inappropriate responses during the survey period. To ensure uniformity in units of measurement and consistent terminology, the researcher and enumerators edited the questionnaires at the survey site.

Key Informant Survey of Rice Breeders

A survey of plant breeders and researchers involved in the national in situ project and rice breeding research in Nepal was carried out in two phases. In the first phase, 16 plant breeders and researchers working on the in situ project were asked to rank lists of farmers' varieties identified in the farm household survey according to their importance for conservation or future use in plant breeding. This survey also enabled the identification of the criteria breeders use to select landraces as potentially useful. Criteria included diversity (expressed as a nonuniform, heterogeneous population), rarity (embodying unique or uncommon traits), and adaptability (exhibiting wide adaptation). In the second phase of the survey, eight plant breeders were asked individually to classify rice landraces according to whether they satisfy each criterion, based on their experiences.

Conceptual Approach

The conceptual approach is based on the theory of the agricultural household (Singh et al. 1986), as applied to analysis of crop biodiversity by Van Dusen (2000; Van Dusen and Taylor 2003). Other related models and applications include those of Brush et al. (1992), Meng (1997), Smale et al. (2001), Benin et al. (2003), and Birol (2004).

In this approach, presented elsewhere in mathematical terms, an agricultural household maximizes utility over a set of consumption items produced on the farm, a set of consumption items purchased on the market, and leisure. The utility a household derives from various consumption combinations and levels depends on the preferences of its members. Preferences in turn depend on various social and demographic characteristics of the household, including its endowments of human capital and other assets, represented by the vector Ω_{HH}.

The amounts the household can produce are constrained by a production technology, given the physical features of the farm (Ω_F). The production technology combines seed and labor with other purchased inputs on the crop area cultivated each season (A). The choice of crop and variety combinations and how much land area to allocate to each determines the levels of farm produce the household expects to harvest and vice versa. The area shares for any given crop or variety can range from zero (when it is not grown) to one (when no other crop or variety is grown).

When these choices are made, expenditures of time and money cannot exceed full income. Full income in any season consists of the net farm earnings (profits) from sales of crop production and income that is exogenous (external) to the season's crop and variety choices, such as stocks carried over, remittances, pensions, and other transfers from the previous season (Y^0). When markets are not functioning well for a crop or its trade is associated with significant costs of transaction (Ω_M), production and consumption decisions cannot be treated separately, and a shadow price for the crop guides decision making rather than its market price. Shadow prices are related to the differential costs of transacting on markets that reflect household-specific characteristics (Ω_{HH}). Previous work in the study area suggests that markets are not complete for rice varieties, and especially landraces (Gauchan et al. 2005).

The random utility model enables statistical interpretation of the variety choice decision with sample data. The household chooses to grow any particular landrace on a portion of rice area if the utility its members expect to derive is greater than for other available alternatives ($U_i > U_j$ for any j not equal to i). Because utility levels (U) cannot be observed, the choices observed in the data reveal the alternatives that provide the greatest utility to households. Variation in these choices is explained systematically by the preferences of households and the constraints they face. Preferences and constraints depend on observable variables related to household, farm, and market characteristics. Drawing data from a random sample of households introduces a stochastic component, providing a statistical context for predicting the probability that a household grows a landrace as a function of the systematic component ($\beta'X$) and random errors (ε):

$$\text{Probability (Landrace } i \text{ chosen)} = \text{Probability } (U_i > U_j)$$
$$= \beta_0 + \beta_H'\Omega_{HH} + \beta_F'\Omega_F + \beta_M'\Omega_M$$
$$+ \beta_y Y^0 + \beta_a A + \varepsilon. \tag{1}$$

Econometric Methods

Equation 1 is the basis of econometric analysis and hypothesis tests. A probit model was used to estimate the regression in LIMDEP (version 7.0), cross-checked in STATA. The econometric tests investigate which explanatory factors specified in the decision-making model significantly alter the predicted probability (the likelihood) that the farm household grows the landraces classified by rice breeders as genetically diverse, rare, or adaptable. Tests are implemented by specifying regressions with different dependent variables (choice criteria) and the same explanatory variables. Signs and significance of regression coefficients are compared.

The dependent variables in the regressions are defined according to the results of the key informant survey (table 16.2).

Explanatory variables and hypothesized effects are shown in table 16.3, grouped according to the sets of observed characteristics that represent the conceptual variables in equation 1.[1] A brief description of these explanatory variables follows.

Household Characteristics

Household characteristics affect the choice among landraces through both preferences and the household-specific costs of market transaction. Age, education, and the gender composition of households influence preferences and habits. Older farmers are more likely to have grown a range of rice landraces and be accustomed to growing them. More education may increase the ability of both production and consumption decision makers

Table 16.2. Definition of dependent variables in the probit regression models.

Diversity	Nonuniform, heterogeneous population	Yes = 1, otherwise = 0	Any landrace satisfying this choice criterion
Rarity	Unique, uncommon traits	Yes = 1, otherwise = 0	Any landrace satisfying this choice criterion
Adaptability	Wide adaptation	Yes = 1, otherwise = 0	Any landrace satisfying this choice criterion

Table 16.3. Definitions of explanatory variables and hypothesized effects on diversity.

Variable Name	Variable Definition	Hypothesized Effect
Household Characteristic		
AGEPDM	Age of production decision maker (years)	(+)
EDUPDM	Education of production decision maker (years)	(+, -)
EDUCDM	Education of consumption decision maker (years)	(+, -)
AAGLABR	Active adults working on farm (number)	(+)
FAADTPCT	Percentage female of actively working adults	(+)
LANIMLV	Value (in Nepalese rupees) of large animals (bullocks, dairy animals)	(+)
TOTEXP	Average monthly household expenditure (in Nepalese rupees) since last harvest preceding this season (exogenous income)	(+, -)
SBRATIO	Ratio of 5-year average of kilograms rice produced to kilograms rice consumed	(+, -)
Farm Physical Characteristic		
IRPCNT	Percentage rice area under irrigation	(+, -)
LNDTYPS	Number of rice land types	(+)
RDPLCULH	Total walking distance (minutes) from house to rice plot, divided by cultivated hectares	(+)
Market Characteristic		
TMKTDS	Total walking distance from house and farm plots to local market (minutes)	(+)
LRSOLD	Landrace grain sold by household in preceding season (kg)	(+)
MVSOLD	Grain of modern variety sold by household in preceding season (kg)	(-)

(typically men and women, respectively) to acquire information and experiment but is often associated with a preference for modern varieties and specialization. More active adult labor allows households to engage in the cultivation of a larger set of rice varieties with differing management requirements. The proportion of active working women may relate positively to growing certain landraces that have unique consumption attributes. An earlier study by the project team revealed a greater role of women on rice seed maintenance and cultivation (Subedi et al. 2000).

Households owning a larger number and value of draft (bullock) and dairy (buffalo, cows) animals are expected to grow more diverse rice varieties because they have better access to inputs and information, more capacity to experiment, and greater demand for fodder. Ownership of bullocks (draft power) also allows timely land preparation, threshing, and transportation of inputs and harvested products. On one hand, external cash income increases farmers' capacity to hire labor and purchase inputs in order to engage in a wider range of activities. On the other hand, it may imply that household members are involved in nonfarm activities and devoting less time to special rice varieties. Farmers producing rice in excess of their expected consumption needs may be better able to maintain landraces, or they may be those specializing in production of modern varieties for the market.

Farm Physical Characteristics

In this labor-intensive farming system characterized by very small farm sizes, fragmentation of plots and the heterogeneity of land types are critical aspects of farm technology. The more distinct the land types on which farmers cultivate rice and the more dispersed the plots, the greater the likelihood they grow landraces to suit certain seasonal or physical niches. Irrigation improves moisture availability and may have either negative or positive effects on the likelihood that they grow specific landraces. Better access to water may increase specialization in a few varieties, making the production process more uniform; it may also enable cultivation of a broader range of varieties with different moisture needs and maturity periods.

Market Characteristics

Market variables affect diversity through the extent to which households trade their rice crop and purchase inputs, foods, and other household needs in the market. The distance of the market from the homestead is a major component of the cost of engaging in market transactions. The more removed a household from a local market center, the more likely it is to rely on its own production to meet its consumption needs. Consumption needs may include a range of food products and fodder. Past sales of grain of landraces is expected to relate positively to incentives for cultivating them. Past grain sales from production of modern varieties may relate to specialization in fewer, uniform modern varieties.[2]

Findings

Descriptive Statistics

Households in the more isolated hillside (Kaski ecosite) are far more likely than those in the plains (Bara ecosite) to grow landraces identified by rice breeders as potentially valuable for their diversity, rarity, or adaptive qualities (table 16.4).

The demographic structure of rice-growing households is similar across the two ecosites with respect to the age and education of the production decision maker, the working adult labor, and numbers and proportions of men and women who are actively engaged in farming. However, in the Bara ecosite women decision makers are significantly less educated. Although the average income levels are similar, livestock asset values are lower in the plains. Households there sell a lot more grain of modern varieties and, by implication, grow more of them. On average, they do better in meeting their rice consumption needs through their own production than do those located in the hills. Differences in grain sales for landraces were not significant, possibly because only one farmer in the Bara ecosite reported a large volume of sales. The physical characteristics of the farms in the two ecosites are similar in terms of number of land types under rice and the percentage irrigated, but rice plots are much less widely dispersed and households are closer to markets on the plains.

Econometric Results

Factors that predict whether private values and social values coincide are shown in table 16.5, according to each choice criterion (diversity, rarity, and adaptability). These are the factors that significantly affect

Table 16.4. Summary statistics for dependent and explanatory variables, Bara and Kaski ecosites.

Variable	Ecosite		
	Bara (N=148)	Kaski (N=159)	All pooled (N=307)
Dependent Variables			
Percentage households growing diverse landraces (+)	2	50.9	27.4
Percentage households growing rare landraces (+)	2.7	20.8	12.1
Percentage households growing adaptable landraces (+)	0.7	74.8	39.1
Explanatory Variables			
AGEPDM	48.27	46.20	47.20
EDUPDM	3.0	3.95	3.52
EDUCDM	0.48**	1.99	1.26
AAGLABR	2.52	2.51	2.52
FAADTPCT	0.27	0.28	0.28
LANIMLV	10,270**	18,490	14,527
TOTEXP	2,483	2,581	2,533
SBRATIO	1.40**	0.76	1.07
IRPCNT	0.42	0.39	0.407
LNDTYPS	1.54	1.49	1.517
RDPLCULH	120*	146	134.58
TMKTDS	163**	340	255.14
LRSOLD	16.89	43.68	30.76
MVSOLD	971**	38	487.8

Note: Pairwise t tests show significant difference of means at $**p<.01$ and $*p<.05$ between Kaski and Bara ecosites with 2-tailed test, equal variance assumed. (+) χ^2 tests show significant difference ($p<.05$) between Bara and Kaski ecosites. See table 16.3 for definitions of explanatory variables.

the likelihood that farmers will grow landraces identified by rice breeders as important.

Among household characteristics, education, labor composition, and livestock assets are statistically significant predictors that households will grow landraces that are considered important for future crop improvement.

Table 16.5. Factors predicting that farmers will grow landraces breeders identify as potentially valuable in two ecosites of Nepal, by choice criterion.

Explanatory Variables	Choice Criterion of Rice Breeders		
	Diversity	Rarity	Adaptability
Constant	−0.6221***	−0.4289***	−2.6499***
Site	0.2792***	0.1074***	1.0596***
AGEPDM	−0.000029	−0.00058	0.000387
EDUPDM	−0.0101	0.00212	0.00931
EDUCDM	0.0218**	−0.00483	−0.00679
AAGLABR	0.04315**	0.01702	0.14948***
FAADTPCT	−0.03892	0.13687*	−0.05048
LANIMLV	0.000005*	−0.0000019	−0.000002
TOTEXP	−0.000023	−0.000018	0.0000003
SBRATIO	−0.09510	−0.02833	0.05185
IRPCNT	0.080216	0.005799	0.1390
LNDTYPS	−0.05990	0.06588***	0.03843
RDPLCULH	0.000029	0.000056	0.001112**
TMKTDS	0.00040**	0.000137**	0.000665*
LRSOLD	0.00021	0.000111*	−0.000094
MVSOLD	−0.00004	−0.000005	−0.0001188
Log likelihood function	−93.79	−75.50	−54.65
Pseudo R-squared	0.478	0.734	0.332

Note: N = 307. The regression model used in all cases is a probit. One-tailed Z tests significant at ***$p < .01$, **$p < .05$, and *$p < .1$. See Table 16.3 for variable definitions. Z statistic is relevant for maximum likelihood estimation. The values reported in the table are marginal effects that are computed as the means of explanatory variables.

Human capital appears to be critical. The more educated the decision maker in rice consumption (typically a woman), the greater the likelihood that a household grows a landrace that is genetically heterogeneous. More adult labor engaged in agriculture has a large effect on the probability that adaptive landraces are grown, also contributing significantly to cultivation of genetically diverse landraces. A higher percentage of women among active adults in the households means that a rare landrace is more likely to be grown. The more endowed with livestock assets (buffalo, cattle, and bullocks), the more likely the household is to grow landraces also selected by rice breeders as diverse or adaptive. External income is of no apparent significance because growing landraces does not cost money. The number of rice land types (diverse farm production niches) increases the chances that a rare landrace is grown, and the dispersion of rice plots relative to the total area cultivated contributes positively to growing adaptive landraces. Location in the hills ecosite and isolation from markets are associated with higher probabilities of growing *any* landrace that is identified as potentially valuable to future crop improvement by rice breeders.

For statistically significant predictors that are common across landrace subsets, the direction of effect is the same, although the magnitude of effect differs (ecosite location, proportion of active adults engaged in farm production, total walking distance to market). Three policy-related factors have non-neutral effects. That is, the statistical significance of their effect depends on the choice criterion of breeders: diversity, rarity, or adaptability. First, women's education and involvement in farm production predict only that the household will grow rare or diverse landraces, and the magnitude of effects differs by choice criterion. Second, past sales of grain from landraces is one policy-relevant factor that is significantly associated with growing rare landraces but not diverse or adaptive landraces. This finding suggests that specialized markets may provide incentives for farmers to continue cultivating rare landraces. Third, the dispersion of farm plots, normalized by farm areas, is a predictor that the household will grow adaptable landraces. Tenure and land use practices are factors that underlie the spatial distribution of plots.

Findings from Related Studies

Findings from related studies are presented in table 16.6, according to the conceptual sets of variables used in this chapter hypothesized to explain variation in the levels of crop diversity maintained by agricultural households or the probability that they will continue to grow landraces. In each case study, the theoretical basis is the model of the agricultural household applied econometrically to data collected in household and plot surveys. Countries and crops include a range of income levels and crops: potatoes in Peru (Brush et al. 1992), wheat in Turkey (Meng 1997), maize, beans, and squash in Mexico (Van Dusen 2000; Smale et al. 2001), cereal crops in Ethiopia (Benin et al. 2003), and home gardens in Hungary (Birol 2004). Signs entered in the table indicate a statistically significant direction of effect, and a zero refers to a regression coefficient that is not statistically significant.

Table 16.6. Comparison of findings from related economic studies about crop genetic diversity or crop species diversity.

Direction of Effects Predicted in Case Studies

Factor Affected by Economic Change and Public Policies	Peru: Potato Landrace Diversity	Turkey: Wheat Landrace Diversity	Mexico: Milpa System, Total Crop and Variety Diversity	Mexico: Maize Landrace Diversity	Ethiopia: Cereal Crop Diversity	Ethiopia: Cereal Variety Diversity	Hungary: Home Gardens, Crop Species Diversity	Hungary: Home Gardens, Landraces
Household characteristics								
Age of head	o	o	+		o	+, o, −	o	+
Education		o	+		o	+, o		
Farm labor supply		−	o		+	+, o	o	+, o
Women's education and participation					o	+, o		
Off-farm income, migration	−	−	−	o				
Assets	−, +, o	−	o	o	+	−, o	+, o	−, o
Farm characteristics								
Fragmentation	+	−	+		+	−, o		
Altitude		+	+					
Soil heterogeneity, multiple slopes		o	+	+	o	+, o, −		
Productivity potential, soil quality			o	o, +	−	+, o, −	+, o	−, o
Market infrastructure	−, o	+	−	−	o	+, o, −	−, o	
Modern varieties grown	−, o	n/a	n/a	n/a	n/a	o	o	o

Sources: Brush et al. (1992), Meng (1997), Van Dusen (2000), Smale et al. (2001), Benin et al. (2003), Birol (2004).
Note: "o" means that the effect was not statistically significant; an empty cell means that the variable was not included in the regression. n/a=not applicable.

Despite the application of related approaches, generalization is not easy. One reason is that in each empirical setting, although the conceptual variables are the same, measurement of the dependent and explanatory variables through survey instruments must be adapted to the study context. In Peru, equations were estimated to predict the adoption of modern varieties and the effect of modern varieties on the number of potato landraces grown; in Turkey, equations were estimated to predict what wheat landraces were grown, and the diversity of wheat landraces was explained conditional on the decision to grow them. In one study in Mexico, the total richness of maize, beans, and squash varieties in the milpa system was explained; in the other, the area share allocation among maize landraces was investigated. The Ethiopia study examined the richness and evenness of cereal crops and their varieties. In Hungary's transitional, high-income economy, several components of agricultural biodiversity were studied, including crop species richness and landrace richness in home gardens.

As is the case for rice in Nepal, the age of the decision maker has no significant effect in lower-income countries of Peru and Ethiopia. In the middle- and high-income countries of Mexico and Hungary, older farmers are more likely to cultivate landraces. In Nepal and Ethiopia, where gender-related variables have been measured, women's education and their involvement in farm production positively influences household levels of crop genetic diversity. Across all countries, the income levels, coefficients of farm labor supply, and off-farm or migrant income indicate that as alternative employment is generated, crop diversity and genetic diversity levels decline at the household level. The predicted effect of wealth is ambiguous; in some cases, farmers who are wealthier in land, livestock, and farm labor are more able to maintain diversity.

Although results are mixed for agroecological factors, fragmentation of land on farms, soil heterogeneity, elevation, and farming on slopes tend to be associated with greater diversity in crops and varieties. Most often, the more developed the local market infrastructure, the lower the household diversity levels. However, in the hillsides of Ethiopia, the proximity of seed or product markets appears to enable the introduction of crops and varieties that complement those maintained by farmers. In Turkey, local markets appear to have encouraged cultivation of diverse wheat landraces; in Nepal, landrace sales were positively related to the likelihood that farmers grow rare landraces.

Conclusion

Farmers determine the survival of crop varieties or the maintenance of specific gene complexes in any given reference area by choosing whether to grow them and in what proportions. The choices they make today affect not only their welfare but also that of future society. With diminishing plant populations, some potentially valuable alleles and gene combinations may be lost. Farmers choose which varieties of a crop to grow according to their private value, and this varies in semisubsistence agriculture according to their characteristics and market conditions as well as the physical features of their farms.

Plant breeders use decision-making criteria when they select materials for breeding or conservation purposes, and these differ from those of farmers and between breeders. For example, they may identify varieties that are genetically diverse, those that have rare traits, or those that express wide adaptation as potentially important for breeding programs and hence for genetic resource conservation. These are best guesses regarding the social value of landraces.

The analysis presented here has focused on policy trade-offs associated with the choice of criteria for conservation. Increasing the likelihood that farmers will maintain varieties that are members of one choice set may decrease the prospects that varieties in other sets continue to be grown. If so, policies designed to attain one objective might have serious consequences for another. Our results show no such conflicts. However, they do suggest that the programs or policies designed to support the continued cultivation of rare landraces are different from those needed for diverse or adaptable landraces. In particular, investment in women's involvement in rice production and niche market development may increase the likelihood that households grow rare landraces.

Regression results and summary statistics suggest how sites and households might be targeted for local conservation of rice biodiversity. Clearly, any rice-growing household in the hill ecosite (Kaski) is more likely to grow genetically diverse, rare, or adaptable landraces. Rice-growing households in the lowland ecosite (Bara) grow and sell more modern varieties of rice. They are better able to satisfy their consumption needs through their own production, even though women decision makers are less educated, households in this location are poorer in assets, and they are no better off in terms of external income.

Not all households in Kaski, and not all landraces in Kaski, are equally promising candidates for conservation. Households with more active adults engaged in agriculture are more likely to maintain landraces of social value, so that increasing opportunities for off-farm employment may have a negative impact on prospects for conservation. Households maintaining socially valued landraces have more heterogeneous farms and are more isolated from markets. The evidence that farmers more likely to grow rare landraces also sell the grain locally suggests that the development of specialized, controlled markets may provide an incentive for maintaining such materials, although the feasibility and costs of implementing such a program are unknown. Finally, targeting may involve other trade-offs in terms of equity considerations. Those most likely to grow socially valuable landraces are also richer in livestock assets, have higher expected production and consumption needs, and are at least as well off in terms of cash income. Although most farmers on the hillsides of Nepal are ranked as poor by global standards, targeting the locations and households more likely to maintain valuable landraces is by no means equivalent to targeting the poor.

Comparing the findings of related economic studies that have been completed or are under way underscores the location specificity of predicted effects of many of the factors, particularly those related to human capital and wealth characteristics of households. Although the market infrastructure and environmental heterogeneity hypotheses are robust, a much more refined comprehension of seed markets and systems as a means of supporting crop diversity management is needed. At least one methodological caveat must also be borne in mind when one interprets the results of case studies such as these. Although we would argue that agroecological considerations, in some cases interacting with market infrastructure development, probably will support the persistence of differences in rice diversity management between the hillsides and plains, the extent to which cross-sectional variation can substitute for temporal variation is limited. Longitudinal data, or data that enable periodic monitoring of both diversity outcomes and underlying processes, would enable stronger conclusions to be drawn and are necessary to establish appropriate incentives for conservation in important centers of diversity.

Acknowledgments

This chapter is based on research conducted as part of the "*In Situ* Conservation of Agrobiodiversity On-Farm Project," Nepal. We are grateful to senior scientists T. Hodgkin, D. Jarvis, P. Eyzaguirre, and B. Sthapit (International Plant Genetic Resources Institute) for their insights and for the contributions of E. Van Dusen, University of California, Berkeley. This research was supported by the International Development Research Centre of Canada, the Swiss Agency for Development and Cooperation, the Netherlands Directorate-General for International Cooperation, and the European Union. The authors are also thankful to members of *In Situ* Project team, Nepal, particularly the field staff in Bara and Kaski ecosites, for their support in the field survey.

Notes

1. Because total farm plot distance was highly correlated with area cultivated, the two variables were combined into one to capture the effect of scattered plots while controlling for total hectares cultivated.

2. Past rather than current sales are used to ensure that independent variables used to explain rice variety choices are not also choice variables. Sales amounts were preferred to a zero–one variable because they express more variation.

References

APSD (Agri-Business Promotion and Statistics Division). 2001. *Statistical Information on Nepalese Agriculture*. Nepal: Ministry of Agriculture, HMG.

Benin, S. B., M. Gebremedhin, M. Smale, J. Pender, and S. Ehui. 2003. Determinants of cereal diversity in communities and on household farms of the northern Ethiopian highlands. *Agriculture and Development Economics Division (ESA) Working Paper* 03-14. Rome: FAO. Available at www.fao.org/es/ESA/wp/ESAWP03_14.pdf.

Birol, E. 2004. *Agri-environmental Policies in a Transitional Economy: Conservation of Agricultural Biodiversity in Hungarian Home Gardens*. PhD thesis, University College London, University of London.

Brush, S. B., J. E. Taylor, and M. R. Bellon. 1992. Biological diversity and technology adoption in Andean potato agriculture. *Journal of Development Economics* 39:365–387.

Gauchan, D., M. Smale, and P. Chaudhary. 2005. Market-based incentives for conserving diversity on farms: The case of rice landraces in central Tarai, Nepal. *Genetic Resources and Crop Evolution* 52:293–303.

Meng, E. 1997. *Land Allocation Decisions and In Situ Conservation of Crop Genetic Resources: The Case of Wheat Landraces in Turkey.* PhD dissertation, University of California, Davis.

Shrestha, G. L. and D. A. Vaughan. 1989. *Wild Rice in Nepal.* Paper presented at the Third Summer Crop Working Group Meeting, National Maize Research Program, Rampur, Chitwan, National Agricultural Research Centre, Nepal.

Simmonds, N. 1979. *Principles of Crop Improvement.* Harlow, UK: Longman.

Singh, I., L. Squire, and J. Strauss, eds. 1986. *Agricultural Household Models: Extensions, Applications, and Policy.* Washington, DC and Baltimore: The World Bank and Johns Hopkins University Press.

Smale, M., M. Bellon, and A. Aguirre. 2001. Maize diversity, variety attributes, and farmers' choices in southeastern Guanajuato, Mexico. *Economic Development and Cultural Change* 50(1):201–225.

Smale, M., E. Meng, J. P. Brennan, and R. Hu. 2003. Determinants of spatial diversity in modern wheat: Examples from Australia and China. *Agricultural Economics* 28(1):13–26.

Subedi, A., D. Gauchan, R. B. Rana, S. N. Vaidya, P. R. Tiwari, and P. Chaudhary. 2000. Gender: Methods for increased access and decision making in Nepal. In D. Jarvis, B. Sthapit, and L. Sears, eds., *Conserving Agricultural Biodiversity In Situ: A Scientific Basis for Sustainable Agriculture,* 78–84. Rome: IPGRI.

Upadhyay, M. P. and S. R. Gupta. 2000. The wild relatives of rice in Nepal. In P. K. Jha, S. B. Karmacharya, S. R. Baral, and P. Lacoul, eds., *Environment and Agriculture: At the Crossroad of the New Millennium,* 182–195. Kathmandu, Nepal: Kathmandu Ecological Society.

Van Dusen, E. 2000. *In Situ Conservation of Crop Genetic Resources in Mexican Milpa Systems.* PhD thesis, University of California, Davis.

Van Dusen, E. and J. E. Taylor. 2003. *Missing Markets and Crop Genetic Resources: Evidence from Mexico.* Berkeley: University of California.

Vaughan, D. and T. T. Chang. 1992. In situ conservation of rice genetic resources. *Economic Botany* 46:369–383.

World Bank. 2003. *World Development Indicators 2003.* Washington, DC: The World Bank.

17 ❦ Economics of Livestock Genetic Resources Conservation and Sustainable Use

State of the Art

A. G. DRUCKER

Livestock supply some 30% of the total human requirements for food and agriculture (FAO 1999), and some 70% of the world's rural poor depend on livestock as a component of their livelihoods (Livestock in Development 1999). Animal genetic resource (AnGR) diversity thus contributes in many ways to human survival and well-being, with differing animal characteristics and hence outputs being tailored to suit a variety of local community needs.

However, an estimated 16% of these uniquely adapted breeds, bred over thousands of years of domestication in a wide range of environments, have been lost since the beginning of the 19th century (Hall and Ruane 1993). Another 32% (22% of mammals and 48% of avian species) are at risk of becoming extinct, and the rate of extinction, currently at two breeds per week, continues to accelerate (FAO 2000). The small genepool of domestic AnGRs (6,000–7,000 breeds of 40 species) means that this loss is of particular concern. Such an irreversible loss of genetic diversity reduces opportunities to improve food security, reduce poverty, and shift toward sustainable agricultural practices.

The large number of AnGRs at risk in developing countries, together with the limited financial resources available for conservation, means that economic analysis can play an important role in ensuring an appropriate focus for conservation efforts (UNEP 1995). Nevertheless, despite the importance of the economics of AnGR conservation and sustainable use, the subject has only recently begun to receive attention. These studies reveal that not can a range of methods be used to value farmer breed and trait preferences, but that they can be of use in designing policies that counter the present trend toward marginalization of indigenous breeds. In particular, it becomes possible to recognize the importance livestock keepers place on adaptive traits and nonincome functions and the need to consider these in breeding program design; to identify breeds that are a priority for participation in cost-efficient diversity-maximizing conservation programs; and to contrast the costs involved with the large benefits that non–livestock keepers place on breed conservation.

This chapter briefly discusses the theoretical background, potential methods, data needs, and difficulties of carrying out such studies before analyzing the results of a range of AnGR economic studies recently carried out in Africa, Latin America, and Europe.

Economics of AnGR Conservation and Sustainable Use

What Can Economics Contribute?

Economic arguments for the conservation and sustainable use of AnGRs can be an effective means of garnering the necessary public and political support, including development of appropriate policies. In this regard, important tasks include determining the economic contribution that AnGRs make to various societies and specific groups within those societies, supporting the assessment of priorities through the identification of cost-effective measures that might be taken to conserve domestic animal diversity, and assisting in the design of economic incentives and institutional arrangements for the promotion of AnGR conservation by individual livestock keepers or communities.

Analytical Framework

AnGR erosion can be understood in terms of the conversion[1] of the existing slate of domestic animals with a selection from a small range of specialized "improved" breeds that are considered to be better able to directly contribute to human welfare. At the same time, economic theory has shown that functioning markets can be a powerful ally in the efficient allocation of resources by reflecting the scarcity of a given resource, through the price mechanism, and thus providing the correct incentives for the resources' use or replacement.

Viewing AnGR loss in these terms led Mendelsohn (2003) to argue that the primary challenge facing AnGR conservation is identifying sound reasons why society should preserve animals that farmers have abandoned. Given that the market will preserve valuable livestock breeds, conservationists must focus on what the market will not do. This includes identifying and quantifying the potential social benefits of AnGRs that have been abandoned by the market. So conservationists first must make a case for why society should be willing to pay to protect apparently unprofitable AnGRs and then must design conservation programs that will effectively protect what society treasures.

Tisdell (2003) also recognizes the importance of market impacts on AnGRs, noting the influence of developed country livestock technologies (e.g., artificial insemination, industrialized intensive animal husbandry) on livestock populations in developing countries, together with the fact that the extension of markets and economic globalization can be expected to accelerate the loss of breeds. Such an outcome can occur through regional specialization, the reduced costs of international breed transfers, the Swanson dominance effect (i.e., breeds used in developed countries tend to replace those in developing countries), specialization by comparative advantage leading to reduced demand for multipurpose breeds, changing consumer tastes and demand, changes in availability and price of feed imports, and the increased scope for controlling the environments in which livestock are reared (Tisdell 2003:367–368). Tisdell (2003:371) cites a Food and Agriculture Organization (FAO undated:45) example from Nigeria to illustrate how the opening of interregional trade and rising tree crop prices have led farmers to abandon their local livestock. The FAO notes that although "this is a perfectly rational medium-term strategy on their [the farmers'] part, it would be short-sighted for the national government to lose the genetic resources these livestock represent because of a temporary pattern in world trade."

The ability of such "free" market forces to provide a socially desirable outcome is further questioned by Pearce and Moran (1994), who argue that the activity of biodiversity (and genetic resource) conservation generates economic values (use and nonuse), which may not be captured in the marketplace because of market, intervention, or global appropriation failures. The result of such failures is a distortion in which the incentives are against genetic resource conservation and in favor of the economic activities that destroy such resources. Smale (2005) recently compiled methods and empirical studies for economic valuation of crop genetic resources on farm and discusses these issues.

For example, economic rationality suggests that decisions such as the replacement of an indigenous breed of livestock with an imported breed will be determined by the relative rates of return of the two options. However, the relevant rates of return are those that accrue to the farmer rather than to the nation or the world as a whole. To the livestock keeper the loss of the indigenous breed appears to be economically rational because returns are higher than those from activities compatible with genetic resource conservation. This is because the latter may consist of nonmarket benefits that accrue to people other than the farmer and because subsidized inputs and services (e.g., artificial insemination, veterinary treatment) may be available for the imported breed.

Swanson (1997) notes that such nonmarket values are likely to be important because biodiversity is not equally distributed among nation states, suggesting that global external values are indeed likely to be significant. The biodiversity problem thus can be conceived of as the set of difficulties that derive from the fact that the conversion process traditionally has been regulated on a globally decentralized basis. Historically, states and individuals have been able to make their own conversion decisions regarding their own lands and resources without regard for the consequences for others. This creates an important regulatory problem because the cost—in terms of the value of lost services— of each successive conversion is not the same. As the conversion process advances, the cost of each successive conversion—in terms of diverse resource services lost to all societies on Earth—escalates rapidly. The absence of any mechanism to bring these costs into the decision-making framework of the converting state or individual is a big part of the biodiversity problem.

Economic analysis therefore is needed to help in understanding the financial incentives livestock keepers face in making the choice between indigenous and imported breeds and the interventions necessary to ensure that the ongoing agricultural development process will be compatible with the conservation and sustainable use of livestock breed diversity.

Methods and Constraints in the Economic Analysis of ANGR Conservation and Sustainable Use

Despite the importance of the economics of ANGR conservation, the subject has received little attention (FAO/ILRI 1999), even though a conceptual framework exists for the valuation of biodiversity in general. There are a number of reasons for this.

Methodological Constraints

First, there are a number of methodological difficulties, many of which have been encountered in valuing plant genetic resources (PGRs). For example, Evenson (1991) argues that the measurement of the benefits of germplasm diversity to crop development is extremely difficult. The genetic resources seldom are traded in markets and often are the product of generations of informal innovations. Therefore identifying the contribution of a particular indigenous breed to the success of an improved variety or breed would be complex. Furthermore, the base materials used for breeding are themselves the result of a production function, and identifying the returns to respective factors (e.g., labor, on-farm technology, intellectual inputs) is likely to be possible only in the most general terms (Evenson 1991; Pearce and Moran 1994).

Nevertheless, a range of analytical techniques for carrying out such an analysis could be adopted from other areas of economics. These were reviewed by Drucker et al. (2001), and the methods were broadly categorized into three groups on the basis of the practical purpose for which they may be conducted. As can be seen in table 17.1, these are: determining the appropriateness of AnGR conservation program costs, determining the actual economic importance of the breed at risk, and priority setting in AnGR breeding programs.

Limited Data Availability

Second, data availability is a constraint. In order to use these methods, it is necessary to perform the following activities:

• Measure breed performance parameters.
• Characterize actual and potential breeding systems.
• Identify uses and livestock-keeper trait preferences (including eliciting the values that livestock keepers place on specific traits and the trade-offs they are willing to make between them) for indigenous breeds under different production systems and the forces influencing such factors and the uptake of alternative breeds.

Table 17.1. AnGR valuation method evaluation.

Valuation Method	Purpose, Objective, or Strength	Actors for Whom Valuation Method Is Most Relevant	Role in Conservation	Type of Data Needed	Data Availability	Conceptual Weakness or Difficulties
Methods for Determining the Appropriateness of AnGR Conservation Program Costs						
Contingent valuation methods	Identify society's willingness to pay (WTP) for the conservation of AnGR, farmer WTP compensation for raising indigenous AnGR instead of exotics or to determine farmer trait value preferences and net returns by breed	Policymakers in charge of conservation	Define upper limit to economically justified conservation program costs	Society preferences expressed in terms of WTP or willingness to accept	Not normally available; survey needed	Response difficulties when used for non-charismatic species or chronic genetic erosion
Production loss averted	Indicate magnitude of potential production losses in the absence of AnGR conservation	Farmers and policymakers in charge of conservation	Justify conservation program costs of at least this magnitude	Estimate of potential production losses (e.g., percentage of herd and market value of animals)	Animal market values available for commercial breeds; potential herd loss must be estimated	Not a consumer/producer surplus measure of value; ignores substitution effects
Opportunity cost	Identify cost of maintaining AnGR diversity	Farmers and policymakers in charge of conservation	Define opportunity cost of AnGR conservation program	Household costs of production and net income	Not normally available; survey needed	

Table 17.1. continues to next page

Table 17.1. *continued*

Valuation Method	Purpose, Objective, or Strength	Actors for Whom Valuation Method Is Most Relevant	Role in Conservation	Type of Data Needed	Data Availability	Conceptual Weakness or Difficulties
Least cost	Identify cost-efficient program for AnGR conservation	Policymakers in charge of conservation; farmers and breeders to some extent	Define minimum cost of conservation program	Household costs of production and profitability	Not normally available; survey needed	
Safe minimum standard	Assess trade-offs involved in maintaining a minimum viable population	Policymakers in charge of conservation	Define opportunity cost of AnGR conservation program	Conservation program costs and benefit differential involved in raising different breeds	Not normally available; survey and modeling needed	Judgment needed as to whether breed substitution will generate utility in excess of the unquantifiable benefits of indigenous breed conservation

Methods for Determining the Actual Economic Importance of the Breed

| Aggregate demand and supply | Identify value of breed to society | Policymakers in charge of conservation and livestock policy; breeders | Value of potential losses associated with AnGR loss | Intertemporal or farm-level data | Available for commercial breeds; not normally available for others; survey needed | Shadow pricing of home labor and forage needed |

Cross-sectional farm and household	Identify value of breed to society	Policymakers in charge of conservation and livestock policy; breeders and farmers	Value of potential losses associated with AnGR loss	Consumer and producer price differences by location	Not normally available; survey needed	Shadow pricing of home labor and forage needed
Market share	Indication of current market value of a given breed	Policymakers in charge of conservation and livestock policy; breeders and farmers	Justify economic importance of given breed	Market value of animal products by breed	Generally available but not always by breed	Not a consumer/producer surplus measure of value; ignores substitution effects
Intellectual property rights and contracts	Market creation and support for fair and equitable sharing of AnGR benefits	Policymakers in charge of conservation; breeders and farmers	Generate funds and incentives for AnGR conservation	Royalty payments or terms of contract	Usually available when such arrangements exist but can be commercial secret	Limited duration of contracts

Methods for Priority Setting in AnGR Breeding Programs

Evaluation of breeding program	Identify net economic benefits of stock improvements	Farmers and breeders	Maximize economic benefits of conserved AnGR	Yield effects and input costs	Available for commercial breeds; not normally available for others; survey or research needed	Difficulty in separating the contribution of genetic resources from other costs of program

Table 17.1. continues to next page

Table 17.1. continued

Valuation Method	Purpose, Objective, or Strength	Actors for Whom Valuation Method Is Most Relevant	Role in Conservation	Type of Data Needed	Data Availability	Conceptual Weakness or Difficulties
Genetic production function	Identify net economic benefits of stock improvements	Farmers and breeders	Maximize expected economic benefits of conserved AnGR	Yield effects and input costs	Available for commercial breeds; not normally available for others; survey or research needed	
Hedonic	Identify trait values	Farmers and breeders; policymakers in charge of conservation	Value potential losses associated with AnGR loss; understand breed preferences	Characteristics of animals and market prices	Available for commercial breeds; not normally available for others; survey or research needed	Not a consumer/producer surplus measure of value; ignores substitution effects
Farm simulation model	Model improved animal characteristics on farm economics	Farmers and breeders	Maximize economic benefits of conserved AnGR	Inputs and outputs; technical coefficients of all main activities	Available for commercial breeds; not normally available for others; survey needed	Correct definition of farm objective function; aggregation for estimating consumer surplus can also be problematic

Source: Drucker et al. (2001).

- Identify factors affecting livestock demand and prices, including the impact of policy-induced changes in agricultural commodity (e.g., forage or crop) prices and external (e.g., veterinary) input costs in the context of different breed use.
- Analyze the potential impact of the uptake of alternative breeds on livelihoods, together with constraints on adoption and potential access and dissemination mechanisms.
- Consider the role of such factors as land tenure, agricultural potential, population density, market access and integration, licensing requirements, tax regimes, credit and extension programs, and education.

Nonmarket Data Needs and Survey Techniques

Third, the important nonmarket contributions (e.g., drought and disease resistance, suitability for traction, cultural and social values, livestock as a means of finance and insurance) of livestock to livelihoods must be incorporated into economic models and analyses because such information is critical to the identification of appropriate breeding program goals and an assessment of the relative profitability of different breed use.

However, despite a wealth of livestock production data at the national level, such information tends to be limited to a number of the principal breeds and largely ignores the important nonmarket contributions. Initiatives such as the FAO Domestic Animal Diversity Information System and the International Livestock Research Institute Domestic Animal Genetic Resource Information System have only recently begun to address this problem. The issue of methodological choice is thus compounded by the lack of data availability and limited potential for acquiring relevant data.

Fourth, the issue of data availability is also closely related to that of data "get-ability." This is because most of the benefits produced by indigenous livestock in marginal production systems are captured by producers rather than consumers. As a consequence, the genetic resources of these breeds have been shaped mostly by producers' preferences. It is therefore to the identification and characterization of these preferences that research must turn in order to identify the implicit value of genetically determined traits as a first approximation to the value of indigenous AnGRs. In marginal production systems the breeding pressure on livestock is directed to creating animals capable of performing satisfactorily on marginal resources. Livestock performance is valued by producers but assessed mostly in nonmarket terms. It is therefore this category of economic agents and nonmarket functions that one needs to be able to study in order to derive economic values (Scarpa et al. 2003a). The question then arises as to how this can best be done.

Need to Use Rural Appraisal Techniques

In the context of the empirical results of biodiversity valuation studies and the difficulties confronted in applying the methods and surveys in rural areas or sectors remote from the market economy, Pearce and Moran (1994:94) note, "One area of further research involves the possible modification of economic techniques for use in conjunction with an established body of participatory and rapid rural appraisal methods."

Rural appraisal methods have been advocated as useful planning tools with livestock keepers (Waters-Bayer and Bayer 1994), ways to improve understanding of livestock keepers' breed interests and their preference for production and functional traits (Steglich and Peters 2002), ways to select genetic traits in cattle improvement programs (Tano et al. 2003), ways to accomplish situation analysis and technology development (Conroy 2003), and ways to facilitate processes of local innovation where the livestock keeper is the key knowledge holder (e.g., forage options in low–external input systems) (Peters et al. 2001).

The key is to match the type of method with the kind of information that is needed. In many cases, the best approach will involve combining several different rural appraisal methods.

Results and Discussion

Notwithstanding the aforementioned constraints on the economic analysis of AnGR conservation and sustainable use, where suitable methods and approaches to attaining the necessary data have been identified, a number of interesting results have been obtained.

Decision Support Tool for Identifying Breed Conservation Priorities

Recognizing the large number of indigenous livestock breeds that are currently threatened and the fact that not all can be saved given limited conservation budgets, Simianer et al. (2003) developed a decision support tool by elaborating a framework for the allocation of a given budget among a set of breeds such that the expected amount of between-breed diversity conserved is maximized. Drawing on Weitzman (1993), they argued that the optimum criterion for a conservation scheme is to maximize the expected total utility of the set of breeds, which is an economically weighted sum of diversity, breed characteristics represented in the set, and the value of the conserved breeds. The method is illustrated with an example of 23 African Zebu and Zenga cattle breeds. The results indicate that conservation funds should be spent on only 3 to 9 (depending on the model considered) of the 23 breeds and that these are not necessarily the most endangered ones. In addition, where the models are sufficiently specified and essential data on key parameters are available, the framework can be used for rational decision making on a global scale.

Stated Preference (Contingent Valuation) Techniques for Nonmarket Valuation

Obtaining the data for use in such decision support tools and elsewhere often entails the development of a number of techniques capable of attributing values to the many unpriced inputs and outputs of household production functions. Tano et al. (2003) and Scarpa et al. (2003a, 2003b) use stated preference choice experiments to value the phenotypic traits expressed in indigenous breeds of livestock. Adaptive traits and nonincome functions are shown to be important components of the total value of the animals to livestock keepers. In West Africa, for example, the most important traits for incorporation into breed improvement program goals were found to be disease resistance, fitness for traction, and reproductive performance. Beef and milk production were less important. The studies also show that not only do these techniques (adapted from other areas of environmental economic analysis) function for AnGR research, but they can be used to investigate values of genetically determined traits currently not widely recognized in livestock populations but desirable candidates for breeding or conservation programs (e.g., disease resistance).

Furthermore, the articles examine how household characteristics determine differences in breed preferences. This additional information can be of use in designing policies that counter the present trend toward marginalization of indigenous breeds. For example, they can be used to target incentives for breed conservation. In the Mexican case, the choice experiment reveals that because the net value backyard producers place on the creole pig is very similar to that of the other breeds, minimal incentives and interventions would be needed to ensure its continued sustainable use.

In a developed country case study, Cicia et al. (2003) show that a dichotomous choice stated preference approach can be used to estimate the benefits of establishing a conservation program for the threatened Italian Pentro horse. A bioeconomic model is used to estimate the costs associated with conservation, and a cost–benefit analysis is subsequently realized. The results not only show a large positive net present value associated with the proposed conservation activity but also show that this approach is a useful decision support tool for policymakers allocating scarce funds to a growing number of animal breeds facing extinction.

Revealed Preference Techniques for Market Valuation

In contrast to the aforementioned preference approaches, Jabbar and Diedhiou (2003) show that a revealed preference hedonic approach can also be used to determine livestock keepers' breeding practices and breed preferences. Analyzing such factors in southwest Nigeria, they confirm a strong trend away from trypanotolerant breeds, especially Muturu, and identify the traits livestock keepers find least desirable in these breeds relative to other Zebu breeds. The results suggest that the best hope for implementing a conservation or sustainable use strategy for breeds at risk such as Muturu is likely to be in other areas of West Africa, such as in southeast Nigeria, where trypanosomosis remains a constraint, where the Muturu is better suited to the farming systems, and where a large market for this breed continues to exist.

Aggregated Productivity Model for Comparative (Indigenous vs. Crossbreed) Performance Evaluation

The secondary importance of meat and milk production traits in many production systems led Ayalew et al. (2003) to argue that conventional productivity evaluation criteria are inadequate to evaluate subsistence livestock production because they fail to capture nonmarketable benefits of the livestock, and the core concept of a single limiting input is inappropriate to subsistence production because multiple limiting inputs (livestock, labor, land) are involved in the production process. Therefore as many of the livestock functions as possible (physical and socioeconomic) should be aggregated into monetary values and related to the resources used, irrespective of whether these "products" are marketed, home consumed, or maintained for later use. A broad evaluation model involving three complementary flock-level productivity indices was developed and applied to evaluate subsistence goat production in the eastern Ethiopian highlands. The results show that indigenous goat flocks generated significantly higher net benefits under improved than under traditional management, which challenges the prevailing notion that indigenous livestock do not adequately respond to improvements in the level of management. Furthermore, it is shown that under the subsistence mode of production considered, the premise that crossbred goats are more productive and beneficial than the indigenous goats is wrong. The model thus provides a more realistic platform on which to propose sound improvement interventions.

Conservation Costs and Benefits (Various Techniques)

Even where the value of indigenous breeds has been recognized and support mechanisms implemented, significant shortcomings can be identified. In an examination of farm animal biodiversity conservation measures and their potential costs in the European Union (EU), Signorello and Pappalardo (2003) report that many breeds at risk of extinction according to the FAO World Watch List are not covered by support payments because they do not appear in countries' rural development plans. Furthermore, where payments are made these do not take into account the different degrees of extinction risk between breeds, and payment levels are in any case inadequate, meaning that it can still be unprofitable to rear indigenous breeds. The EU AnGR conservation support measures thus urgently need to be reviewed if they are to meet their goals.

Incentives for indigenous breed conservation are inadequate despite the fact that conservation costs are shown to be small by Drucker (in press), who draws on the safe minimum standards (SMSs) literature and adapts Crowards's (1998) minimax payoff matrix to consider breeds rather than species. The basic framework considers that the uncertain benefits of indigenous livestock breed conservation can be maintained as long as a minimum viable population (the SMS, in this case the FAO measure of "not at risk," which is equivalent to approximately 1,000 animals) of the breed is also maintained. The costs of implementing SMSs are made up of the opportunity cost differential (if any exists) of maintaining the indigenous breed rather than an exotic or crossbreed. In addition, the administrative and technical support costs of the conservation program must be accounted for. Empirical cost estimates are then obtained using data from the three AnGR economic case studies mentioned earlier (i.e., EU, Italy, and Mexico). The findings support the hypothesis that the costs of implementing SMSs are low (as shown in table 17.2, depending on the breed, costs range from approximately 3,000 to 426,000 Euros annually), both when compared with the size of subsidies currently being provided to the livestock sector (less than 1% of the total subsidy) and with regard to the benefits of conservation (benefit–cost ratio of more than 2.9). Encouragingly, the costs are lowest in developing countries, given that 70% of the livestock breeds existing today are to be found there where the risk of loss is highest (Rege and Gibson 2003). Costs are particularly low where such conservation approaches are applied in communities that still favor the maintenance of the local breed. Therefore SMS approaches must be implemented with government or nongovernment support in collaboration with those communities.

The SMS approach is thus shown to have a role to play in AnGR conservation, but more extensive quantification of the components needed to determine SMS costs must be undertaken before it can be applied in practice. Such economic valuation must cover the full range of breeds or species being considered and ensure that as many as possible of the elements making up their total economic value are accounted for.

Drucker and Anderson provide additional data supporting the hypothesis that AnGR conservation costs are likely to be small compared with the benefits. In an unpublished paper that shows how data obtained through the use of rural appraisal methods[2] can be applied to some of the valuation methods reviewed in Drucker et al. (2001), they show that the conservation costs are several orders of magnitude smaller[3] (table 17.3).

Furthermore, the low estimated annual cost for the indigenous breed pig conservation and sustainable use program suggests that the least-cost approach (Brush and Meng 1996) does indeed provide a useful framework in which households or villages where conservation costs would be minimal can be costed into a conservation program. A very strong economic argument for implementing a conservation and sustainable use program therefore can be made, and such as program must be undertaken urgently if the breed, currently classified as critical on the FAO scale of risk, is not to become extinct.

The size of the net benefits identified also raises the question of whether the indigenous breed is, as predicted by theory, in fact being lost because, from the farmer's private perspective, it is less profitable than other breeds. Although certain household types (e.g., larger, better-off ones) did express trait value preferences that support this theory, most households did not.[4]

Table 17.2. Annual costs (in Euros) of achieving a safe minimum standard (SMS) for selected livestock breeds.

Case Study	PB_{exotic}	EB_{ind}	Opportunity Cost Differential for 1,000 Animals	C_{ind}	Annual Cost of SMS*
Creole pigs, Mexico[a]	24.1	21.5	2,600	200–1,100	<2,800–3,700
Various, Sicily, Italy[b]	29 (Comisana sheep)	−12 (Bar-baresca sheep)	41,000	Administrative and technical support costs not detailed; assumed to be 5% of total and thus ranges from € 2,000 to 20,000/yr	43,000
	15 (Maltese goat)	−27 (Gir-gentana goat)	42,000		44,000
	201 (Aveglinese horse)	−58 (Ragusana horse)	259,000		272,000
	306 (Bruna cattle)	−53 (Modicana and Cinisara cattle)	359,000		377,000
	224 (Landrace pig)	−182 (Nera Siciliana pig)	406,000		426,000
Pentro horse, Molise, Italy[c]	20.8[†]	<−158.4[‡]	<179,200	Unknown percentage of administrative and technical support costs included as production costs in EB_{ind} column	179,200

Sources: In Drucker (2005) and adapted from [a]Scarpa et al. (2003a), [b]Signorello and Pappalardo (2003), and [c]Cicia et al. (2003). C_{ind} = the cost of conservation of the indigenous breed under the SMS (includes administrative and technical support costs); EB_{ind} = expected benefits of continuing to use the indigenous breed; PB_{exotic} = private benefits of livestock breed substitution based on the use of exotics.

*Annual cost of SMS = 1,000(PB_{exotic} − EB_{ind}) + C_{ind}.

[†]Data not strictly comparable with the other studies. Refers to forgone income.

[‡]Data not strictly comparable with the other studies. Refers to "new sales" minus "production costs," and opportunity cost differential for 1,000 animals = 1,000(PB_{exotic} − EB_{ind}).

Table 17.3. Summary of results of valuation techniques using rural appraisal data related to the Yucatec (Mexico) creole pig.

	Conservation and Sustainable Use Benefits (¢)	Conservation Costs (¢)
Market share	0.54 million	
Production loss averted (Yucatan state only)	1.21 million	
Contingent valuation (urban consumer taste test)	1.43 million	
Contingent valuation (producer choice experiment) and least cost approach		<2,800–3,700

Source: Drucker and Anderson, unpublished.
Note: Original US$ values converted to Euros at a rate of Euro 1 = US$1.10.

Yet backyard indigenous pig production has declined across all households. It therefore appears that the purebred population has fallen to such a low level that such factors as the lack of availability of indigenous breeding stock, rather than farmer net returns per se, are determining breed choice.

At the level of society, the large size of the net benefits of a conservation and sustainable use plan suggest that a number of very significant market failures must be addressed if the benefit values (e.g., indigenous breed pigs as a reservoir of disease resistance or in terms of their existence value to urban consumers) are to be harnessed for conservation purposes. In addition, the market distortions introduced by subsidizing exotic breed production in the commercial sector are considerable, and the levels of subsidy are several orders of magnitude greater than the costs of indigenous breed conservation. A genetic resource important in the maintenance of subsistence farmer livelihoods thus is being lost for the lack of minimal funds, while large and AnGR diversity-threatening subsidies are provided to commercial farmers.

Conclusion

The findings described in this chapter (based on a variety of species, breeds, production systems, locations, and analytical approaches) show that methods for the economic analysis of AnGR conservation and sustainable use do in fact exist and, particularly when used in conjunction with rural appraisal methods, can reveal useful estimates of the values that are placed on market, nonmarket, and potential breed attributes.

Such information on livestock keeper knowledge about breed characteristics and management needs, as well as livestock keeper preferences for different traits, constitutes critical input for breeding and conservation strategies. In addition, information regarding the performance and potential of indigenous breeds under improved management, breed conservation priorities, and the relative size of the costs and benefits resulting from conservation and sustainable use programs can be obtained. Such data are crucial for understanding the type and net costs of the interventions necessary to promote the conservation and sustainable use of AnGRs.

The challenge is to now apply work of this type in contexts where the results can be used to benefit livestock keepers and support the work of national researchers and policymakers.

Notes

1. Conversion or replacement can occur not only through substitution but also through crossbreeding and eliminating livestock due to production system changes.

2. These included semistructured interviews, direct observation, inventories, timelines, seasonal calendars, wealth ranking, preference ranking, and pairwise rankings. Selections of such tools were applied in focus groups, at household, commercial farm, and market levels, with key informants (e.g., local pig breeders, butchers, consumers, livestock association personnel) and were also applied longitudinally by monitoring of selected households over 12-month periods.

3. Similarly large net benefits to conservation were identified by Signorello and Pappalardo (2003) in the case of the Italian Pentro horse, suggesting that this is not an isolated finding.

4. Furthermore, even in these larger, better-off households, the crossbred is preferred over the exotic. Thus there remains the issue of how to maintain a purebred line that can be used in crossbreeding.

References

Ayalew, W., J. King, E. Bruns, and B. Rischkowsky. 2003. Economic evaluation of smallholder subsistence livestock production: Lessons from an Ethiopian goat development program. *Ecological Economics* 45(3):331–339.

Brush, S. and E. Meng. 1996. *Farmers' Valuation and Conservation of Crop Genetic Resources.* Paper prepared for the Symposium on the Economics of Valuation and Conservation of Genetic Resources for Agriculture, Centre for International Studies on Economic Growth, Tor Vergata University, Rome, May 13–15.

Cicia, G., E. D'Ercole, and D. Marino. 2003. Valuing farm animal genetic resources by means of contingent valuation and a bio-economic model: The case of the Pentro horse. *Ecological Economics* 45(3):445–459.

Conroy, C. 2003. *Participatory Livestock Research: A Guide.* London: Intermediate Technology Publications.

Crowards, T. 1998. Safe minimum standards: Costs and opportunities. *Ecological Economics* 25:303–314.

Drucker, A. G. In press. The role of safe minimum standards in the conservation of livestock biodiversity. *International Journal of Agricultural Sustainability.*

Drucker, A., V. Gomez, and S. Anderson. 2001. The economic valuation of farm animal genetic resources: A survey of available methods. *Ecological Economics* 36(1):1–18.

Evenson, R. 1991. Genetic resources: Assessing economic value. In J. Vincent, E. Crawford, and J. Hoehn, eds., *Valuing Environmental Benefits in Developing Economies.* Proceedings of a seminar series held February–May 1990 at Michigan State University, Special Report No. 29.

FAO (Food and Agriculture Organization of the United Nations). Undated. *Extensive Pastoral Livestock Systems: Issues and Options for the Future.* Available at www.fao-kyokai.or.jp/edocuments/document2.html.

FAO (Food and Agriculture Organization of the United Nations). 1999. *The Global Strategy for the Management of Farm Animal Genetic Resources.* Rome: FAO.

FAO (Food and Agriculture Organization of the United Nations). 2000. *World Watch List for Domestic Animal Diversity,* 3rd ed. Rome: FAO.

FAO/ILRI (Food and Agriculture Organization of the United Nations/International Livestock Research Institute). 1999. *Economic Valuation of Animal Genetic Resources.* Proceedings of an FAO/ILRI workshop. Rome, March 15–17. Rome: FAO.

Hall, S. J. G. and J. Ruane. 1993. Livestock breeds and their conservation: Global review. *Conservation Biology* 7(4):815–825.

Jabbar, M. and M. Diedhiou. 2003. Does breed matter to cattle farmers and buyers? Evidence from West Africa. *Ecological Economics* 45(3):461–472.

Livestock in Development. 1999. *Livestock in Poverty-Focused Development.* Crewkerne, UK: Livestock in Development.

Mendelsohn, R. 2003. The challenge of conserving indigenous domesticated animals. *Ecological Economics* 45(3):501–510.

Pearce, D. and D. Moran. 1994. *The Economic Value of Biodiversity.* London: Earthscan.

Peters, M., P. Horne, A. Schmidt, F. Holmann, P. C. Kerridge, S. A. Tarawali, R. Schultze-Kraft, C. E. Lascano, P. Argel, W. Stür, S. Fujisaka, K. Müller-Sämann, and C. Wortmann. 2001. *The Role of Forages in Reducing Poverty and Degradation of Natural Resources in Tropical Production Systems.* AgREN Network Paper 117, July. London: ODI.

Rege, J. E. O. and J. P. Gibson. 2003. Animal genetic resources and economic development: Issues in relation to economic valuation. *Ecological Economics* 45(3):319–330.

Scarpa, R., A. Drucker, S. Anderson, N. Ferraes-Ehuan, V. Gomez, C. Risopatron, and O. Rubio-Leonel. 2003a. Valuing animal genetic resources in peasant economies: The case of the Box Keken creole pig in Yucatan. *Ecological Economics* 45(3):427–443.

Scarpa, R., P. Kristjanson, E. Ruto, M. Radeny, A. Drucker, and J. E. O. Rege. 2003b. Valuing indigenous farm animal genetic resources in Africa: A comparison of stated and revealed preference estimates. *Ecological Economics* 45(3):409–426.

Signorello, G. and G. Pappalardo. 2003. Domestic animal biodiversity conservation: A case study of rural development plans in the European Union. *Ecological Economics* 45(3):487–499.

Simianer, H., S. Marti, J. Gibson, O. Hanotte, and J. E. O. Rege. 2003. An approach to the optimal allocation of conservation funds to minimize loss of genetic diversity between livestock breeds. *Ecological Economics* 45(3):377–392.

Smale, M., ed. 2005. *Valuing Crop Biodiversity: On-Farm Genetic Resources and Economic Change.* Wallingford, UK: CABI Publishing.

Steglich, M. and K. J. Peters. 2002. *Agro-Pastoralists' Trait Preferences in N'dama Cattle: Participatory Methods to Assess Breeding Objectives.* 7th World Congress on Genetics Applied to Livestock Production, August 19–23, 2002, Montpellier, France.

Swanson, T. 1997. *Global Action for Biodiversity.* London: Earthscan.

Tano, K., M. Faminow, M. Kamuanga, and B. Swallow. 2003. Using conjoint analysis to estimate farmers' preferences for cattle traits in West Africa. *Ecological Economics* 45(3):393–407.

Tisdell, C. 2003. Socioeconomic causes of loss of animal diversity genetic: Analysis and assessment. *Ecological Economics* 45(3):365–376.

UNEP (United Nations Environment Programme). 1995. *Global Biodiversity Assessment.* Cambridge, UK: Cambridge University Press.

Waters-Bayer, A. and W. Bayer. 1994. *Planning with Pastoralists: PRA and More, a Review of Methods Focused on Africa.* Eschborn, Germany: GTZ.

Weitzman, M. 1993. What to preserve? An application of diversity theory to crane conservation. *Quarterly Journal of Economics* February:157–183.

18 ⚘ Ecological and Economic Roles of Biodiversity in Agroecosystems

M. CERONI, S. LIU, AND R. COSTANZA

As ecosystems become less diverse as a consequence of land conversion and intensification, there is a shared concern over the functioning of these systems and their ability to provide a continuous flow of services to human societies (Ehrlich and Wilson 1991). The ecological consequences of biodiversity loss on ecosystem functioning have been investigated for more than a decade, but only recently has interest developed around the consequences of agricultural biodiversity loss on the functions of agroecosystems. Agricultural intensification has led to a widespread decline in agricultural biodiversity measured across many different levels, from a reduction in the number of crop and livestock varieties, to decreasing soil community diversity, to the local extinction of a number of natural enemy species.

Each time species go locally extinct, energy, and nutrient pathways are lost with consequent alteration of ecosystem efficiency and of the ability of communities to respond to environmental fluctuations. Monocultural agroecosystems typically display low resilience to perturbations such as drought, flooding, pest outbreaks, and invasive species and to uncertainties related to market fluctuations. Large inputs of energy are then needed in the form of fertilizers, pesticides, herbicides, and irrigation.

Multifunctional and sustainable agriculture, where production is achieved with respect for ecosystem functions and processes and with reduced impacts to other systems, is expected to produce a whole array of ecosystem services besides edible and fiber biomass production, such as soil erosion control, carbon sequestration, nutrient cycling, wildlife refugia, and sources of spiritual and cultural enjoyment.

Ecosystem functioning refers to the rates and magnitudes of ecosystem processes, such as primary production, decomposition, and nutrient cycling. Ecosystem services are the functions that directly or indirectly affect human welfare. Whereas well-established measures of ecosystem functioning exist, such as mineralization rates and organic matter production, it is difficult to translate what ecologists measure into ecosystem services. Because ecosystem services represent anthropocentric properties of the ecosystems, the notion of value is inherently part of their definition. For this reason ecosystem services often are measured in economic terms rather than in ecological terms of energy and material flux (see Costanza et al. 1997). Although local and global economies depend heavily on ecosystem services, these have been traditionally ignored by commercial markets and therefore have been given little weight in policy decisions.

This is well exemplified by the study conducted by Costanza and colleagues (1997) on the economic value of ecosystem services at the global level. The study estimated the total economic value of ecosystem services for the globe based on economic valuations of ecosystem services for each of 16 biomes (communities of plants and animals that are well adapted to different climatic regions of the earth, such as deserts, grasslands, or temperate forests).

The authors found that estimates of global economic activities, such as the yearly global gross national product, failed to account for the substantial economic contribution of ecosystem services from the different world's biomes. Whereas global gross national product was estimated to be around US$18 trillion per year, the economic value of ecosystem services ranged between US$16 and 54 trillion per year, with an average of US$33 trillion per year (in 1994 U.S. dollars). It is not well understood from this study to what extent different agricultural ecosystems contributed to the total value of ecosystem services. Croplands, with a total value of US$128 billion per year (0.38% of total estimated value), seem to contribute little to the global flow of ecosystem services beyond food production (table 18.1). However, this result is mainly a consequence of the limited information available on ecosystem services in food production systems and of the assumption that croplands do not provide habitat for wildlife, nor do they represent a valuable source for recreation. When grass and rangeland systems are included, most of which are assumed to be subject to various levels of grazing for farming purposes, the total value of annual ecosystem services from agricultural lands jumps to US$1.03 trillion (3.1% of total estimated value). Croplands and grass and rangelands together contribute mainly to

Table 18.1. Summary of average global value of annual ecosystem services (in US$/ha/yr).

Biome	Area (ha × 10⁶)	Gas Regulation	Climate Regulation	Disturbance Regulation	Water Regulation	Water Supply	Erosion Control	Soil Formation	Nutrient Cycling	Waste Treatment	Pollination	Biological Control	Habitat or Refugia	Food Production	Raw Materials	Genetic Resources	Recreation	Cultural	Total Value per ha*	Total Global Flow Value†
Marine	36,302																		577	20,949
Open ocean	33,200	38							118			5		15	0		76		252	8,381
Coastal	3,102			88					3,677			38	8	93	4		82	62	4,052	12,568
Estuaries	180			567					21,100			78	131	521	25		381	29	22,832	4,110
Seagrass and algae	200								19,002					2					19,004	3,801
Coral reefs	62			2,750						58		5	7	220	27		3,008	1	6,075	375
Shelf	2,660								1,431			39	2	68			70		1,610	4,283
Terrestrial	15,323																		804	12,319
Forest	4,855		141	2	2	3	96	10	361	87		2		43	138	16	66	2	969	4,706
Tropical	1,900		223	5	6	8	245	10	922	87				32	315	41	112	2	2,007	3,813
Temperate and boreal	2,955		88		0			10		87	4			50	25		36	2	302	894
Grasslands and rangelands	3,898	7			3		29	1		87	25	23		67	0	2	2		232	906

																	Total value per hectare[a]	Total global flow value[b]
Wetlands	330	133		4,539	15	3,800			4,177				304	256	106	574	881	14,785	4,879
Tidal marsh and mangroves	165	265		1,839					6,696				169	466	162	658		9,990	1,648
Swamps and floodplains	165			7,240	30	7,600			1,659				439	47	49	491	1,761	19,580	3,231
Lakes and rivers	200			665	5,445	2,171			665					41		230		8,498	1,700
Deserts	1,925																		
Tundra	743																		
Ice and rock	1,640																		
Crop-land	1,400										14	24		54				92	128
Urban	332																		
Total	51,625	1,341	684	1,779	1,115	1,692	576	53	17,075	2,277	117	417	124	1,386	721	79	815	3,015	33,268

Source: Adapted from Costanza et al. (1997).

Note: Rows and column totals are in $ × 10⁹/yr. Shaded cells indicate services that do not occur or are known to be negligible. Open cells indicate lack of available information.

[a] Total value per hectare in $/ha/yr.

[b] Total global flow value in $ × 10⁹/yr.

food production (US$336 billion), followed by biological control (US$121 billion) and pollination services (US$117 billion). The main services contributed by the grass and rangeland component of agricultural lands are waste treatment (US$339 billion) and erosion control (US$113 billion). Given the large scale of this study and the broad categories used to identify the main biomes, these figures do not capture the role of the different agricultural land uses (e.g., shrimp farming, aquaculture, flooded fields, and agroforestry), inevitably underestimating the contribution of the world's agroecosystems.

No matter what the final figures of the total contribution of agroecosystems to human welfare would be, agricultural biodiversity is what supports the ecosystem services that our societies depend on. Yet estimating the specific economic contributions of agricultural biodiversity and biodiversity in general to the value of ecosystem services is a formidable challenge (see Turner et al. 2003 and Smale 2005).

For the sake of economic valuation of biodiversity, a distinction can be made between biological resources and biological diversity (OECD 2002). Biological *resources* are elements of ecosystems, such as genes or species, which are of direct importance to human economies. Biological *diversity* is considered to be of value to human societies as the source of the variety of species' ecological interactions, physiological tolerances, structural arrangements in space, and genetic structures that in the end determine ecosystem functioning.

The importance of economic valuation of biodiversity is recognized by the Convention on Biological Diversity (CBD). CBD's Conference of the Parties Decision IV/10 recognizes that "economic valuation of biodiversity and biological resources is an important tool for well-targeted and calibrated incentive measures."

Most studies on biodiversity valuation have assessed the direct value of *biological resources* (i.e., the value that is more readily captured by commercial markets), focusing in particular on plant or crop and animal genetic resources or the direct use of plant species for medicinal or ornamental use (for the direct value genetic resources in crop improvement, see reviews in Alston et al. 1998; Evenson and Gollin 2003). The nonmarket values of genetic resources have been assessed in a very few cases, including livestock genetic resources (Drucker, chapter 17) and, most recently, components of agricultural biodiversity in home gardens (Birol 2004; Birol et al. 2004). Two collections of studies about valuing crop genetic resources conserved in banks (Koo et al. 2004) and the biological diversity of crop plants on farms (Smale 2005; see also chapter 16), both using primary data, have been published recently. These studies are based on detailed field research and have advance methods for valuing some components, or entry points, of biological diversity.

Almost no information exists on the economic value of most components of biological diversity to human societies and, particularly, their indirect value. For example, the diversity in species or functional groups in an ecological community is of value to our society to the extent that it matters to the provision of the services we benefit from, such as nutrient cycling, biomass production, and stability of biomass production. But proving that community diversity does actually matter is extremely difficult, and even more difficult is to identify general ecological rules that can fit the broad purposes of economic valuation. In this chapter we report results from empirical ecological studies that measured the relationship between diversity and ecosystem functions (mostly in agricultural systems), under the assumption that measures of ecosystem functions provide a useful indication of the direction and intensity of the flow of ecosystem services without necessarily translating directly into ecosystem services. We focus primarily on the role of agricultural biological diversity (instead of biological resources). Besides providing evidence from empirical ecological studies, each section briefly addresses how ecological knowledge of agrobiodiversity can be applied to inform economic valuation. Valuation methods for biodiversity and ecosystem services have been extensively reviewed recently (Wilson 1988; Orians et al. 1990; Drucker et al. 2001; Nunes and van den Bergh 2001), so methodological considerations are not part of our discussion. We begin the chapter with an overview of the main concepts and findings from a decade of biodiversity and ecosystem functioning literature. We then discuss how agrobiodiversity relates to stability and resilience in agricultural systems. The role of habitat heterogeneity to support wild species is then examined, followed by a section on agrobiodiversity at the landscape scale. We conclude with observations on research needs in assessing the relationship between agrobiodiversity and ecosystem services and implications for agrobiodiversity economic valuation studies.

Diversity of Producers and Biomass Production

Over the last decade, the most influential empirical research on the links between biodiversity and ecosystem function has been the series of experiments manipulating plant species diversity and functional group richness in grasslands (e.g., Naeem et al. 1994; Tilman et al. 1996, 2002; Hector et al. 1999) and in aquatic microbial microcosms (reviewed by Petchey et al. 2002).

Because recent publications cover biodiversity functioning research extensively (Chapin et al. 2000; Loreau et al. 2001, 2002; Kinzig et al. 2002; see also chapters 9 and 10), we only briefly review the central issues.

Empirical and theoretical studies in many cases have confirmed associations between biodiversity and ecosystem functioning, but many relationships, from insignificant to significant, from positive to negative, have been identified depending on the scale of the investigation (Naeem 2001). Many factors, such as site fertility, disturbance, habitat size, climate (Wardle et al. 1997), the presence or absence of trophic groups (Mulder et al. 1999; Naeem et al. 2000), and the functional composition of species (Hooper and Vitousek 1997; Tilman et al. 1997a), can determine the relationship between biodiversity and ecosystem function.

Several studies found significant positive correlations between species richness and plant biomass accumulation (reviewed by Schmid et al. 2002). The mechanisms behind these correlations were long debated around two main hypotheses, although alternative explanations also have been discussed (reviewed by Eviner and Chapin 2003). Aarssen (1997), Huston (1997), and Tilman et al. (1997b) suggested that the often-observed increase in primary productivity in more diverse plots may have reflected a sampling effect. A community with a higher number of species inherently has a higher probability of including species with superior traits. Another explanation of diversity effects on ecosystem functioning is niche complementarity (Naeem et al. 1995; Tilman et al. 1997a). Higher species diversity in a community increases the range of ecological traits—and consequently the variety of niches available—leading to a more efficient resource use in a variable environment. Recently, the debate appears to have been reconciled (Loreau et al. 2002; Naeem 2002). Niche complementarity and sampling effects seem to play different roles in different phases of the experimental manipulations: Initially, a rapid growth response that seems compatible with the sampling mechanism is observed, with the best diversity plots reaching a productivity almost equal to that of the best monocultures. After two or more years, a longer-term response shows the best diversity plots producing higher yields than the best monocultures, a pattern that can be explained by interspecific competition resulting from niche differentiation (Pacala and Tilman 2002).

One general conclusion seemed to emerge among the contrasting results and interpretations that a decade of diversity functioning research has generated: The species' role in the functioning of these experimental communities can vary widely. Some species might be indispensable in maintaining the functioning of an ecosystem, as in the case of keystone species (Paine 1966) or ecosystem engineers (Jones et al. 1994; Wright et al. 2002). Some other species may even appear redundant in their ecological functions and may be easily replaced by other species with no appreciable consequences for ecosystem functioning, should they go locally extinct (Walker 1992; Gitay et al. 1996; Naeem 1998).

As discussed also in chapter 10, one of the limitations of biodiversity function studies is that they have been performed in small, controlled patches that are far from mimicking the conditions of natural or even managed ecosystems. For example, it is hard to extrapolate the implications of this type of research for agricultural systems, where the number of crop species used typically is low and rotation cycles govern the temporal dynamics of the system.

Very few experiments have manipulated species richness in agricultural systems to assess the effects on biomass production. Results from a study on hay fields in southern Britain show that restoration of species richness in fields that were previously impoverished in species had a positive effect on hay production. Bullock and colleagues (2001) reported a 60% yield increase in species-rich treatments in hay meadow restoration experiments at seven sites across southern Britain. At each site two seed mixes (species poor, with 6 ± 17 species, and species rich, with 25 ± 41 species) were applied in a randomized block experiment. Hay yield was higher in the species-rich treatment from the second year onward, by up to 60% (figure 18.1). Comparing the two treatments in all sites, there was a simple linear relationship between the difference in species number and the amount of increase in hay production. Fodder quality was the same in both treatments. This suggests that farmers can maximize high-quality herbage production in resown grasslands by maximizing biodiversity. The results of this study are particularly remarkable if we think that there is a common misconception among farmers that every effort to increase biodiversity results in lower food production.

The only apparent shortcoming in this study was the higher cost of the high-diversity seed mix; a higher increase in yield would be needed to offset these additional costs. The ecological mechanisms behind the observed patterns seem to be a result of species number differences between

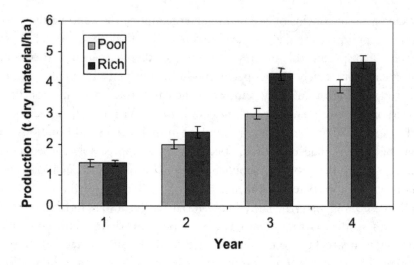

FIGURE 18.1. Biodiversity treatment effects on hay production in different years (mean across plots and sites is displayed ± 1 *SE*). The species-rich treatment had higher dry matter yield from the second year onward (adapted from Bullock et al. 2001).

treatment and control plots, but the authors warn that because species number and composition were not varied independently (as done by Hector et al. 1999), compositional differences also might have contributed to yield differences.

Economic Considerations

In this study case, the economic contribution of species richness to hay production is straightforward to assess as the difference between production outcomes under the two different richness treatments. Valuations of this kind could be used to develop incentives to farmers to promote higher plant diversity in hayfield systems.

In most cases, though, assessing the economic contribution of crop species richness to other ecosystem services such as nitrogen cycling or CO_2 regulation is not as straightforward. In the best-case scenario, even assuming that the ecological causalities between agrobiodiversity and ecosystem functions have been clearly identified, economic assessments would rarely reach a validity that goes beyond the scale of the studied site.

Attempts are being made to assess the specific ecological and economic contributions of species richness to net primary productivity and nutrient cycling in natural and seminatural environments at a regional scale based on multiple regression models (Costanza et al. unpublished).

Diversity of Consumers and Decomposers

Most studies have focused on the role of the diversity of primary producers in providing fundamental ecosystem services. However, very little is known about the factors influencing ecosystem services provided by higher trophic levels in natural food webs. A recent study of 19 plant–herbivore–parasitoid food webs (Montoya et al. 2003) showed that differences in food web structure and the richness of herbivores influence parasitism rates on hosts, promoting the service supplied by natural enemies. One main result of this study was that parasitoids function better in simple food webs than in complex ones, indicating that species richness per se might not be a key factor in the provision of higher-level ecosystem services when more complex, multitrophic communities are investigated.

As Brown et al. (chapter 9) noted, most evidence suggests that in soils there is no predictable relationship between species diversity and specific soil functions, making it difficult to foretell the consequences of decreased soil species richness (Mikola and Setälä 1998). In many cases, soil ecosystem function seems to be controlled by individual traits of dominant species and by the complexity of biotic interactions that occur between components of soil food webs (Cragg and Bardgett 2001).

Higher functional diversity in microbial communities has been associated with higher efficiency in resource use. For example, a 21-year study comparing biodynamic, organic, and conventional farming systems in central Europe (Mäder et al. 2002) shows that more diverse microbial communities, typical of organically managed soils, transformed carbon from organic debris into biomass at lower energy costs.

Economic Considerations

In systems where the role of one individual species determines the rate of a given set of ecological processes and the flux of a given ecosystem service, that species could be valued independently. However, this is rarely the case. Complex ecological interactions normally make it difficult to disentangle the role of particular species and the effect of diversity per se in supporting certain ecosystem functions. For these reasons, ecological economists have tended to value biodiversity indirectly by valuing the services biodiversity supports. For example, Walker and Young (1986) estimate that soil erosion was responsible for revenue loss from agriculture in the Palouse region, northern Idaho and western Washington, in the range of US$10 to US$15 per hectare. This estimate is an aggregated indicator of the ecological functions responsible for erosion control in agroecosystems of that particular region.

Diversity and Resilience in Agroecosystems

Most studies on biodiversity and ecosystem functioning have been conducted in stable conditions. Agroecosystems typically are subject to cyclical perturbations of variable intensity as a consequence of agricultural practices and to unpredictable events such as pest outbreaks and drought. However, the relationship between diversity and ecosystem function might change in a fluctuating environment (see chapters 13 and 14).

There is a general agreement that a major role of biodiversity in relation to ecosystem services is insurance against environmental change (e.g., Holling et al. 1995; Perrings 1995). A higher number of functionally similar species ensures that when environmental conditions have turned against the dominant species, other species can readily substitute for their functions, thereby maintaining the stability of the ecosystem (Yachi and Loreau 1999) and enhancing ecosystem reliability (i.e., the probability that a system will provide a consistent level of performance over a given unit of time) (Naeem and Li 1997).

For example, diversity of pollinators is essential to food production systems, not only because pollen limitation to seed and fruit set is widespread (Burd 1994) but, most importantly, in the face of the ongoing trends of pollinator disruptions (Nabhan and Buchmann 1997; Kremen and Ricketts 2000; Cane and Tepedino 2001; see also chapter 8). Kremen et al. (2002) found that a diversity of pollinators was a determinant for sustaining pollination services in conventional (versus organic) farms in California because of annual variation in composition of the pollinator community.

Redundancy in soil microbial communities seems to be very common and crucial in maintaining soil resilience to perturbations (see chapter 9). For example, experimental reductions of soil biodiversity through fumigation techniques show that soils with the highest biodiversity are more resistant to stress than soils with impaired biodiversity (Griffiths et al. 2000).

Studies conducted in extreme regions of the world, such as the Dry Valley in Antarctica, where soil communities are much less diverse, provide unique experimental sites to address the role of food web complexity in soil function. Nematode communities in this region, comprising three species at the most, typically lack redundancy and are particularly sensitive to environmental change (Freckman and Virginia 1997).

Agrobiodiversity at the genetic level also provides an insurance value in the face of changing environmental conditions. Chapters 2 through 6 describe empirical evidence of how in food production systems, genetic diversity ensures adaptability and evolution by providing the raw material for desirable genetic traits in crops and livestock. In chapter 15, Johns demonstrates how agricultural diversity and the knowledge imbedded in its management are essential for dietary diversity and human health.

Ecosystems that are capable of absorbing a higher degree of perturbation before their functioning is significantly altered (i.e., are more ecologically resilient, *sensu* Holling 1973) can provide ecosystem services more consistently. Planting of varietal mixtures with differing levels of pest resistance has proved to be a successful strategy to fight fungal pathogens (see also chapters 11 and 12 and Zhu et al. 2002).

Resilience in industrial monocultures is achieved through use of external inputs such as chemical fertilizers, pesticides, and fossil fuels. As noted in chapters 12, 13, 14, 16, and 17, in less intensive systems agricultural biodiversity may provide a buffer to unpredictable environmental and market fluctuations. Several scientists have urged recognition of the indissoluble link between ecological and sociological resilience in managed systems (Scoones 1999; Folke et al. 2003; Milestad and Hadatsch 2003). In fact, systems may be ecologically resilient but socially vulnerable or socially resilient but environmentally degrading (Folke et al. 2003). Agricultural systems can then be thought of as social-ecological systems that behave as complex adaptive systems, in which the managers are integral components of the system (Conway 1987). In chapter 13 the term *agrodiversity* is used to interrelate agrobiodiversity, management diversity, and biophysical diversity into organizational diversity. To be resilient to natural and market fluctuations, agroecosystems should withstand disturbance, be able to reorganize after disturbance, and have the ability to learn and adapt in the face of change (Walker et al. 2002). Exponents of the Resilience Alliance argue that resilience is something that can and should be managed to "prevent the system from moving to undesired system configurations in the face of external stresses and disturbances" and to "nurture and preserve the elements that enable the system to renew and reorganize itself following a massive change" (Walker et al. 2002). Both ecological components and human capabilities can play an important role in resilience management. For example, the insurance value of agricultural biodiversity has a recognized role in protecting ecosystem resilience (Heywood 1995). Furthermore, agricultural systems with high levels of social and human assets are more flexible and more capable of incorporating innovations in the face of uncertainty (Pretty and Ward 2001).

Economic Considerations

Identifying and measuring the insurance value of biodiversity is a far from trivial exercise. For example, what premium would be paid to preserve resilience in a given system? One option would be to consider the cost of maintaining a nonresilient system. In agroecosystems this premium would be equivalent to the entire costs of maintaining intensive agricultural practices through the use of external inputs, including costs of pesticides and chemical fertilizers. As noted earlier in this chapter and in chapter 8, diversity of pollinators is needed to maintain the resilience of production systems in the face of declining pollinators. Southwick and Southwick (1992) calculated for each of 62 U.S. crops the extent to which wild pollinators could replace honeybee functions, should they decline to the degree predicted by their model. In the absence of compensation from wild pollinators, alfalfa yield losses were estimated to be 70% of total production, equivalent to US$315 million a year.

Maintaining or enhancing the insurance function of species and genetic diversity might come with a cost to other functions that are relevant to human welfare, such as food and fiber production. For example, Heisey et al. (1997) assessed the yield losses associated with switching to a more genetically diverse portfolio of wheat varieties in Pakistan at tens of millions of U.S. dollars per year. Widawsky and Rozelle (1998), Di Falco and Perrings (2003), Meng et al. (2003), and Smale et al. (1998) found both positive and negative associations between crop variety diversity, crop productivity, and yield variability, depending on the cropping system context. Whereas the insurance value of genetic diversity in food production systems has been assessed at least in some cases (e.g., see the studies assessing costs of conservation programs for genetic resources reviewed by Drucker et al. 2001), there are no studies addressing the insurance value of a diversified portfolio of functions and phenotypic traits provided by crop species, soil organisms, or natural enemies. The difficulties in determining the insurance value are related to the intangible nature of this service and the inability to account for future benefits adequately. In addition, the outcomes of a valuation study might vary according to the perceived level of collapse threat.

Agricultural Habitats and Landscape Diversity

Various studies show that agricultural landscape diversity can reduce yield losses to pests by affecting populations of both herbivorous insects and natural enemies (see Andow 1991 for a review). For example, healthier populations of predator carabid beetles can be found in more heterogeneous farm systems (where heterogeneity is measured as perimeter-to-area ratio) and in systems with higher crop species diversity (Ostman et al. 2001).

The composition and spatial arrangement of perennial and annual crops in the agricultural landscape can also be crucial for the long-term predator–population dynamics (Bommarco 1998; Thies and Tscharntke 1999).

In other cases polycultures do not seem to provide any advantage to natural enemy populations when compared with monocultures (Tonhasca and Stinner 1991).

Inconsistent results in experiments that have manipulated landscape structure and vegetational diversity might reflect the variation related to the different spatial scale of the experimental vegetation plots. A comprehensive meta-analysis of the literature results in this field over a period of 18 years shows that in experiments performed in small plots, spatial heterogeneity tends to have a large negative effect on herbivores, intermediate-sized plots show an intermediate effect, and the largest plots exhibit a negligible effect (Bommarco and Banks 2003).

Finding general patterns in the relationship between landscape diversity and species diversity becomes even more complicated when diversity across multiple taxa is investigated (Tews et al. 2004 and references therein; see also chapters 13 and 14). This relationship specifically depends on at least three factors: the species groups studied, the measurement of landscape diversity, and the temporal and spatial scales.

More diverse agricultural landscapes provide important habitats not only for natural enemies but also for pollinators, enhancing the provision of pollination services (see also chapter 8). A study on the effects of agricultural landscape structure on bees found that species richness and abundance of solitary wild bees were positively correlated with the percentage of seminatural habitats, an indicator of landscape diversity (Steffan-Dewenter et al. 2002). The correlation depended on spatial scale and species group. For example, whereas solitary wild bees responded to landscape complexity at the small scales, honeybees were correlated with landscape structural characteristics only at large scales. In other cases, the availability of suitable foraging habitats matters more than landscape heterogeneity in determining the species richness of pollinators (Steffan-Dewenter 2003).

Bird and mammal species richness also can be enhanced by agricultural landscape diversity. A recent review (Benton et al. 2003) provides ample evidence that habitat heterogeneity matters to farmland biodiversity from the individual field to the whole landscape. For example, seed-eating birds seemed to occur in higher numbers in pastoral areas containing small patches of arable land than in pure grassland landscapes (Robinson et al. 2001). Some bird species specifically depend on the open habitats provided by farming systems in Africa (Söderström et al. 2003), as in Europe (Pain and Pienkowski 1997) and Central America (Daily et al. 2001).

Agroforestry patches can harbor a number of wild species similar to or higher than that of original forest patches. For example, Ricketts et al. (2001) found no significant difference in the abundance and richness of moth species between forest and agricultural fragments composed of coffee monocultures, shade-grown coffee, pasture, and mixed farms. Polycultural coffee plantations designed to mimic natural systems in various cases show species richness equal to or greater than that of adjacent natural forest patches (figure 18.2) (Perfecto et al. 1997; Daily et al. 2003). A decline in species diversity in agroforests can be observed with increasing distance from the forest patches (Ricketts et al. 2001; Armbrecht and Perfecto 2003), although this result is not consistent across studies (e.g., Daily et al. 2003). In Central and South America, shaded coffee plantations that include leguminous, fruit, fuelwood, and fodder trees are reported to contain more than 100 plant species per field and support up to 180 bird species (Michon and de Foresta 1990; Altieri 1991; Thrupp 1997).

Noncultivated areas (e.g., riparian buffers, windbreaks, or border plantings), improved fallows, and woody vegetation play an important role in maintaining biodiversity of weeds, insects, arthropods, and birds (Benton et al. 2003 and references therein; McNeely and Scherr 2003). Hedgerows and woody vegetation, while providing habitats for wild biodiversity, may enhance other ecosystem services such as soil stabilization, soil erosion control, and carbon sequestration.

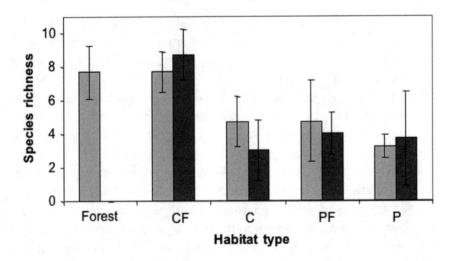

FIGURE 18.2. Mammal species richness by habitat type and distance class from an extensive forest patch (mean±1 SE). Shaded bars represent sites in and near (<1 km) the forest; black bars represent sites far (5–7 km) from the forest. Species richness varied significantly between habitat types but not with distance from extensive forest. Small forest remnants contiguous with coffee plantations (CF) did not differ from more extensive forest in species richness and were richer than coffee plantations (C), pastures with adjacent forest remnant (PF), and pastures (P) (adapted from Daily et al. 2003).

Economic Considerations

Once a strong link between agricultural habitat diversity and wild species diversity has been documented, the value of agrobiodiversity to wildlife habitat protection can be assessed through the expenditures associated with the enjoyment of a biologically richer environment. Alternatively, assessments can include the costs of protecting the diversity of habitats that agrobiodiversity provides. For example, citizens in the Netherlands were willing to pay between 16 and 45 guilders per household per year (corresponding to $10.80 and $30.35 in 2003 U.S. dollars) to fund management practices that would enhance wildlife habitat in the Dutch meadow region (cited in Nunes and van den Bergh 2001).

Recreational and Cultural Roles of Agricultural Biodiversity

A variety of different agricultural land uses can promote scenic beauty, with positive effects on the economy of local communities. For example, it is known that aesthetic properties are associated with heterogeneity in the landscape (Stein et al. 1999). Entire communities in the Tuscany region, in Italy, benefit from a rural tourism economy that is based on the diversity of agricultural patches ranging from vineyards, wheat fields, pasture lands, and orchards to olive tree cultivations. Similarly, the Montado, in the Alentejo region of southern Portugal, is a highly diverse agricultural landscape. Cork and helm oaks are grown in varying densities, combined with a rotation of crops, fallows, and pastures, providing natural, scenic, and recreational value (Pinto-Correia 2000). Another example of an agriculturally rich region is the Pinar del Rio province in Cuba, where a healthy agrotourism industry relies on different natural attractions interspersed in a mosaic of agricultural lands, integrating tobacco fields, sugarcane cultivations, and fruit trees (Honey 1999). Various European countries and states in the United States have policies to preserve the traditional character of agricultural landscapes. For example, Switzerland subsidizes farmers in mountain areas to maintain a mix of agricultural and natural land covers because of the recreational value of these heterogeneous systems (McNeely and Sherr 2003). Conservation organizations such as the Land Trust in the United States often use the purchase of development rights as a way to maintain the rural, multiuse character of agricultural landscapes, which is perceived as a source of recreational activities and cultural enjoyment.

Agricultural biodiversity is a crucial source of nonmaterial well-being that derives from nutrition traditions, dietary diversity, and longstanding knowledge (chapter 15). Plant and animal diversity in small-scale farming often can serve the purpose of personal enjoyment or the fulfillment of family or clan tradition or may meet spiritual needs. For example, the variety of domesticated plants and livestock breeds in various regions of the world have provided raw materials for artistic expression in textiles and other crafts for centuries. As another example, home gardens are cultivated not only for food production but also with ornamental and aesthetic values in mind (Kumar and Nair 2004).

Economic Considerations

A comprehensive assessment of the value of landscape agricultural diversity for recreational purposes has not been conducted. However, data sources abound for recreational expenditures in regions that comprise a variety of agricultural land uses (e.g., Fleischer and Tsur 2000). Alternatively, the value of agricultural landscape heterogeneity might be assessed by surveys to estimate the economic value that visitors would place on the maintenance of the landscape. For example, Drake (1992) found that Swedish citizens were willing to pay US$130/ha each year to preserve agricultural land against conversion into forest, a value that was higher than the return from agricultural production in most regions of Sweden.

Whereas ecologists have identified measures of ecosystem functions (such as biomass for primary productivity or mineralization rates for nitrogen cycling), there are no corresponding quantities that can be used as measures of social function related to agricultural diversity. In many rural societies the cultural value of certain plant species resides beyond any notion of monetary measure. It may be argued that intrinsic values for these plant uses cannot be measured. These are cases in which monetary valuations of biodiversity services may be inappropriate. Alternative valuation methods that are relevant to policy and decision making must be developed for these kinds of contributions. An initial step in this direction is represented by a recent study assessing the historical and cultural value of livestock diversity in Italy (Gandini and Villa 2003). The authors qualitatively evaluated nine local cattle breeds based on their value to folklore, gastronomy, handicrafts, and the maintenance of local traditions.

Conclusion

The services that agricultural biodiversity provides are critical to the functioning of food support systems. They contribute to human welfare, both directly and indirectly, and therefore represent part of the total economic value of the planet.

There is a general agreement that the management of agricultural biodiversity can provide ways to increase food production while beneficially affecting other ecosystem services. Multifunctional and sustainable agriculture are expected to produce higher flows of ecosystem services, but the extent of these contributions and their economic value has yet to be quantified.

The positive results from studies of multifunctional agricultural systems often are overlooked because these results normally are achieved at a small scale and are difficult to document. Nonetheless, small-scale farming is the predominant form of farming in many regions of the world and is projected to remain so in marginal areas where little investment in new agricultural technologies is expected to occur (Wood et al. 2000). Identifying alternative experimental models may be crucial if more conclusive understanding of the relationship between agrobiodiversity and ecosystem functions and services is to be achieved. For example, it is well understood that large-scale experiments in agriculture (involving hundred of small farmers) might take place only as a result of a strong political will and where economic benefits for the farmers involved are clearly prospected, as in the case of the use of mixed rice varieties in the Yunnan province of China (Zhu and colleagues, chapter 12).

Often, however, the benefits to small farmers of experimenting or adopting new practices to maintain agrobiodiversity on their land might not be immediately available or apparent. This is especially the case for the values of agrobiodiversity that are not directly traceable in the marketplace (chapter 16). These include the insurance value against risk and uncertainty, the value of supporting relevant ecosystem services, and the cultural and aesthetic functions. A full assessment of these values (that includes monetary as well as ecological evaluations) is key to encouraging decision makers to invest in programs for the active protection and maintenance of agrobiodiversity. In particular, economic valuations of nonmarket benefits of agrobiodiversity can be used to identify incentives for farmers to adopt innovative cultivation methods that might be beneficial for agrobiodiversity but might not be economically viable.

In general, current valuation methods must be supported by a better understanding of the relationships between agrobiodiversity and ecosystem functions and by the identification of the functions that are irreplaceable.

Recent developments in the field of ecosystem service valuation show geographic information system–based spatial representation of valuation data as a valuable visualization tool to facilitate management planning and to identify target areas for conservation. For example, in a study commissioned by the Audubon society in Massachusetts, researchers M. Wilson and A. Troy were able to visualize nonmarket values of ecosystem services at the watershed level (Breunig 2003).

So far, valuation studies conducted at a regional scale do not differentiate between the various agricultural land uses, making it difficult to assess the economic value of ecosystem services provided by agricultural ecosystems at the larger scale.

Whenever used to inform and redesign policy, economic valuation studies of agrobiodiversity should be regarded as indicative estimates, recognizing the uncertainties about the actual contributions of diversity at various levels of ecological organization.

References

Aarssen, L. W. 1997. High productivity in grassland ecosystems: Effected by species diversity or productive species? *Oikos* 80:183–184.

Alston, J. M., G. W. Norton, and P. G. Pardey. 1998. *Science Under Scarcity: Principles and Practice for Agricultural Research Evaluation and Priority Setting.* Wallingford, UK: CAB International.

Altieri, M. A. 1991. How best can we use biodiversity in agroecosystems? *Outlook on Agriculture* 20:15–23.

Andow, D. A. 1991. Vegetational diversity and arthropod population response. *Annual Review of Entomology* 36:561–586.

Armbrecht, I. and I. Perfecto. 2003. Litter-twig dwelling ant species richness and predation potential within a forest fragment and neighboring coffee plantations of contrasting habitat quality in Mexico. *Agriculture, Ecosystems and Environment* 97:107–115.

Benton, T. G., J. A. Vickery, and J. D. Wilson. 2003. Farmland biodiversity: Is habitat heterogeneity the key? *Trends in Ecology and Evolution* 18:182–188.

Birol, E. 2004. *Valuing Agricultural Biodiversity on Home Gardens in Hungary: An Application of Stated and Revealed Preference Methods.* PhD thesis, University College London, University of London.

Birol, E., M. Smale, and Á. Gyovai. 2004. *Agri-environmental Policies in a Transitional Economy: The Value of Agricultural Biodiversity in Hungarian Home Gardens.* Environment and Production Technology Division Discussion Paper No. 117. Washington, DC: International Food Policy Research Institute.

Bommarco, R. 1998. Reproduction and energy reserves of a predatory carabid beetle relative to agroecosystem complexity. *Ecological Applications* 8:846–853.

Bommarco, R. and J. E. Banks. 2003. Scale as modifier in vegetation diversity experiments: Effects on herbivores and predators. *Oikos* 102:440–448.

Breunig, K. 2003. *Losing Ground: At What Cost? Changes in Land Uses and Their Impact on Habitat, Biodiversity, and Ecosystem Services in Massachusetts.* Lincoln: Massachusetts Audubon Summary Report.

Bullock, J. M., R. F. Pywell, M. J. W. Burke, and K. J. Walker. 2001. Restoration of biodiversity enhances agricultural production. *Ecology Letters* 4:185–189.

Burd, M. 1994. Bateman's principle and plant reproduction: The role of pollen limitation in fruit and seed set. *Botanical Review* 60:81–109.

Cane, J. H. and V. J. Tepedino. 2001. Causes and extent of declines among native North American invertebrate pollinators: Detection, evidence, and consequences. *Conservation Ecology* 5:1. Available at www.consecol.org/vol5/iss1/art1.

Chapin, F. S. III, E. S. Zavaleta, V. T. Eviner, R. L. Naylor, P. M. Vitousek, H. L. Reynolds, D. U. Hooper, S. Lavorel, O. E. Sala, S. E. Hobbie, M. C. Mack, and S. Diaz. 2000. Consequences of changing biodiversity. *Nature* 405:234–242.

Conway, G. 1987. The properties of agroecosystems. *Agricultural Systems* 24:95–117.

Costanza, R., R. d'Arge, R. de Groot, S. Farber, M. Grasso, B. Hannon, S. Naeem, K. Limburg, J. Paruelo, R. V. O'Neill, R. Raskin, P. Sutton, and M. Van den Belt. 1997. The value of the world's ecosystem services and natural capital. *Nature* 387:253–260.

Cragg, R. G. and R. D. Bardgett. 2001. How changes in soil faunal diversity and composition within a trophic group influence decomposition processes. *Soil Biology and Biochemistry* 33:2073–2081.

Daily, G. C., G. Ceballos, J. Pacheco, G. Suzan, and A. Sanchez-Azofeifa. 2003. Countryside biogeography of neotropical mammals: Conservation opportunities in agricultural landscapes of Costa Rica. *Conservation Biology* 17:1814–1826.

Daily, G. C., P. E. Ehrlich, and G. A. Sanchez-Azofeifa. 2001. Countryside biogeography: Use of human-dominated habitats by the avifauna of southern Costa Rica. *Ecological Applications* 11:1–13.

Di Falco, S. and C. Perrings. 2003. Crop genetic diversity, productivity and stability of agroecosystems. A theoretical and empirical investigation. *Scottish Journal of Political Economy* 50:207–216.

Drake, L. 1992. The non-market value of the Swedish agricultural landscape. *European Review of Agricultural Economics* 19:351–364.

Drucker, A., V. Gomez, and S. Anderson. 2001. The economic valuation of farm animal genetic resources: A survey of available methods. *Ecological Economics* 36:1–18.

Ehrlich, P. R. and E. O. Wilson. 1991. Biodiversity studies: Science and policy. *Science* 253:758–762.

Evenson, R. E. and D. Gollin. 2003. Assessing the impact of the Green Revolution, 1960 to 2000. *Science* 300:758–762.

Eviner, V. T. and F. S. Chapin III. 2003. Biogeochemical interactions and biodiversity. In J. M. Melillo, C. B. Field, and B. Moldan, eds., *Interactions of the Major Biogeochemical Cycles. Global Change and Human Impacts*, 151–173. Washington, DC: Island Press.

Fleischer, A. and Y. Tsur. 2000. Measuring the recreational value of agricultural landscape. *European Review of Agricultural Economics* 27:385–398.

Folke, C., J. Colding, and F. Berkes. 2003. Synthesis: Building resilience and adaptive capacity in social-ecological systems. In F. Berkes, J. Colding, and C. Folke, eds., *Navigating Social-Ecological Systems: Building Resilience for Complexity and Change*, 352–387. Cambridge, UK: Cambridge University Press.

Freckman, D. W. and R. A. Virginia. 1997. Low-diversity Antarctic soil nematode communities: Distribution and response to disturbance. *Ecology* 78:363–369.

Gandini, G. C. and E. Villa. 2003. Analysis of the cultural value of local livestock breeds: A methodology. *Journal of Animal Breeding and Genetics* 120:1–11.

Gitay, H., J. B. Wilson, and W. G. Lee. 1996. Species redundancy: A redundant concept? *Journal of Ecology* 84:121–124.

Griffiths, B. S., K. Ritz, R. D. Bardgett, R. Cook, S. Christensen, F. Ekelund, S. J. Sorensen, E. Baath, J. Bloem, P. C. de Ruiter, J. Dolfing, and B. Nicolardot. 2000. Ecosystem response of pasture soil communities to fumigation-induced microbial diversity reductions: An examination of the biodiversity–ecosystem function relationship. *Oikos* 90:279–294.

Hector, A., B. Schmid, C. Beierkuhnlein, M. C. Caldiera, M. Diemer, P. G. Dimitrakopoulos, J. A. Finn, H. Freitas, P. S. Giller, J. Good, R. Harris, P. Higberg, K. Huss-Danell, J. Joshi, A. Jumpponen, C. Korner, P. W. Leadly, M. Loreau, A. Minns, C. P. H. Mulder, G. O. O'Donovan, S. J. Otway, J. S. Pereira, A. Prinz, D. J. Read, M. Scherer-Lorenzen, E.-D. Schulze, A.-S. Siamantziouras, D. E. M. Spehn, A. C. Terry, A. Y. Troumbis, F. I. Woodward, S. Yachi, and J. H. Lawton. 1999. Plant diversity and productivity experiments in European grasslands. *Science* 286:1123–1127.

Heisey, P., M. Smale, D. Byerlee, and E. Souza. 1997. Wheat rusts and the cost of genetic diversity in the Punjab of Pakistan. *American Journal of Agricultural Economics* 79:727–737.

Heywood, V. H. 1995. *Global Biodiversity Assessment*. United Nations Environmental Programme (UNEP). Cambridge, UK: Cambridge University Press.

Holling, C. S. 1973. Resilience and stability of ecological systems. *Annual Review of Ecology and Systematics* 4:1–23.

Holling, C. S., D. S. Schindler, B. W. Walker, and J. Roughgarden. 1995. Biodiversity in the functioning of ecosystems: An ecological synthesis. In C. Perrings, K.-G. Mäler, C. Folke, C. S. Holling, and B.-O. Jansson, eds., *Biodiversity Loss: Economic and Ecological Issues*, 44–83. Cambridge, UK: Cambridge University Press.

Honey, M. 1999. *Ecotourism and Sustainable Development: Who Owns the Paradise?* Washington, DC: Island Press.

Hooper, D. U. and P. M. Vitousek. 1997. The effects of plant composition and diversity on ecosystem processes. *Science* 277:1302–1305.

Huston, M. A. 1997. Hidden treatments in ecological experiments: Re-evaluating the ecosystem function of biodiversity. *Oecologia* 110:449–460.

Jones, C. G., J. H. Lawton, and M. Shachak. 1994. Organisms as ecosystem engineers. *Oikos* 69:373–386.

Kinzig, A. P., D. Pimentel, and D. Tilman, eds. 2002. *The Functional Consequences of Biodiversity. Empirical Progress and Theoretical Extensions.* Princeton, NJ: Princeton University Press.

Koo, B., P. G. Pardey, and B. D. Wright. 2004. *Saving Seeds: The Economics of Conserving Crop Genetic Resources Ex Situ in the Future Harvest Centres of the CGIAR.* Wallingford, UK: CABI Publishing.

Kremen, C. and T. Ricketts. 2000. Global perspectives on pollination disruptions. *Conservation Biology* 14:1226–1228.

Kremen, C., N. M. Williams, and R. W. Thorp. 2002. Crop pollination from native bees at risk from agricultural intensification. *Proceedings of the National Academy of Sciences* 99:16812–16816.

Kumar, B. M. and P. K. R. Nair. 2004. The enigma of tropical homegardens. *Agroforestry Systems* 61:135–152.

Loreau, M., S. Naeem, and P. Inchausti, eds. 2002. *Biodiversity and Ecosystem Functioning: Synthesis and Perspectives.* Oxford, UK: Oxford University Press.

Loreau, M., S. Naeem, P. Inchausti, J. Bengtsson, J. P. Grime, A. Hector, D. U. Hooper, M. A. Huston, D. Raffaelli, B. Schmid, D. Tilman, and D. A. Wardle. 2001. Ecology—Biodiversity and ecosystem functioning: Current knowledge and future challenges. *Science* 294:804–808.

Mäder, P., A. Fließbach, D. Dubois, L. Gunst, P. Fried, and U. Niggli. 2002. Soil fertility and biodiversity in organic farming. *Science* 296:1694–1697.

McNeely, J. A. and S. J. Scherr. 2003. *Ecoagriculture: Strategies to Feed the World and Save Biodiversity.* Washington, DC: Island Press.

Meng, E. C. H., M. Smale, S. Rozelle, H. Ruifa, and J. Huang. 2003. Wheat genetic diversity in China: Measurement and cost. In S. Rozelle and D. A. Sumner, eds., *Agricultural Trade and Policy in China: Issues, Analysis and Implications.* Burlington, VT: Ashgate.

Michon, G. and H. de Foresta. 1990. Complex agroforestry systems and the conservation of biological diversity. Agroforests in Indonesia: The link between two worlds. In *Proceedings of the International Conference on Tropical Biodiversity,* Kuala Lumpur, Malaysia. Kuala Lumpur: United Selangor Press.

Mikola, J. and H. Setälä. 1998. No evidence of trophic cascades in experimental microbial-based soil food web. *Ecology* 79:153–164.

Milestad, R. and S. Hadatsch. 2003. Organic farming and social-ecological resilience: The alpine valleys of Sölktäler, Austria. *Conservation Ecology* 8:3. Available at www.consecol.org/vol8/iss1/art3.

Montoya, J. M., M. A. Rodriguez, and B. A. Hawkins. 2003. Food web complexity and higher-level ecosystem services. *Ecology Letters* 6:587–593.

Mulder, C. P. H., J. Koricheva, K. Huss-Danell, P. Högberg, and J. Joshi. 1999. Insects affect relationships between plant species richness and ecosystem processes. *Ecology Letters* 2:237–246.

Nabhan, G. P. and S. Buchmann. 1997. Services provided by pollinators. In G. C. Daily, ed., *Nature's Services. Societal Dependence on Natural Ecosystems*, 133–150. Washington, DC: Island Press.

Naeem, S. 1998. Species redundancy and ecosystem reliability. *Conservation Biology* 12:39–45.

Naeem, S. 2001. Experimental validity and ecological scale as tools for evaluating research programs. In R. H. Gardner, W. M. Kemp, V. S. Kennedy, and J. E. Petersen, eds., *Scaling Relationships in Experimental Ecology*, 223–250. New York: Columbia University Press.

Naeem, S. 2002. Ecosystem consequences of biodiversity loss: The evolution of a paradigm. *Ecology* 83: 1537–1552.

Naeem, S. and S. Li. 1997. Biodiversity enhances ecosystem reliability. *Nature* 390:507–509.

Naeem, S., D. Hahn, and G. Shuurman. 2000. Producer–decomposer codependency modulates biodiversity effects. *Nature* 403:762–764.

Naeem, S., L. J. Thompson, S. P. Lawler, J. H. Lawton, and R. M. Woodfin. 1994. Declining biodiversity can alter the performance of ecosystems. *Nature* 368:734–737.

Naeem, S., L. J. Thompson, S. P. Lawler, J. H. Lawton, and R. M. Woodfin. 1995. Empirical evidence that declining species diversity may alter the performance of terrestrial ecosystems. *Philosophical Transactions of the Royal Society of London Series B: Biological Sciences* 347:249–262.

Nunes, P. A. L. D. and J. C. J. M. van den Bergh. 2001. Economic valuation of biodiversity: Sense or nonsense? *Ecological Economics* 39:203–222.

OECD (Organization for Economic Cooperation and Development). 2002. *Handbook of Biodiversity Valuation. A Guide for Policy Makers*. Paris: OECD.

Orians, G. H., G. M. Brown, W. E. Kunin, and J. E. Swierzbinski, eds. 1990. *Preservation and Valuation of Biological Resources*, 203–226. Seattle: University of Washington Press.

Ostman. O., B. Ekbom, J. Bengtsson, and A. C. Weibull. 2001. Landscape complexity and farming practice influence the condition of polyphagous carabid beetles. *Ecological Applications* 11:480–488.

Pacala, S. W. and D. Tilman. 2002. The transition from sampling to complementarity. In A. P. Kinzig, D. Pimentel, and D. Tilman, eds., *The Functional Consequences of Biodiversity. Empirical Progress and Theoretical Extensions*, 151–166. Princeton, NJ: Princeton University Press.

Pain, D. J. and M. W. Pienkowski. 1997. *Farming and Birds in Europe: The Common Agricultural Policy and Its Implications for Bird Conservation.* Cambridge, UK: Academic Press.

Paine, R. T. 1966. Food web complexity and species diversity. *American Naturalist* 100:65–75.

Perfecto, I., J. Vandermeer, P. Hanson, and V. Cartin. 1997. Arthropod biodiversity loss and the transformation of a tropical agro-ecosystem. *Biodiversity and Conservation* 6:935–945.

Perrings, C. 1995. Biodiversity conservation as insurance. In T. Swanson, ed., *Economics and Ecology of Biodiversity Decline,* 69–77. Cambridge, UK: Cambridge University Press.

Petchey, O. L., P. J. Morin, F. D. Hulot, M. Loreau, J. McGrady-Steed, and S. Naeem. 2002. Contributions of aquatic model systems to our understanding of biodiversity and ecosystem functioning. In M. Loreau, S. Naeem, and P. Inchausti, eds., *Biodiversity and Ecosystem Functioning: Synthesis and Perspectives,* 127–138. Oxford, UK: Oxford University Press.

Pinto-Correia, T. 2000. Future development in Portuguese rural areas: How to manage agricultural support for landscape conservation? *Landscape and Urban Planning* 50:95–106.

Pretty, J. and H. Ward. 2001. Social capital and the environment. *World Development* 29:209–227.

Ricketts, T. H., G. C. Daily, P. R. Ehrlich, and J. P. Fay. 2001. Countryside biogeography of moths in a fragmented landscape: Biodiversity in native and agricultural habitats. *Conservation Biology* 15:378–388.

Robinson, R. A., J. D. Wilson, and H. Q. P. Crick. 2001. The importance of arable habitat for farmland birds in grassland landscapes. *Journal of Applied Ecology* 38:1059–1069.

Schmid, B., J. Joshi, and F. Schläpfer. 2002. Empirical evidence for biodiversity–ecosystem functioning relationships. In A. P. Kinzig, S. W. Pacala, and D. Tilman, eds., *Functional Consequences of Biodiversity. Empirical Progress and Theoretical Extensions,* 120–150. Princeton, NJ: Princeton University Press.

Scoones, I. 1999. New ecology and the social sciences: What prospects for a fruitful engagement? *Annual Review of Anthropology* 28:479–507.

Smale, M., ed. 2005. *Valuing Crop Biodiversity: On-Farm Genetic Resources and Economic Change.* Wallingford, UK: CABI Publishing.

Smale, M., J. Hartell, P. W. Heisey, and B. Senauer. 1998. The contribution of genetic resources and diversity to wheat production in the Punjab of Pakistan. *American Journal of Agricultural Economics* 80:482–493.

Söderström, B., S. Kiema, and R. S. Reid. 2003. Intensified agricultural land-use and bird conservation in Burkina Faso. *Agriculture, Ecosystems and Environment* 99:113–124.

Southwick, E. E., and L. Southwick. 1992. Estimating the economic value of honeybees (Hymenoptera, Apidae) as agricultural pollinators in the United States. *Journal of Economic Entomology* 85:621–633.

Steffan-Dewenter, I. 2003. Importance of habitat area and landscape context for species richness of bees and wasps in fragmented orchard meadows. *Conservation Biology* 17:1036–1044.

Steffan-Dewenter, I., U. Munzenberg, C. Burger, C. Thies, and T. Tscharntke. 2002. Scale-dependent effects of landscape context on three pollinator guilds. *Ecology* 83:1421–1432.

Stein, T. V., D. H. Anderson, and T. Kelly. 1999. Using stakeholders' values to apply ecosystem management in an upper midwest landscape. *Environmental Management* 24:399–413.

Tews, J., U. Brose, V. Grimm, K. Tielborger, M. C. Wichmann, M. Schwager, and F. Jeltsch. 2004. Animal species diversity driven by habitat heterogeneity/diversity: The importance of keystone structures. *Journal of Biogeography* 31:79–92.

Thies, C. and T. Tscharntke. 1999. Landscape structure and biological control in agroecosystems. *Science* 285:893–895.

Thrupp, L. A. 1997. *Linking Biodiversity and Agriculture: Challenges and Opportunities for Sustainable Food Security*. Washington, DC: World Resources Institute.

Tilman, D., J. Knops, D. Wedin, and P. Reich. 1997a. The influence of functional diversity and composition on ecosystem processes. *Science* 277:1300–1302.

Tilman, D., J. Knops, D. Wedin, and P. Reich. 2002. Experimental and observational studies of diversity, productivity, and stability. In A. P. Kinzig, D. Pimentel, and D. Tilman, eds., *The Functional Consequences of Biodiversity. Empirical Progress and Theoretical Extensions*, 42–70. Princeton, NJ: Princeton University Press.

Tilman, D., C. L. Lehman, and K. T. Thompson. 1997b. Plant diversity and ecosystem productivity: Theoretical considerations. *Proceedings of the National Academy of Science* 94:1857–1861.

Tilman, D., D. Wedin, and J. Knops. 1996. Productivity and sustainability influenced by biodiversity in grassland ecosystems. *Nature* 379:718–720.

Tonhasca, A. and B. R. Stinner. 1991. Effects of strip intercropping and no-tillage on some pests and beneficial invertebrates of corn in Ohio. *Environmental Entomology* 20:1251–1258.

Turner R. K., J. Paavola, P. Cooper, S. Farber, V. Jessamy, and S. Georgiou. 2003. Valuing nature: Lessons learned and future research directions. *Ecological Economics* 46:493–510.

Walker, B. H. 1992. Biological diversity and ecological redundancy. *Conservation Biology* 6:18–23.

Walker, B., S. Carpenter, A. Anderies, N. Abel, C. Cumming, M. Janssen, L. Lebel, J. Norberg, G. D. Peterson, and R. Pritchard. 2002. Resilience management in social-ecological systems: A working hypothesis for a participatory approach. *Conservation Ecology* 6:14. Available at www.consecol.org/vol6/iss1/art14.

Walker, D. J. and D. L. Young. 1986. The effect of technical progress erosion damage and economic incentives for soil conservation. *Land Economics* 62:83–93.

Wardle, D. A., O. Zackrisson, G. Hörnberg, and C. Gallet. 1997. The influence of island area on ecosystem properties. *Science* 277:1296–1299.

Widawsky, D. and S. Rozelle. 1998. Varietal diversity and yield variability in Chinese rice production. In M. Smale, ed., *Farmers, Gene Banks, and Crop Breeding*, 159–187. Boston: Kluwer.

Wilson, E. O., ed. 1988. *Biodiversity*. Washington, DC: National Academy Press.

Wood, S., K. Sebastian, and S. J. Scherr. 2000. *Pilot Analysis of Global Ecosystems: Agroecosystems*. Washington, DC: International Food Policy Research Institute and World Resources Institute.

Wright, J. P., C. G. Jones, and A. S. Flecker. 2002. An ecosystem engineer, the beaver, increases species richness at the landscape scale. *Oecologia* 132:96–101.

Yachi, S. and M. Loreau. 1999. Biodiversity and ecosystem functioning in a fluctuating environment: The insurance hypothesis. *Proceedings of the National Academy of Science* 96:1463–1468.

Zhu, Y. Y., H. R. Chen, J. H. Fan, Y. Y. Wang, Y. Li, J. B. Chen, J. X. Fan, S. S. Yang, L. P. Hu, H. Leung, T. W. Mew, P. S. Teng, Z. H. Wang, and C. C. Mundt. 2002. Genetic diversity and disease control in rice. *Nature* 406:718–722.

Index

Abiotic stress, 28*f*, 29
Accessions, agromorphologically classified, 52*t*, 53
Afghanistan, phylogeny of wild barley, 17
Africa: banana leaf spot, 302–04*b*; banana mutation, 94; beans, floury leaf spot, 306; beans, resistance to angular leaf spot, 304, 305–06*b*, 306; cacao export industry creation, 349; cassava mosaic virus, 294; castration of livestock, 160*b*; cataracts, 395; cattle, Boran, 153–55*b*; cattle, replacement by camels, 164–65; cattle, study of genetic influences, 130; cattle, trypanotolerant, 151*b*, 438; cattle biodiversity management, 157–60*b*; cattle breed improvement program goals, 437; cattle resource optimization, 134; crop failures, 365–72, 367–68*b*, 369*b*, 371*t*; diet, leafy vegetables in, 395; dietary diversity's benefits, 383; dietary energy overconsumption, 391; farmers' decision making, 374*b*; farmers' predictions of rain, 375–76*bt*; FCA method use, 58; food available for consumption, 388, 388*f*, 391, 392*f*; Guinea, 354; honeybees and beekeeping, 204–05; land use stages, 344, 345–47*b*; leafy vegetables in traditional diet, 395; loss of AnGRs, 428; mammalian and avian breeds, risk status, 161; pollinator habitat destruction, 209; pollinators, wild habitat for, 210–11*b*;

rice, weeds, and generalist herbivores, 278; termites, 257; tsetse belt drift, 162; *See also* Burkina Faso; Ethiopia; Ghana; Kenya; Mali; Morocco; Nigeria; Uganda
Africanized bees (killer bees), 205, 212
Aggregate demand and supply method (AnGR conservation), 432*t*
Aggregated productivity model (AnGR conservation), 438–39
Agricultural biodiversity. *See* Biodiversity in agricultural ecosystems
Agricultural ecosystems. *See* Ecosystem engineering; Ecosystem functioning
Agricultural ecosystem services, 6–8, 447
Agricultural intensification, 234–35, 238*f*, 446; *See also* Monocultures
Agricultural technology, degradation resulting from, 340
Agrodiversity, 6, 347, 457; *See also* Biodiversity in agricultural ecosystems
Agroforestry: *banana emcapoeirada* system, 370–71, 371*t*; China, reforestation system, 368–69, 369*b*, 377–78; land use stages and field types, 346*b*; species richness in patches, 460, 461*f*; swidden-fallow systems, 351–54, 355–56*b*, 356–57
Alemayehu, F., 308
Alfalfa: farmers' names for, 48–49; outbreeding, 55; yield losses, 458
Algae biome's average annual global value, 448*t*

Allard, A. W., 299
Allogenic engineers, 229*b*
Amazon floodplain, 353
Amphibians: consumption frequency in Vietnam, 191*b*; number of species collected in Asia, 184*t*; uses of various aquatic organisms, 187*t*
Andes: castration of livestock, 160*b*; faba beans, resistance to chocolate spot, 296–97
AnGR (Animal Genetic Resources): accelerating losses, 426, 429; changes and threats to, 161–62; common groupings, 145; global program for comprehensive management, 119; information systems, 123–24, 125–27*b*; valuation methods, 156; *See also* Livestock genetic diversity; Livestock, economic analysis for conservation and sustainable use
AnGR management, indigenous: biodiversity management in Nigeria, 157–60*b*; breed maintenance, 160–62; castration of livestock, 160*b*; cattle, ratings of Gambian breeds, 152*b*; cattle, replacement by camels, 164–65*b*; cattle, value of Ethiopian Boran, 153–55*b*; change and threats to AnGRs, 162–63; conclusions and challenges, 170–71, 173–74; dynamicism and influences, 150; environmental changes, 163–64; participatory methods to assess breeding objectives, 151–52*b*; reasons for keeping species and breeds, 150, 152, 155–56; risk status of mammalian and avian breeds, 162*t*; sheep, genetic improvement project, 166–68*b*; social and economic changes, 166, 169–70
Animal welfare movements, 147
Antarctica, nematode communities, 457
Anthophiles, 201–02, 213–14
Anthracnose pathogens, 295–96*b*, 299, 305*b*
Anticancer agents, 384–85*t*, 386
Antioxidants, 384, 385–86*t*, 386, 389, 395
Ants, 226*t*, 228*t*, 229
Apples, 214, 215–16*b*, 216*b*
Aquatic biodiversity in rice-based ecosystems: conclusions and challenges, 196; consumption frequency of fish in Vietnam, 191*b*; consumption frequency per household, 193*b*; contribution to food security, 195*b*; ecological

functions, 184, 186–88; fertilizers and pesticides as threats, 183; fisheries to aquaculture, 183–84; highlights of 20th Session of International Rice Commission, 189*t*; lack of information, 183; number of species collected in Asia, 184*t*; nutrition and aquatic resources, 190–95*b*; productive ecosystems under threat, 189, 195–96; recent studies, 188, 189*b*; underestimation of production, 181, 183; uses of various organisms, 187*t*
Arab horse breeders, 146
Argentina: introduced European bumblebees, 213; maize, 24*t*, 24
Arias, L., 34–68
Arthropod communities. *See* Pest management
Ascochyta blight, 297, 298*b*
Asia and Pacific: buffalo and yak, 143*f*; castration of livestock, 142, 160*b*; crop rotations, 366; declining fish catches, 189; introduced European bumblebees, 213; land use stages, 344; number of aquatic species collected, 184*t*; Philippine rice, 51, 322*b*; risk status of mammalian and avian breeds, 161*t*; rubber systems, 373; rubber tree leaf blight, 294; studies of aquatic species' capture and uses, 188; swamps, conversion to productive plots, 356; swiddens as cyclic agroforestry, 352; *See also* Bangladesh; Cambodia (Kampuchea); China; Himalayas; India; Indonesia; Laos; Malaysia; Nepal; Pakistan; Rice landraces and choices of Nepalese breeders and farmers; Thailand; Vietnam
Asian goats, 134
Asian rice, 407
Asiatic hive bee, 205, 211–12
Asses, 144*t*
Associated biodiversity, 5
Australia: conservation of wild species, 26; introduced bees, 212–13; land clearing in cereal belt, 23
Autogenic engineers, 229*b*
Avian influenza, 169
Ayalew, W., 117–35

Bacteria: ecosystem functions performed by, 228*t*; number of described species, 226*t*; *See also* Crop disease management

Bajracharya, J., 34–68
Balma, D., 34–68, 77–110
Banana: in agroforests, 346b; *banana emcapoeirada* agroforestry system managing Mokko disease, 370–71, 371t; black sigatoka fungus, 294, 302b; leaf spot, 302–04b; mutation, 94; seed flows, 95, 96b; Uganda, 36, 53, 59; within-variety variation, 80–81
Bangladesh: honeybees and beekeeping, 205; livelihood strategies from uncultivated food sources, 350b
Baobab, 393
Barley: adaptedness, 19; diversification concept, 321b; leaf rust, 308; resistance to barley yellow dwarf virus, 297; resistance to scald, 299; topsoil carbon loss/gain, 244f
Bartley, D., 181–96
Bats, 203t
Bayesian clustering, 130
Beans: in agroforests, 346b; angular leaf spot, 304, 305–06b, 306; bean fly maggots, 321b; bean rust, 295–96b; counts of named varieties, 43; diversification concept, 321b; floury leaf spot, 306; gendered seed flows, 91; Hungary, 36, 45, 47, 88t, 88; hurricane reconfiguration of crop, 100; in Mayan milpas, 69; in planted or managed forest, 345–46b; seed losses to pests, 107–08; seed transactions, 89; *See also* Milpas
Bees: burrowing, 232; diseases of, 211–13; Meliponine (stingless) bees, 205; as most valuable pollinators, 202; types of bees and commercial crops, 203t; *See also* Pollinator services
Beetles, 203t, 203–04, 459
Benin, use of FCA method, 58
Bennack, D. E., 224–61
Bilberry, 385t
Biocontrol agents, insects as, 214
Biodiversity in agricultural ecosystems: agricultural, 4; associated, 5; complementarity of elements, 271–73; controversy over causes, 270–71; cultural significance, 1; definition, 5; future of, 9–10; importance, 1; initiatives addressing, 2–4; as insurance against disasters, 456; insuring resilience, 458; at landscape level, 8; managed, 4; measuring, 14–18; multiple dimensions, 4–6; nonmarket

values not assessed, 450–51; planned, 4; producers and biomass production, 451–55; species hypothesis, 273; understanding conservation value, 377–78; value of, 463–64; wild, 4; *See also* Aquatic biodiversity in rice-based ecosystems; Diet, human health, and agrobiodiversity; Economic roles of biodiversity in agricultural ecosystems; Ecosystem functioning; Genetic diversity; Landscape-level biodiversity management
Bioindicators: pollinators as, 217–18; *See also* Indicators of biodiversity
Biological control, global contribution by biome, 448–49t
Biological diversity, 450
Biological resources, 450
Biomass production, 451–55
Biomes, 447, 448–49t
Biota. *See* Soil biota
Biotic stress, 28f, 29
Birds, seed-eating, 460
Blueberries: physiologic effect and active constituents, 385t; pollination in Canada, 206, 208–09, 218
Bolivia: maize, 24t; seed migration, 87
Boreal biome's average annual global value, 448t
Boserup, E., 365
Bottlenecks, genetic, in seed systems, 82, 83
Bouhassan, A., et al., 298b
Boundaries, edges, and hedgerows: in agroforests, 346b; biodiversity in, 357, 460; earthworms in tea plantation trenches, 257; effects of different agricultural management practices, 246f; as marginal for agricultural production, 8; usefulness overlooked, 378
Brazil: cashew pollination, 207–08b; cassava, 87; conservation agricultural practices, 251, 253; farmers' management of banana Mokko disease, 370–71, 371t; soil structure and pasture degradation, 258–59; swidden-fallow agroforestry systems, 354
Breeders (seeds, crop plants), 108, 408, 422; *See also* Rice landraces and choices of Nepalese breeders and farmers
Breeding systems and programs: for crop disease management, 307–09, 309b; evaluation method for AnGRs, 433t; pyramid breeding, 312; relationship to seed stocks, 82

Breeds, livestock, 145, 148, 150, 173–74; *See also* AnGR (Animal Genetic Resources); Livestock genetic diversity

Breeds at risk, 174

Britain: animal breeding, 146; hay production, 453, 454*f*

Broccoli, 385*t*

Brookfield, H., 338–58

Brown, A. H. D., 13–31, 77–110, 292–312

Brown, G. G., 224–61

Brussaard, L., 224–61

Buddenhagen, I. W., 297

Buffalo, 142, 143*f*, 144*t*

Bumblebees, 202, 203*t*, 213

Bunning, S., 224–61

Burdon, J. J., 307

Burgos-May, L. A., 34–68

Burkina Faso: breeding systems for various crop plants, 98*b*, 99*b*; criteria for judging breeding animals, 155–56; millet, 78, 102; soil improvement, 257–58; sorghum, 42*t*, 42–43, 77; study of farmers' decision making, 374*b*

Buzz pollinators, 202, 210*b*

Cabbage, in planted or managed forest, 345–46*b*

Cacao: in blended cash and subsistence crops, 349; destruction of pollinator habitat, 209; Ghanaian crop rotations, 365–66, 367–68*b*

Calamities in agroecosystems. *See* Disasters to ecosystems

Camacho-Villa, T. C., 34–68

Cambodia (Kampuchea): aquatic habitat for endangered species, 188; declining fish catches, 189; honeybees and beekeeping, 205; number of aquatic species collected, 184*t*; number of fish species collected, 185–86*t*; studies of aquatic species' capture and uses, 188

Camelid species, 142, 143*f*, 144*t*, 163, 164–65*b*

Canada: blueberry pollination, 206, 208–09, 218; destruction of pollinator habitat, 209; pesticide use, 206, 208–09

Capture system (rice-fish systems), 183, 196*n1*

Carbon in soil, 242–43, 244*f*

Caribbean: risk status of mammalian and avian breeds, 161; *See also* Cuba

Carpenter bees, 202, 203*t*, 210*b*

Cashew pollination in Brazil, 207–08*b*

Cassava: Brazil, 87; counts of named varieties, 43; mosaic virus, 294

Cataracts, 395

Cattle: African, 130, 134; biodiversity management in Nigeria, 157–60*b*; breed improvement program goals, 437; Bruna, 441*t*; Criollo breeds threatened, 169; criteria for judging breeding animals, 155–56; and draft animals and rare rice varieties, 418; effects of environmental changes, 162–63; Ethiopian Boran, 153–55*b*; in feedlots, 146; Holstein-Friesian, 148; Italian breeds, 463; milk yield in, 132; Murnau-Werdenfelser, 148; Muturu, 438; Namibian, 148; ownership and rice varieties, 415; pastured in fallow fields, 347*b*, 351; population share and number of breeds, 144*t*; replacement by camels in Nigeria, 164–65*b*; trypanotolerant, 151*b*, 438

Cattle Breeds: An Encyclopedia (Felius 1995), 123

CBD (Convention on Biological Diversity): Conference of the Parties, 2, 236*b*, 450; International Initiative for the Conservation and Sustainable Use of Soil Biodiversity, 235, 236*b*; livestock genetic diversity, 119

Centipedes, number of described species, 226*t*

Centro Internacional de Agricultura Tropical (CIAT), 3

Cereal crops: economic factors related to crop diversity, 420*t*, 421; studies of divergence, 18*t*

Ceroni, M., 446–64

Chavez-Servia, J. L., 34–68

Chickens, 142, 143*f*, 144*t*, 368

Chile: introduced European bumblebees, 213; maize, 24*t*

Chili peppers, gendered seed flows, 91

China: Dayi's mixed-tree plantation, 348–49; declining fish catches, 189; deterioration of rice ecosystems, 320; hand pollination of apple blooms, 216*b*; number of aquatic species collected, 184*t*; reforestation system, 368–69, 369*b*, 377–78; studies of aquatic species' capture and uses, 188; swidden-fallow agroforestry systems, 354; varietal rice mixture plantings, 321–22*b*; *See also* Rice crop variety diversification for disease control

Diversity effects: controversy over causes, 270–71; niche complementarity, 452

DNA: microarrays (DNA chips), 17; mitochondrial, 127–28; polymorphisms, 127, 323*t*, 324; slippage, 129; transgenic, 94

Domestic Animal Diversity Information System (DAD-IS), 125–26*b*, 149*b*, 435

Domestic Animal Genetic Resources Information System (DAGRIS), 126–27*b*, 435

Donkeys, 163

Draft animals, 415

Drainage, effects of agricultural management practices, 246*f*

Drucker, A. G., 426–43

Ducks, 144*t*

Durum wheat: Morocco, 47–48; sources of seed, 77

Dyck, P. L., 308

Dyer, L. A., 279

Earthworms: conversion to pasture, 258–59; disregarded in many societies, 240–41; ecosystem functions performed by, 228*t*, 229, 231; as indicator species, 243; number of described species, 226*t*; in tea plantation trenches, 257

Economic issues: economic change as threats to animal populations, 168–70; hedonic price models, 156; role of soil biodiversity, 233–34; See also Livestock, economic analysis for conservation and sustainable use; Markets

Economic roles of biodiversity in agricultural ecosystems: conclusions and challenges, 463–64; consumers and decomposers, 455–56; dependence on biodiversity, 450; diversity and resilience, 9, 456–59; existence value, 7; ignored by commercial markets, 447; importance recognized by CBD's Conference of the Parties, 450; lack of information, 451; option value, 6–7; penalties of extinctions, 446; problem of validity beyond scale of studied site, 454–55; producers and biomass production, 451–55; recreational and cultural roles, 461–63; wildlife habitat protection, 461

Economic value of biodiversity: varying expressions of, 6–8; See also Crop genetic diversity

Ecosystem engineering, 229*b*, 256, 257

Ecosystem functioning: conclusions and challenges, 8, 463–64; definition, 447; diversity and resilience in agroecosystems, 456–59; diversity of consumers and decomposers, 455–56; diversity of producers and biomass production, 451–55; habitats and landscape diversity, 459–61, 461*f*; limitations of studies, 453; measuring, 451; need for clarification, 260; recreational and cultural roles, 461–63; as research goal, 270–74

Ecosystem services, 6–8, 447

Ecosystem stress, pollinators as bioindicators for, 218

Ecotoxicology, 243

Ecuador: deployment of disease-resistant varieties, 295–96*b*; quinoa, 294

Edge-cropping. *See* Boundaries, edges, and hedgerows

Enclosure of common lands, 350*b*

Ensete, 42

Environmental changes: contamination compromising nutritional status, 389; landraces' adaptation to pathogens, 296–300; as potentially disastrous, need for resilience, 9; as threats to animal populations, 162–63

Epidemics of disease, 390–91

Erosion control, global contribution by biome, 448–49*t*

Ethiopia: barley, leaf rust, 308; economic factors related to crop diversity, 420*t*, 421; farmers' naming of varieties, 42; goats, 439; resistance to barley yellow dwarf virus, 297; sorghum, 19, 20, 30; wheat, 42

Europe: characterizing landscape-level biodiversity, 341–44; destruction of pollinator habitat, 209; EU's common agricultural policy, 340; EU's conservation support measures, 439; honeybee declines, 3; honeybee diseases, 211; horses, chickens, and geese, 142, 143*f*; landscape, percentages of natural and managed, 340; mix of land covers, 462; mosaic of habitat, 342–43; multiuse landscape, 462; risk status of mammalian and avian breeds, 161; wildlife habitat protection, 461; *See also* Hungary; Turkey

European Association for Animal Production (EAAP), 125*b*
European Farm Animal Biodiversity Information System, 126*b*
Evolution of biodiversity: as central reference point, 1; coevolutionary linkages, 25; for continued adaptation, 8; time dimension from DNA data, 16, 19; types of forces, 28*f*, 28–29
Evolution of biodiversity in seed systems: conclusions and challenges, 108–10; forces acting upon, 81; migration and selection, 91–93; migration scale, 85–87; mutation, 94–95; population size, bottlenecks, and genetic drift, 81–85; recombination, 93–94; seed and pollen exchange, 85; seed replacement and seed sources, 87–89; seed selection practices, 101–06; selection by farmers, 95, 99; selection by natural disasters and catastrophes, 99–101; storage and selection practices, 106–08; supply networks, 90–91
Evolution of pathogens, 300–301, 302–04*b*, 312; *See also* Crop disease management
Exotic breeds, 169, 174
Experimental studies: criticism of, 270–71; recommendations for, 274
Ex situ conservation: links between in situ and ex situ activities, 22*t*, 27–28
Extinctions: among mammalian and avian breeds, 2–3; breeds at risk, 174; creating through crosses, 148; lack of documentation and risk of, 156; mammalian and avian breeds at risk, 117; probabilities for closely related breeds, 133–34; threat to AnGRs, 161–62

Faba beans: change in population size, 83, 84*b*; farmers' designations for varieties, 43, 44–45*t*, 45, 46*t*, 47*f*, 51, 53; farmers' emphasis on various traits, 36, 41*t*, 41; geographic association with diversity, 65, 67; germplasm enhancement, 309*b*; resistance to chocolate spot, 296–97; seed insecurity, 100–101; seed selection practices, 106; varieties grown in seasons, 58
Fabirama, study of farmers' decision making, 374*b*
Fairhead, J., 354

Fallows, 347*b*, 347
FAO (Food and Agricultural Organization of the United Nations): Domestic Animal Diversity Information Systems, 435; Farmer Field School approach, 249; Food Balance Sheets, 388; Global Pollinator Project, 3; International Rice Commission, 189*t*; Land and Water Digital Media Series, 249; on loss of Nigerian AnGRs, 428; Programme for Community Integrated Pest Management, 245; report of substantial losses of biodiversity, 2; Soil Biodiversity Initiative, 236*b*; visual soil assessment toolbox, 245
Far East. *See* Asia and Pacific
Farm animal genetic resources (AnGRs): *See also* AnGR (Animal Genetic Resources); AnGR management, indigenous; Livestock genetic diversity
Farmer Field School approach, 249
Farmer-named varieties: agronomic traits as indicators of diversity, 43–49; alfalfa, 48–49; conclusions and challenges, 67–68; as indicators of diversity, 34–43; possibilities for crop diversity estimates, 34–35; problems of consistency and comparability, 34, 54–55; representatives of local varieties, 65, 67; spatial diversity patterns and variety names, 55–65; units of diversity (FUDs) management, 53–65; variability and genetic distinctiveness, 53; variety and genetic distinctiveness, 49–53
Farmers and pastoralists: as agrobiodiversity custodians and creators, 10; blending cash and subsistence crops, 349, 351; at center of host–pathogen–environment triangle, 293; collaboration with institutions, 28; crop disease management, 301, 302–04*b*, 304, 309–11; importance of choices made by, 1, 6, 7, 24, 338, 339, 348, 407–08, 422; interactions between components, 8–9; as landscape-level managers, 347–49, 350*b*, 351; long-term approaches, 356; mixed rice planting, training for, 334; nodal for seed exchange networks, 90; production optimization through genetic diversity, 305*b*; resilience and adapability, 9; skills undervalued, 348; soil biological management practices, 245–46, 246*f*,

Genetic distinctiveness: acceptable variability in farmers' names, 53; relationship to farmers' names for crop plant varieties, 49–53

Genetic diversity: as essential, 13; as goal of conservation, 132–33; history of farmers' appreciation and use of, 14; importance of, 30–31; links between in situ and ex situ activities, 27–28; molecular diversity, 14–18; wild species and crop relatives, 25–27; *See also* Crop genetic diversity

Genetic diversity management: conclusions and challenges, 30–31; inclusion of wild crop relatives, 25; indicators for, 20; measurement of genetic diversity as key, 30

Genetic drift: in livestock, 130; in seeds, 82–83; variation in, 132

Geneticists. *See* Breeders (seeds, crop plants)

Genetic markers, studies of divergence, 18t

Genetic production function method (AnGR breeding program valuation), 434t

Genetic resources, global contribution by biome, 448–49t

Genetic vulnerability, 313n3

Genomics, functional, 17–18

Gentry, G., 279

Ghana: cacao, corn and cassava crop failures, 365–72, 367–68b, 369b, 371t; creation of major cacao export industry, 349; farmers predicting rain from environmental signs, 375–76bt

Gibson, J. P., 117–35

GISs (geographic information systems), 23, 24, 122

Global Environment Facility (GEF), 3

Global Environment Facility Below Ground Biodiversity (BGBD) Project, 3

Globalization: accelerating loss of breeds, 428; exotic and local breeds, 169; need for knowledge, 9

Global Pollinator Project, 3

Global Strategy for the Management of Animal Genetic Resources (AnGR). *See* AnGR (Animal Genetic Resources)

Goats, 144t, 347b, 439, 441t

Goldstein, B., 274–75

Grain landraces, studies of divergence, 18t

Grasslands: biome's average annual global value, 448t, 450; maximizing biodiversity, 453

Greeks, animal breeding, 146

Green Revolution: challenging farmers' skills, 348; as HEIA (high–external input agriculture), 237; negative impact on biodiversity, 9

Ground-nesting bees, 210b

Groundnut: breeding systems in Burkina Faso, 98b; study of farmers' decision making, 374b

Grupo Vicente Guerrero for soil management, 251

Guinea, 354

Habitat destruction, threat to pollinator services, 209, 211

Habitat or refugia, global contribution by biome, 448–49t

Halwart, M., 181–96

Hanotte, O., 117–35

Harlan, J. R., 297

Hawkins, B. A., et al., 279

Hay production, 453, 454f

Health issues. *See* Diet, human health, and agrobiodiversity

Hedgerows. *See* Boundaries, edges, and hedgerows

Hedonic method (AnGR breeding program valuation), 434t, 438

Hedonic price models, 156

HEIA (high–external input agriculture), 234, 237, 238f, 241–42

Herbicides: risk of neglectful or ignorant use, 234; as threat to rice-based ecosystems, 192b

Himalayas: apples, 214, 215–16b; cliff bee declines, 3

Hodgkin, T., 13–31, 77–110

Hoffman, I., 141–75

Holstein-Friesian cattle, 148

Home gardens: becoming village plots, 69n1; cash crop seedlings in, 349; economic factors related to crop diversity, 420t, 421; as land use stage, 346b; women making decisions for, 62b

Honeybees: beekeeping, 204–05; as bioindicators, 217–18; declines, 3; habitats and scale, 459–60; as pollinators, 202, 203t

Horses, 142, 143f, 144t, 438, 441t

Hue, N. N., 34–68

Human health issues. *See* Diet, human health, and agrobiodiversity

Hungary: beans, 36, 45, 47; bean seed replacement practices, 88, 88t;

economic factors related to crop diversity, 420t, 421; informal seed systems, 91; maize, 82

Hybrid systems, 373

Ice biome's average annual global value, 449t

Imbruce, V., 292–312

Immunity and infection, 389

Imported breeds, 174

Inbreeders (selfers), 55

India: honeybees and beekeeping, 205; rice, 29, 51; tea production and soil management, 256–57; wheat rust, 310

Indicators of biodiversity: agromorphological traits, 35–43, 45, 51, 53, 54, 65; agronomic traits, 43–49; challenges of developing, 341; combination approach, 357; in developing countries, 342; distinctiveness, 17, 35, 49–53, 67; indicator species, 341, 342; landrace names as, 34; links between in situ and ex situ activities, 22t; for managing genetic diversity, 20; maps and GISs, 23; for measurement of soil biodiversity, 260; molecular, 14–18, 19, 34, 53, 54; for monitoring crop plants, 21t; physical characteristics of farm space and management, 342; PLEC project, 342–44; richness and evenness, 15, 19, 23, 24, 35, 43, 55, 65, 68; sampling, 341; security of traditional knowledge, 25; of soil quality, 241f, 242–45, 243b, 244f; standard inventories, 341; stratified sampling, 341; See also Farmer-named varieties

Indicator species, 341, 342

Indigenous breeds, 174

Indigenous peoples. See Farmers and pastoralists

Indonesia: conversion of swidden farming, 356–57; introduced Asiatic hive bee, 213; rice insect community dynamics, 277; swamps, conversion to productive plots, 356

Information systems, for livestock development, 123, 124

Inoculations of beneficial soil microorganisms, 254, 255b

Insecticides. See Pesticides/insecticides

Insects: beetles, 203t, 203–04, 459; as biocontrol agents, 214; consumption frequency in Vietnam, 191b; ecosystem functions performed by, 228t; as food

for chickens, 368; Japanese beetles, 214; mites as parasites of honeybees, 211; mosquito larvae eaten by fish, 186–87; number of species collected in Asia, 184t; root herbivorous, 226t; seed losses to pests, 107–08; tsetse fly (trypanosomiasis), 151b, 162–63; uses of various aquatic organisms, 187t; weevils, flies, and midges, 107, 203t, 204; See also Pest management; Pollinator services

In situ conservation: indicators for managing genetic diversity, 20; links between in situ and ex situ activities, 22t, 27–28

Insurance hypothesis of species diversity, 273, 456, 464

Intellectual property rights and contracts method (AnGR conservation), 433t

International Bee Research Institute, 214

International Center for Research in Agroforestry (ICRAF), 373

International Center for Tropical Agriculture, Tropical Soil Biology and Fertility Institute, 249

International Centre for Integrated Mountain Development (ICIMOD), 215b

International Livestock Research Institute (ILRI), 126–27b, 435

International Plant Genetic Resources Institute (IPGRI), 3, 20, 68, 79

International Rice Commission, 189t

Intraspecific diversity, 394

Introduced breeds, 174

Iran, phylogeny of wild barley, 17

Iraq, phylogeny of wild barley, 17

Ireland, potato famine, 294

Irrigation, effects of different agricultural management practices, 246f, 248t

Isopods, number of described species, 226t

Israel: introduced European bumblebees, 213; phylogeny of wild barley, 17

Italy: annual costs of safe minimum standards for livestock, 441t; local cattle breeds, 463; Pentro horse, 438, 441t; rural tourism, 462

Janzen, D., 209, 211

Japan, introduced European bumblebees, 213

Japanese beetles, 214

Jarvis, D. I., 1–10, 34–68, 77–110, 292–312

Johns, T., 382–99

accelerating erosion, 141; nonmarket contributions, 435–36; phenotypic characterization, 118, 119–22; population share and number of breeds, 144t; reproductive rate, 147; species domestication and distribution, 142, 143t; *See also* AnGR (Animal Genetic Resources); AnGR management, indigenous

Livestock keepers. *See* AnGR management, indigenous; Farmers and pastoralists

Livestock production systems (LPSs), 143, 145

Livestock Revolution, 141

Local breeds: genetic adaptedness, 174; kept by small farmers, 147; loss through crossbreeding, 161, 169–70; quantitative valuation difficulties, 156

Local crop cultivar genetic diversity, as advantage in reducing susceptibility to pathogens. *See* Crop disease management

Local crop varieties: agromorphological traits, 36, 47, 53; assessing importance, 58; consistency of naming, 43; distinctiveness, 35; diversity of uses, 30; as representative of diversity, 65, 67; *See also* Farmer-named varieties; Landraces

Lope, D., 34–68

Loreau-Hector equation, 271

Loss of biodiversity: genetic drift and bottlenecks, 83–85; globalization accelerating, 428; increasing actions, 3; livestock (AnGRs), 117, 161–62, 162, 428; multiple aspects, 446; recognition of increased rates of, 2–3; understanding of mechanisms as imperative, 274

Lutein, 395

Macrobiota of soil, 232

Macrofauna of soil, 226t, 256–58

Maize. *See* Corn

Malaysia: addition of pollinator habitat, 209; honeybees and beekeeping, 205; poultry relocation to remote areas, 169; use of FCA method, 59

Mali: benefits of dietary diversity, 383; sorghum, 99

Malnutrition. *See* Diet, human health, and agrobiodiversity

Managed biodiversity, 4

"Managing Biodiversity in Agricultural Ecosystems" symposium (Montreal 2001), 3

Mangrove biome's average annual global value, 449t

Mapping of landrace cultivation, 23–24

Mar, I., 34–68, 77–110

Margins of fields. *See* Boundaries, edges, and hedgerows

Marine biome's average annual global value, 448t

Markets: direct value of biological resources, 450; failure of incentives, 428–29; failure to capture economic value, 7; genetic resources not traded in, 430; ignoring ecosystem services, 447; incomplete for rice varieties, 412; price mechanisms, 427

Market share method (AnGR conservation), 433t

Masai, 396

Measuring biodiversity. *See* indicators of biodiversity

Medicinal plants, 369b

Megabiota of soil, 232

Meliponine (stingless) bees, 205

Melon, bitter, 385t

Menalled, F. D., et al,, 275

Mesobiota of soil, 232

Mesofauna of soil, 226t

Metapopulations, 79–80, 109, 339

Mexico: creole pigs, 437–38, 441t, 442t; diversity distribution and production spaces, 59, 60–62b; economic factors related to crop diversity, 420t, 421; maize, 24t, 36, 54, 78, 83, 94; Sierra de Manantlan Biosphere Reserve, 26; soil management by Grupo Vicente Guerrero, 251, 252–53b; sources of maize seed, 77, 93; *See also* Yucatán, Mexico

Microarthropods, 232

Microbes. *See* Soil biota

Microbiota of soil, 232

Microfauna of soil, 226t, 232

Microflora of soil, 232

Microorganisms, 254–56

Microorganisms of soil, 226t

Microsatellite loci, 127, 128–30

Middle Ages, animal breeding, 146

Middle East: honeybees and beekeeping, 204–05; introduced European bumblebees, 213; phylogeny of wild barley, 17

Midges, 203t

Migration: scale of, 85–87; seed-mediated, 85; and seed selection, 91–93; variation in extent, 89

Millet: Burkina Faso, 78, 98b, 99b; study of farmers' decision making, 374b

Millipedes, number of described species, 226t

Milpas, 69, 89, 91, 420t, 421

Mites, 211, 226t

Mitochondrial DNA, 127–28

Mixture planting: expanding scale, 331, 333, 333t; genetic diversity of varieties, 322–25; rice blast, genetic diversity, 325; rice blast, lodging, and yields, 331, 332t

Molecular biology, 14

Molecular diversity: considerations for genetic diversity, 14–15; functional genomics, 17–18; phylogeny and coalescence, 16–17; single nucleotide polymorphisms (SNPs), 15–16

Molecular marker diversity: in conservation decisions, 131–34; estimates of, 127–31

Mollusks: consumption frequency in Vietnam, 191b; number of described species, 226t; number of species collected in Asia, 184t; uses of various aquatic organisms, 187t

Monocultures: agricultural intensification, 238f; commercial agriculture and poor knowledge of soil biota, 241; crop disease and diversity hazard hypothesis, 292, 293; definition, 313n2; effects of different agricultural management practices, 246f; Ghanaian farmers change to agrodiversity, 366; as HEIA (high–external input agriculture), 238f; loss of uncultivated food sources, 350b; low resilience to perturbations, 446; outperformed by diversity plots, 452; outperformed by interplanted mixture farms, 321b; problem of increased weed management and pests, 278; resilience through external inputs, 457, 458

Montáñez, A., 224–61

Montreal symposium "Managing Biodiversity in Agricultural Ecosystems" (2001), 3

Moran, D., 436

Morning glory genome, 95

Morocco: durum wheat, 47–48; faba bean, 36, 41, 41t; faba bean germplasm enhancement, 309b; faba bean landraces, 20; faba beans, 43, 44–45t, 45, 46t, 47f, 51, 53, 58, 65, 67, 83, 84b; faba beans, resistance to chocolate spot, 297, 298b; seed insecurity, 100–101; seed selection practices, 106; sources of duram wheat seed, 77; varieties as percentage of area, 58; women's knowledge of seed selection, 98b

Morphological characters, studies of divergence, 18t

Moths, 107

Mozambique, use of FCA method, 59

Mulches, 246f, 257, 258

Multivariate analysis, 130

Murnau-Werdenfelser cattle, 148

Mushrooms, 368

Mutation, within seed systems, 94–95

Muturu cattle, 438

Mycorrhizae inoculation, 254–55

Namibian cattle, 148

Native breeds, 174

Native pollinators, 205–06

Natural enemy diversity and pest control, 278–82, 283–84

Natural selection, and farmer selection within seed systems, 95, 99

Near East, risk status of mammalian and avian breeds, 161

Negassa, M., 308

Neglected and underutilized species (NUSs), 393

Nematodes: in Antarctica, 457; ecosystem functions performed by, 228t, 229; number of described species, 226t

Nepal: attitude toward seed replacement, 89; as center of diversity for Asian rice, 407; crop cultivar amounts and distribution, 55, 56, 57f; informal seed exchange, 88–89; maintenance of rice seed quality, 102; maintenance of seed quality, 103–06t; rice, 22, 23t, 29, 49, 50–51f, 51, 78; rice grain yield, 63–64b; rice seed selection practices, 102, 106f; sources of rice seed, 77, 90–91; sponge gourd, 80, 81; taro, 36, 37–39t; use of FCA method, 59; See also Rice landraces and choices of Nepalese breeders and farmers

Netherlands, wildlife habitat protection, 461

New Zealand, introduced European bumblebees, 213

Niche complementarity, 452

Niche expression, 273

Nigeria: cattle, replacement by camels, 164–65b; cattle biodiversity management case study, 157–60b; loss of AnGRs, 428; pearl millet, 54, 55

Nitrogen fixation, 254, 255*b*, 368*b*

North America: blueberry pollination, 206, 208–09, 218; destruction of pollinator habitat, 209; honeybee declines, 3; pesticide use, 206, 208–09; risk status of mammalian and avian breeds, 161

No-tillage systems. *See* Tillage and no-tillage systems

Nucleotide diversity studies, 15*t*

Nutrient cycling, global contribution by biome, 448–49*t*

Nutrition: and aquatic resources, 190–95*b*; *See also* Diet, human health, and agrobiodiversity

Oats, 385*t*

Observational studies, criticism of, 271

Oca, 87

Ocean biome's average annual global value, 448*t*

Ocean shelf biome's average annual global value, 448*t*

Ochoa, J., 292–312

Oklahoma State University Breeds of Livestock, 127*b*

Okra, 98*b*, 374*b*

Olive oil, 385*t*

Operational Programme on Agricultural Biodiversity

Opportunity cost (AnGR conservation), 431*t*

Orchard bees, 202, 203*t*

Organic amendments, effects of different agricultural management practices, 246*f*, 247*t*

Organic matter management, 250–51

Organizational diversity, 347–48

Outbreeders (outcrossers), 55, 93

Outpollinators, 93

Pacific ocean nations. *See* Asia and Pacific

Padoch, C., 1–10, 338–58

Pakistan: wheat rust, 310; wheat yield losses, 458

Papua New Guinea: cash crops on subsistence farms, 349; introduced Asiatic hive bee, 213

Parasites as threat to pollinators, 211–12

Parasitoids, 284, 455

Parleviet, J. E., 308

Participatory assessment, by livestock keepers, 151–52*b*, 156

Participatory plant breeding (PPB), 22, 29, 30

Pastoralists. *See* Farmers and pastoralists

Pasture: grazing by livestock in, 347*b*; topsoil carbon loss/gain, 244*f*

Pathogens: as threat to pollinators, 211–12; *See also* Crop disease management; Pest management

Pauropoda, number of described species, 226*t*

PCR amplification for RGA analysis (rice), 323

Peanuts: resistance to rust and leaf spot, 297; seed losses to pests, 107

Pearce, D., 436

Pearl millet: Nigeria, 54, 55; seed selection practices, 102

Pedigrees for animals, 146, 150

Pentro horse, 438, 441*t*

People, Land Management and Environmental Change (PLEC) Project: field selection and sampling, 357; finding need for flexible methodology, 353; focus, 363; locations, 363; studies of diversity strategies, 342–44; undertaking studies, 3

Perturbations of ecosystems. *See* Disasters to ecosystems

Peru: economic factors related to crop diversity, 420*t*, 421; floodplain farmers and aquatic and terrestrial phase management, 353; resistance to peanut rust and leaf spot, 297; seed exchange and migration, 85–86, 86*f*, 87; seed storage, 107; sheep, genetic improvement project, 166–68*b*; villagers' adaptations to river-course changes, 371–72

Pesticides/insecticides: effects of different agricultural management practices, 246*f*, 247*t*; excessive use, 320; forcing recolonization, 277–78; impact on natural enemy diversity, 283; loss of uncultivated food sources, 350*b*; pollinators as bioindicators for, 217; risk of neglectful or ignorant use, 234; threat to pollinators, 206, 208–09; as threat to rice-based ecosystems, 183, 192*b*, 195

Pest management: alternative food supplies, 5; arthropod communities, 274–78, 276*f*; conclusions and challenges, 285; crop genetic diversity, 6; life span of herbivorous insects, 282;

Pest management (*continued*)
    as linked study of assembly and function, 274; natural, mosquito larvae eaten by fish, 186–87; natural enemy diversity and pest control, 278–82; pest occurrence, 282–84; for seed storage, 107–08; species richness, 279–80, 280f
PGRs (plant genetic resources), 382
Phenotypic characterization, of livestock, 119–22
Philippines: rice, 51; rice varietal mixtures, 322b
Phylogeny and coalescence, molecular diversity, 16–17
Pigs: commercial production and food security, 169; creole, 437–38, 441t, 442t; high-input intensive pig farming, 146, 147; landrace, 135–36n1; number of breeds in world's regions, 144t
Pinedo-Vasquez. M., 362–79
Planned biodiversity, 4
Plants, aquatic: number of species collected in Asia, 184t; uses of various organisms, 187t
PLEC project. *See* People, Land Management and Environmental Change (PLEC) Project
Plot size, 459
Pollen-mediated gene flow, 85, 93–94
Pollinator services: in agroecosystems, 202–04, 203t; alternative food supplies for, 283; apple valleys of Hindu Kush, 214, 215–16b; blueberry pollination in Canada, 206, 208–09, 218; for commercial crops, 202–04, 203t; conclusions and challenges, 219; cropping systems, sustainability and biodiversity, 214; crucial to sustaining ecosystems, 200, 201f; declines in every continent, 3, 200; description and definitions, 201–02; diversity as key, 456, 458; global contribution by biome, 448–49t; habitats and scale, 459–60; hand pollination of apple blooms, 216b; honeybees and beekeeping, 204–05; native pollinators, 205–06; need for knowledge, 200, 214, 216f; other anthophiles, 213–14; other managed pollinators, 206; pollinators as bioindicators, 217–18; protection and promotion of native pollinators, 213; reward sought by pollinators, 204; threat of competitive interactions, 212–13; threat of habitat destruction, 209, 211; threat of parasites and pathogens, 211–12; threat of pesticides, 206, 208–09; variety of anthophiles, 202; wild habitat for pollinators, 210–11b

Pollution: effects of different agricultural management practices, 246; environmental contamination compromising nutritional status, 389; from industrial growth and urban waste, 390; nitrogen and phosphorus, 147; pollinators as bioindicators for, 217–18
Population genetics, 109
Population of breeds, 175
Population size of seed stocks, 81–82
Population structure, 79–81
Portugal, multiuse landscape, 462
Potato, study of farmers' decision making, 374b
Potatoes: breeding systems in Burkina Faso, 98b; counts of named varieties, 43; economic factors related to crop diversity, 420t, 421; intraspecific diversity, 394; migration scale, 87; in planted or managed forest, 345–46b
Potato famine, Ireland, 294
Poultry, 142, 144t, 146, 147, 149b, 169
Prickly pear leaf, 385t
Processed foods, 390–91; *See also* Diet, human health, and agrobiodiversity
Production loss averted (AnGR conservation cost), 431t
Programme for Community Integrated Pest Management, 245
Protozoa, number of described species, 226t
Provisioning ecosystem services, 8
Psyllium, 385t
Pyndji, M. M., 304
Pyramid breeding, 312

Qualset, C. O., 297
Quinoa: downy mildew, 295b; Ecuador, 294

Rana, R., 77–110
Rangelands biome's average annual global value, 448t, 450
Rao, V. R., 34–68
Raw materials, global contribution by biome, 448–49t
Recombination of genes, 93–94
Recreation cultural, global contribution by biome, 448–49t

Reptiles: consumption frequency in Vietnam, 191*b*; number of species collected in Asia, 184*t*; uses of various aquatic organisms, 187*t*

Rerkasem, K., 362–79

Research challenges and development opportunities, crop genetic diversity, 28*f*, 29–30

Resilience, 9, 456–59

Resilience Alliance, 457

Resistance gene analogue (RGA) analysis (rice), 322–23, 323*t*, 324

Resistance gene sequence analysis (rice), 322

Resistance structure, crop disease, 292

Rhizobiaceae bacteria, 254

Rice: Asian, 407; brown planthopper problems, 278, 283; conversion from swidden farming to irrigated cropping, 356–57; effects of increased cropping cycles, 277; India, 29, 51; as key commodity, 183; maintenance of seed quality, 102; Nepal, 22, 23*t*, 29, 49, 50–51*f*, 51, 55, 56, 57*f*, 78; Nepal, grain yield, 63–64*b*; number and distribution of landraces, 23*t*; Philippines, 51; reduction in populations, 82; rice blast, 297, 320; rice neck blast, 329; seed selection practices, 102, 106*f*; seed supply networks, 90–91; sources of seed, 77; tungro virus, 322*b*; varietal mixtures, 322*b*; varieties and agronomic traits, 327*t*; weeds and generalist herbivores, 278; worldwide cultivation, 181, 182*f*; *See also* Aquatic biodiversity in rice-based ecosystems

Rice crop variety diversification for disease control: composition of rice blast isolates, 326; conclusions and challenges, 335; expanding scale of mixed-variety planting, 331, 333, 333*t*; genetic diversity of rice blast in mixtures, 325; genetic diversity of varieties in mixture planting, 322–25; growing procedures, 333–34; humidity and surface area effects on yield, 326–28, 328*t*; plant silica content, 328–31, 330*t*; rice blast, lodging, and yields in mixture fields, 331, 332*t*; training, 334; varietal combinations, 333

Rice-fish systems, 183; *See also* Aquatic biodiversity in rice-based ecosystems

Rice landraces and choices of Nepalese breeders and farmers: conceptual

approach, 411–12; conclusions and challenges, 422–23; descriptive statistics, 416, 417*t*; econometric methods, 413, 413*t*, 414*t*; econometric results, 416–19, 418*t*; farm physical characteristics, 415; findings from related studies, 419, 420*t*, 421; household characteristics, 413, 415; key informant survey of rice breeders, 411; market characteristics, 415–16; overview an evolutionary forces, 408–09; policy trade-offs, 408–09, 422; sample survey of rice-growing households, 410–11; study sites, 409–10, 410*t*, 422

Rijal, D. K., 34–68

River biome's average annual global value, 449*t*

Rock biome's average annual global value, 449*t*

Romans, animal breeding, 146

Rotation of crops: adaptation by Chinese smallholders, 368; in agricultural intensification, 238*f*; constraints, effects on biota and function, 248*t*; for crop disease management, 310–11; effects of different agricultural management practices, 246*f*, 248*t*; in Ghanaian villages, 366, 367–68*b*; by Hmong farmers, 366; improving soil biological activity, 253, 256; increases in soil species diversity, 246*f*; of local crop cultivars, 311; in planted or managed forest, 345–46*b*; Thailand, 366

Rubber leaf blight, 294

Rubber systems, 373

Sadiki, M., 34–68, 77–110, 292–312

Safe minimum standard valuation method (AnGR conservation), 432*t*, 439–40, 441*t*

Sample complementarity, 27

Sampling effects, 274, 452

Sawadogo, M., 34–68, 77–110

Seagrass biome's average annual global value, 448*t*

Seed lots, 79, 80

Seed-mediated migration, 85

Seed systems: conclusions and challenges, 108–10; disasters and catastrophes, 99–101; diversity comparisons, landraces and frequencies, 84*b*; dynamism in, 85; farmer selection and natural selection, 95, 99; farmers'

Seed systems (*continued*)
informal sources, 77–78; features, 78–79; importance to crop genetic diversity, 78; migration, seed and pollen exchange, 85; migration and selection, 91–93; mutation, 94–95; on-farm seed storage and selection, 106–08; overview of evolutionary forces, 81; population size, bottlenecks, and genetic drift, 81–85; population structure and breeding systems, 79–81; recombination, 93–94; scale of migration, 85–87; seed lots, 79, 80; seed replacement and sources, 87–89; seed selection practices, 101–06; studies of, 78; supply networks, 90–91

Selfers (inbreeders), 55

Self-pollinators, 55, 80, 93

Senegal, food available for consumption, 391, 392*f*

Sheep, 144*t*, 166–68*b*, 441*t*

Sierra de Manantlan Biosphere Reserve (Mexico), 26

Silica content (rice varieties), 330*t*

Single nucleotide polymorphisms (SNPs), 15–16

Smale, M., 407–23

Smallholders. *See* Farmers and pastoralists

Snails. *See* Mollusks

SNPs (single nucleotide polymorphisms), 15–16

Social and economic changes, as threats to animal populations, 168–70

Soil biodiversity: accidents, 258–59; adaptive management, 249–50; agricultural intensification, 234–35, 238*f*; agricultural practices, 235, 237; biological management of soil fertility, 237; conclusions and challenges, 259–60; dimensions of, 224; direct biological management technologies, 253–58; economic benefits, 233–34; identification and use of soil quality indicators, 241*f*, 242–45, 243*b*, 244*f*; implementing integrated management, 250; indirect biological management, 251, 252–53*b*, 253; interventions for soil biological management, 238–40, 239*b*; land use trends and threats, 234–35, 237; limited knowledge, 235, 260; organically managed, 455; organic matter management, 250–51; overcoming limitations, 245–46, 246*f*, 247–48*t*, 249; recognizing importance of soil

biota, 240–42, 241*f*; redundancy as crucial, 456; seven-step process to soil health and conservation, 241*f*

Soil biota: beneficial macrofauna, 256–58; beneficial soil microorganisms, 254–56; body size, 231–32; bottom-up control (feedback), 231; as disparate assemblages of organisms, 224–25; ecosystem functions, scale effects, and regulatory hierarchies, 227–31; ecosystem functions performed by different members, 228*t*, 233; effects of different agricultural management practices, 246*f*, 247–48*t*; feeding behavior, 232–33; functional classifications, 231; hierarchical organization of function, 230*f*, 230–31; number of described species, 226*t*, 226–27; as physically complex, 225; physical structures and functional domains, 233; as poorly understood, 225–27, 241, 260; recognizing importance, 240–42, 241*f*

Soil engineering, 229, 229*b*

Soil formation, global contribution by biome, 448–49*t*

Sorghum: adaptedness, 19; breeding and storage systems in Burkina Faso, 98*b*, 99*b*; Burkina Faso, 42*t*, 42–43; counts of named varieties, 43; earlier maturation, 99; Ethiopia, 30; outcrossing, 55; sources of seed, 77; study of farmers' decision making, 374*b*

South America. *See* Latin America

Soybeans: nucleotide diversity studies, 15*t*; phylogeny, 17; physiologic effect and active constituents, 385*t*

Sponge gourd, Nepal, 80, 81

Springtails, number of described species, 226*t*

Squash: diversity distribution and production spaces, 59, 60–62*b*; gendered seed flows, 91; in Mayan milpas, 69; seed transactions, 89; *See also* Milpas

Sri Lanka: honeybees and beekeeping, 205; use of FCA method, 59

Standard inventories, 341

Stated preference techniques for valuation of AnGR conservation programs, 437–38

State of the World Plant Genetic Resources for Food and Agriculture (PGRFA), 2–3

Sthapit, B. R., 34–68

Storage of seeds. *See* Seed systems